Applications of Hypothesis Testing for Environmental Science

Applications of Hypothesis Testing for Environmental Science

Abbas F. M. Alkarkhi
University of Kuala Lumpur (Unikl)–Malaysia

Elsevier
Radarweg 29, PO Box 211, 1000 AE Amsterdam, Netherlands
The Boulevard, Langford Lane, Kidlington, Oxford OX5 1GB, United Kingdom
50 Hampshire Street, 5th Floor, Cambridge, MA 02139, United States

Notices

Knowledge and best practice in this field are constantly changing. As new research and experience broaden our understanding, changes in research methods, professional practices, or medical treatment may become necessary.

Practitioners and researchers must always rely on their own experience and knowledge in evaluating and using any information, methods, compounds, or experiments described herein. In using such information or methods they should be mindful of their own safety and the safety of others, including parties for whom they have a professional responsibility.

To the fullest extent of the law, neither the Publisher nor the authors, contributors, or editors, assume any liability for any injury and/or damage to persons or property as a matter of products liability, negligence or otherwise, or from any use or operation of any methods, products, instructions, or ideas contained in the material herein.

British Library Cataloguing-in-Publication Data
A catalogue record for this book is available from the British Library

Library of Congress Cataloging-in-Publication Data
A catalog record for this book is available from the Library of Congress

ISBN: 978-0-12-824301-5

Publisher: Joseph Hayton
Acquisitions Editor: Marisa LaFleur
Editorial Project Manager: Grace Lander
Production Project Manager: Sruthi Satheesh
Cover Designer: Christian Bilbow

Typeset by MPS Limited, Chennai, India

Dedication

To the memory of my parents (deceased)

To my beloved children Atheer, Hibah, and Farah

To my beloved wife
Abbas F. M. Alkarkhi

Contents

Preface

In this book applications of hypothesis testing for environmental science are described in an easy and enjoyable style to help and assist graduate and postgraduate students and researchers working in the fields of environmental science, environmental engineering, and related fields. The lack of knowledge of many researchers about the concept and application of hypothesis testing in the environmental field, including how to state the hypothesis, how to choose the most suitable test, and how to interpret and draw conclusions motivated me to write this book and introduce hypothesis testing for one and two parameters covering three procedures for hypothesis testing (critical value, *P*-value, and confidence interval) and to connect the theory to the environmental field to meet the needs of researchers and enable them to carry out various testing procedures and draw conclusions.

The concept of hypothesis testing, the type of test, interpretation of the results, and drawing conclusions are carried out in a step-by-step, easy, and clear manner to enable researchers to understand it clearly and use it in their research. The book focuses on the application of hypothesis testing for one and two population parameters including *Z*-test, *t*-test, chi-square test, and *F*-test, which are frequently used by researchers in the environmental field. Most of the tutorials in the book use real data obtained through over 18 years of research in this area.

Finally, I wish to thank my beloved family for their continuous support. I also would like to express my gratitude to University of Kuala Lumpur (Unikl) for its support.

Abbas F. M. Alkarkhi

June 2020

Introduction to statistical hypothesis testing

1

Abstract

Chapter 1 is the introductory chapter that presents the philosophy of hypothesis testing and the concept of null and alternative hypotheses, rejection region, nonrejection region, significance level, and critical values. Furthermore, the general procedure for performing hypothesis testing is given along with the two types of errors that could occur while performing the procedure. Types of hypothesis testing including one-tailed (right-tailed and left-tailed) and two-tailed tests are delivered in a simple and easy to understand way. This chapter includes examples obtained from the field of environmental science to show the application of hypothesis testing in the field of environmental science and engineering.

Keywords: Null and alternative hypotheses; critical value; rejection and nonrejection regions; significance level; types I and II errors

Learning outcomes

After completing this chapter, readers will be able to:

- Understand the philosophy behind hypothesis testing idea;
- Know the general procedure for performing hypothesis testing;
- Know how to define the two hypotheses (null and alternative);
- Describe the steps on how to set up the null and alternative hypotheses in terms of mathematical forms;
- Understand the difference between rejection and nonrejection regions;
- Know how to obtain critical values from a probability distribution table;
- Know the level of significance and how to use it;
- Know how to use the sample data in calculating the test statistic;
- Describe the two types or errors (type I and type II errors);
- Understand the difference between two-tailed and one-tailed tests;
- Understand the procedure for applying hypothesis testing in the field of environmental science or engineering;
- Explain the outputs and match the results to environmental issues;
- Know how to extract helpful inferences regarding the problem under study.

1.1 Introduction

Researchers usually carry out various experiments using statistical techniques for collecting, analyzing, summarizing, and draw conclusions to gather

Applications of Hypothesis Testing for Environmental Science. DOI: https://doi.org/10.1016/B978-0-12-824301-5.00006-X

information regarding the behavior of populations under investigation and help in making judgments regarding the purpose of study. Statistical hypothesis testing is used to make judgments regarding any claim or issue using the sample data to obtain the correct decision that guides researchers to accomplish the research and achieve the objectives. Gathering information can be done using a representative sample chosen from the same population under investigation. Hypothesis testing is an important method to examine various claims or statements regarding different issues to support or deny the claims in order to plan for the next step.

Hypothesis testing concerning population proportions, population means, and population variances is of the subject of this text. A *Z*-test is used to carry out hypothesis testing regarding population proportions. Furthermore, a *Z*-test can be used for population means when the sample size is large, and a *t*-test is used when the sample size is small. A chi-square test and *F*-test are used to test a hypothesis concerning one and two variances.

As an example, consider a researcher who wants to test a claim regarding the concentration of total suspended solids (TSSs) in the surface water of a river, the claim says that the mean concentration of TSSs is 465 (mg/L). A sample should be selected from the river and tested for the concentration of TSSs and then the results summarized and analyzed to finally decide whether to support the claim or to reject it.

1.2 What is hypothesis testing?

Hypothesis testing is a statistical technique used to make judgments regarding claims or statements related to populations. The opinion and definition of hypothesis testing with the general procedure of conducting the analysis are clarified before being applied to environmental data.

Any claim or statement made on population parameters (mean, standard deviation, and proportion) is called a hypothesis. A sample should be chosen from the population of interest to understand the behavior of the issue under investigation and make a judgment based on the collected sample information.

We can identify two types of hypotheses: the null hypothesis and the alternative hypothesis. The two hypotheses are mutually exclusive (they cannot occur together).

The null hypothesis is a statistical hypothesis that says no real difference exists between a population parameter (such as proportion, average, and standard deviation) and some claimed value (given value). The symbol for null hypothesis is H_0.

The alternative hypothesis is a statistical hypothesis that represents the opposite of the null hypothesis. An alternative hypothesis says that a real difference exists between a population parameter and a given value. The symbol for alternative hypothesis is H_1.

Three steps can be employed to specify the null and alternative hypotheses mathematically as given below.

1. Find out the hypothesis concerning the issue under investigation.
2. Translate the words of the null hypothesis into a mathematical expression.
3. Translate the words of the alternative hypothesis into a mathematical expression.

Example 1.1: Specify the two hypotheses for the two-sided test: An environmentalist wishes to investigate a claim regarding the mean concentration of cadmium (Cd) of surface water. The environmentalist wants to examine the claim that the mean concentration of cadmium of surface water is 0.17 mg/L.

The three steps for specifying the null and alternative hypotheses can be used to write the two hypotheses as shown below.

1. *Find out the hypothesis concerning the issue under investigation*
   One can see that the hypothesis is given in words; thus we need to figure out the hypothesis. The hypothesis says, "the mean concentration of cadmium of surface water is 0.17 mg/L."
2. *Translate the words of the null hypothesis into a mathematical expression*
   The next step in specifying the null and alternative hypotheses is to translate the words of the null hypothesis into a mathematical form. The null hypothesis states that the mean concentration of cadmium of surface water is exactly 0.17 mg/L. This statement can be translated into a mathematical form as given below:

   $H_0{:}\mu = 0.17$ (Claim)

   where μ refers to the population mean.

   One can see that the equality sign is under the null hypothesis.
3. *Translate the words of the alternative hypothesis into a mathematical expression*
   The null hypothesis states that the mean concentration of cadmium of surface water equals 0.17 mg/L. The opposite of the null hypothesis represents the alternative hypothesis, which is not equal. Thus, if the mean concentration of cadmium of surface water is not equal to 0.17 mg/L, then the mean concentration of cadmium of surface water will be either greater than or lesser than 0.17 mg/L. We can recognize two directions for the alternative hypothesis, the first direction is greater than 0.17 mg/L and the second direction is less than 0.17 mg/L. The two directions can be represented in a mathematical symbol as $\neq$. The alternative hypothesis in terms of a mathematical form is

   $H_1{:}\mu \neq 0.17$

   One can see that the null and alternative hypotheses are mutually exclusive (they cannot occur at the same time).

Example 1.2: Specify the two hypotheses for a right-tailed test: An environmentalist wishes to investigate a claim regarding the mean concentration of cadmium (Cd) of surface water. The environmentalist wants to examine the claim that the mean concentration of cadmium of surface water is greater than 0.17 mg/L.

The three steps for specifying the null and alternative hypotheses can be used to write the two hypotheses as shown below.

1. *Find out the hypothesis concerning the issue under investigation*

   One can see that the hypothesis is given in words; thus we need to figure out the hypothesis. The claim says, "the mean concentration of cadmium of surface water is greater than 0.17 mg/L."

2. *Translate the words of the null hypothesis into a mathematical expression*

   The next step in specifying the null and alternative hypotheses is to translate the words of the null hypothesis into a mathematical form. The claim states that the mean concentration of cadmium of surface water is greater than 0.17 mg/L. Thus, if the mean concentration of cadmium is not greater than 0.17 mg/L, then the mean concentration of cadmium of surface water will be either equal to 0.17 mg/L or less than 0.17 mg/L. Thus we can use the two directions to represent the null hypothesis as $\leq$. This statement can be translated into a mathematical form as given below:

   $$H_0: \mu \leq 0.17$$

   One can see that the equality sign is under the null hypothesis.

3. *Translate the words of the alternative hypothesis into a mathematical expression*

   The null hypothesis states that the mean concentration of cadmium in surface water is less than or equal to 0.17 mg/L. The opposite of the null hypothesis represents the alternative hypothesis; thus if the mean concentration of cadmium of surface water is not less than or equal to 0.17 mg/L, then the mean concentration of cadmium of surface water will be greater than 0.17 mg/L. Thus we can recognize one direction for the alternative hypothesis which is greater than 0.17 mg/L. This direction can be represented in a mathematical symbol as $>$. The alternative hypothesis in terms of a mathematical form is:

   $$H_1: \mu > 0.17$$

   One can see that the null and alternative hypotheses are mutually exclusive (they cannot occur at the same time).

Example 1.3: Specify the two hypotheses for a left-tailed test: An environmentalist wishes to investigate a claim regarding the mean concentration of cadmium (Cd) of surface water. The environmentalist wants to examine the claim that the mean concentration of cadmium of surface water is less than 0.17 mg/L.

The three steps for specifying the null and alternative hypotheses can be used to write the two hypotheses as shown below.

1. *Find out the hypothesis concerning the issue under investigation*

   One can see that the hypothesis is given in words; thus we need to figure out the hypothesis. The hypothesis says, "the mean concentration of cadmium of surface water is less than 0.17 mg/L."

2. *Translate the words of the null hypothesis into a mathematical expression*

   The next step in specifying the null and alternative hypotheses is to translate the words of the null hypothesis into a mathematical form. The claim states that the mean concentration of cadmium of surface water is less than 0.17 mg/L. Thus, if the mean concentration of cadmium is not less than 0.17 mg/L, then the mean concentration of cadmium of surface water will be either equal to 0.17 mg/L or greater than 0.17 mg/L. Thus we can use the two directions to represent the null hypothesis as $\geq$. This statement can be translated into a mathematical form as given below:

$$H_0{:}\mu \geq 0.17$$

One can see that the equality sign is under the null hypothesis.

3. *Translate the words of the alternative hypothesis into a mathematical expression*
   The null hypothesis states that the mean concentration of cadmium in surface water is greater than or equal to 0.17 mg/L. The opposite of the null hypothesis represents the alternative hypothesis; thus if the mean concentration of cadmium of surface water is not more than or equal to 0.17 mg/L, then the mean concentration of cadmium of surface water will be less than 0.17 mg/L. Thus we can recognize one direction for the alternative hypothesis which is less than 0.17 mg/L. This direction can be represented in a mathematical symbol as $<$. The alternative hypothesis in terms of a mathematical form is:

$$H_1{:}\mu < 0.17$$

One can see that the null and alternative hypotheses are mutually exclusive (they cannot occur at the same time).

We can summarize the three situations of statistical hypotheses including one situation for the two-tailed test and two situations for the one-tailed test.

1. The two-tailed test will have the two hypotheses as given in Eq. (1.1).

$$\begin{aligned} H_0&{:}\mu = c \\ H_1&{:}\mu \neq c \end{aligned} \tag{1.1}$$

2. The right-tailed test is a one-tailed test, and the two hypotheses for the right-tailed test are given in Eq. (1.2).

$$\begin{aligned} H_0&{:}\mu \leq c \\ H_1&{:}\mu > c \end{aligned} \tag{1.2}$$

3. The left-tailed test is a one-tailed test, and the two hypotheses for the left-tailed test are given in Eq. (1.3).

$$\begin{aligned} H_0&{:}\mu \geq c \\ H_1&{:}\mu < c \end{aligned} \tag{1.3}$$

where c is a given value.

Researchers usually employ one of the hypotheses in their research to match the purpose of the study.

Note

- One can see that the equality sign $(=, \geq, \leq)$ must be placed with the null hypothesis H_0.
- We always include $<$, $>$, or $\neq$ in the alternative hypothesis H_1.
- One can see that the two hypotheses are mutually exclusive.

1.3 The general procedure for performing statistical hypothesis testing

The general procedure for conducting hypothesis testing can be summarized by the steps given below.

Step 1: Specify the null and alternative hypotheses.

Step 2: Select the significance level (α) for the study.

Step 3: Use the sample information to calculate the test statistic value.

Step 4: Identify the critical and noncritical regions for the study.

Step 5: Make a decision and interpret the results.

We will discuss the general procedure step-by-step supported by examples where necessary.

Step 1: Specify the null and alternative hypotheses

We have studied this step earlier on how to write and specify the null and alternative hypotheses.

Step 2: Select the significance level (α) for the study

The significance level, usually called alpha (α), is also called the level of significance. Consider that the null hypothesis is true (H_0), then the probability of rejecting the null hypothesis (H_0) is called the alpha or significance level. The value of alpha refers to the significance of the results, and we usually select the value of alpha (α) to be 0.05 or 0.01.

Step 3: Use the sample information to calculate the test statistic value

The sample information (data) is used to calculate a value called the test statistic. This value is calculated using a specific mathematical formula for each distribution. Researchers use the test statistic value to make a decision to reject the null hypothesis or not.

Step 4: Identify the critical and noncritical regions for the study

The critical and noncritical regions are also called rejection and nonrejection regions; we should deliver the concept of critical values, and critical and noncritical regions.

The range of values of the test statistic that would reject the null hypothesis is called the critical region or the rejection region for a hypothesis testing.

The range of values of the test statistic that would not reject the null hypothesis is called the noncritical region or the nonrejection region for a hypothesis testing.

The value that is calculated from a probability distribution related to the problem under study is called the critical value. This value is used in hypothesis testing to differentiate the critical region (where H_0 should be rejected) from the noncritical region. We can calculate critical values for each distribution based on the significance level.

We can recognize the critical and noncritical regions by employing three simple steps given below:

1. *Specify the alternative hypothesis;*
2. *Select the appropriate significance level;*
3. *Extract the correct critical value.*

We should make a decision regarding the null hypothesis of the study whether to reject H_0 or not. Then we need to interpret the decision and draw a conclusion regarding the study under investigation.

Examples about the mean value will be shown to identify the critical and noncritical regions. We consider the variable under study to be normally distributed.

Example 1.4: Identify the critical and noncritical regions for the two-tailed test: An environmentalist wishes to identify the critical and noncritical regions for a claim regarding the mean concentration of cadmium (Cd) of surface water. The environmentalist wants to examine the claim that the mean concentration of cadmium of surface water is 0.17 mg/L. Use a significance level of $\alpha = 0.05$ to identify the two regions. Assume that the data are normally distributed.

We can use the three steps presented earlier to identify the critical and noncritical regions for the two-tailed test.

1. *Specify the alternative hypothesis*

   The key to specifying the critical and noncritical regions is to know the alternative hypothesis. The null and alternative hypotheses are

   $$H_0: \mu = 0.17$$
   $$H_1: \mu \neq 0.17$$

2. *Select the appropriate significance level*

   The second step in the procedure for identifying the critical and noncritical regions is to select the appropriate significance level. The appropriate significance level is selected to be 0.05 ($\alpha = 0.05$). Because the test is a two-tailed test, we should divide alpha by 2 to identify two equally critical regions on both tails.

   $$\frac{\alpha}{2} = \left(\frac{0.05}{2}\right) = 0.025.$$

   An area of 0.025 $\left(\frac{\alpha}{2}\right)$ will be on each tail to represent the critical (rejection) region.

3. *Extract the correct critical value*

   The extracted critical value e for the two-tailed test with $\alpha = 0.05$ is ± 1.96 (see standard normal Table A). The critical and noncritical regions for this study which represent the two-tailed test are presented in Fig. 1.1. The shaded area in Fig. 1.1 is called the critical (rejection) region.

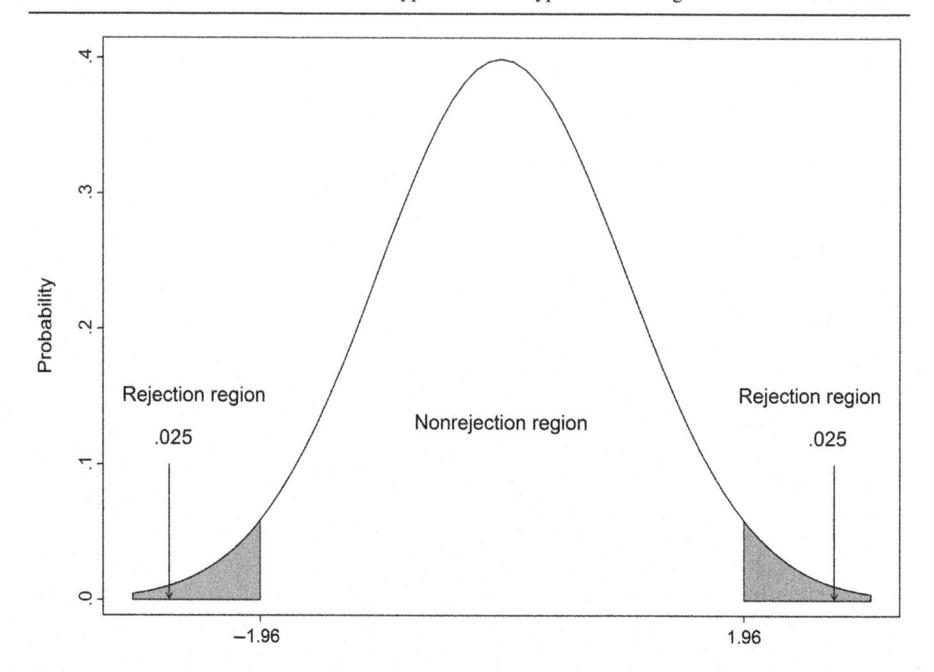

Figure 1.1 The critical and noncritical regions for $H_1{:}\mu \neq 0.17$.

Example 1.5: Identify the critical and noncritical regions for the right-tailed test: An environmentalist wishes to specify the critical and noncritical regions for a claim regarding the mean concentration of cadmium (Cd) of surface water. The environmentalist wants to examine the claim that the mean concentration of cadmium of surface water is greater than 0.17 mg/L. Use a significance level of $\alpha = 0.05$ to specify the two regions. Assume that the data are normally distributed.

We can use the three steps presented earlier to identify the critical and noncritical regions for the right-tailed test.

1. *Specify the alternative hypothesis*

   The key to specifying the critical and noncritical regions is to know the alternative hypothesis. The null and alternative hypotheses are

   $H_0{:}\mu \leq 0.17$
   $H_1{:}\mu > 0.17$

2. *Select the appropriate significance level*

   The second step in the procedure for identifying the critical and noncritical regions is to select the appropriate significance level. The appropriate significance level is selected to be 0.05 ($\alpha = 0.05$). Because the test is a one-tailed test (right-tailed), we identify the critical region (rejection) of $\alpha = 0.05$ on the right tail.
3. *Extract the correct critical value*

   The extracted critical value for the one-tailed test with $\alpha = 0.05$ is ± 1.645 (see standard normal Table A). The critical and noncritical regions for this study which represent a one-tailed test are presented in Fig. 1.2. The shaded area in Fig. 1.2 is called the critical (rejection) region.

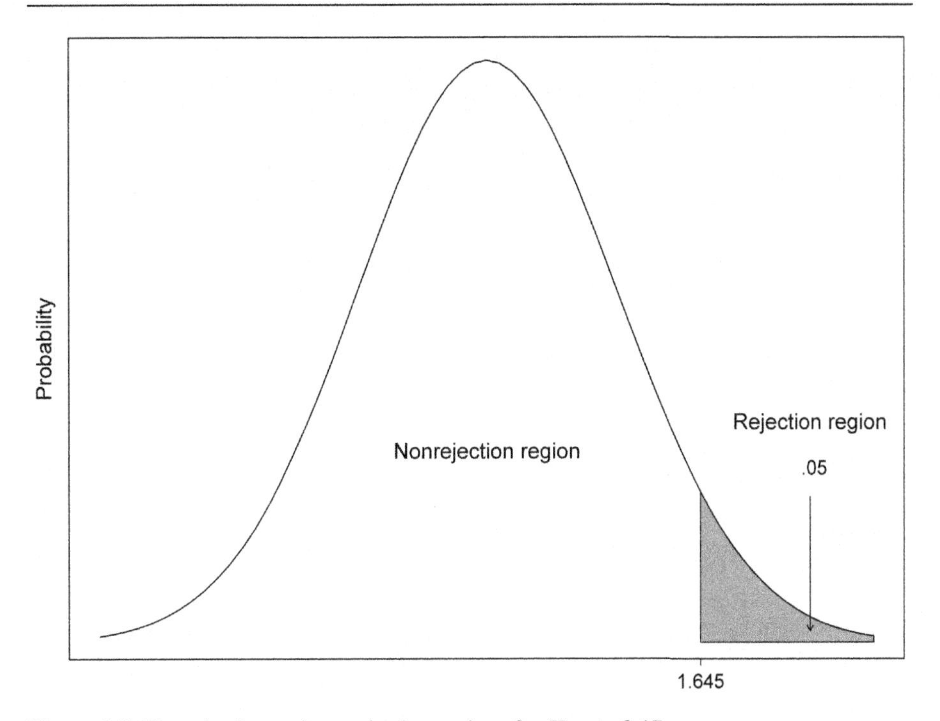

Figure 1.2 The rejection and nonrejection regions for $H_1: \mu > 0.17$.

Example 1.6: Identify the critical and noncritical regions for the right-tailed test: An environmentalist wishes to specify the critical and noncritical regions for a claim regarding the mean concentration of cadmium (Cd) of surface water. The environmentalist wants to examine the claim that the mean concentration of cadmium of surface water is less than 0.17 mg/L. Use a significance level of $\alpha = 0.05$ to identify the two regions. Assume that the data are normally distributed.

We can use the three steps presented earlier to specify the critical and noncritical regions for the right-tailed test.

1. *Specify the alternative hypothesis*

   The key to identifying the critical and noncritical regions is to know the alternative hypothesis. The null and alternative hypotheses are

   $H_0: \mu \geq 0.17$
   $H_1: \mu < 0.17$

2. *Select the appropriate significance level*

   The second step in the procedure for identifying the critical and noncritical regions is to select the appropriate significance level. The appropriate significance level is selected to be 0.05 ($\alpha = 0.05$). Because the test is a one-tailed test (left-tailed), we identify the critical region (rejection) of $\alpha = 0.05$ on the left tail.

3. *Extract the correct critical value*

   The extracted critical value for the one-tailed test with $\alpha = 0.05$ is $\pm$ 1.645 (see standard normal Table A). The critical and noncritical regions for this study which represent the one-tailed test are presented in Fig. 1.3. The shaded area in Fig. 1.3 is called the critical (rejection) region.

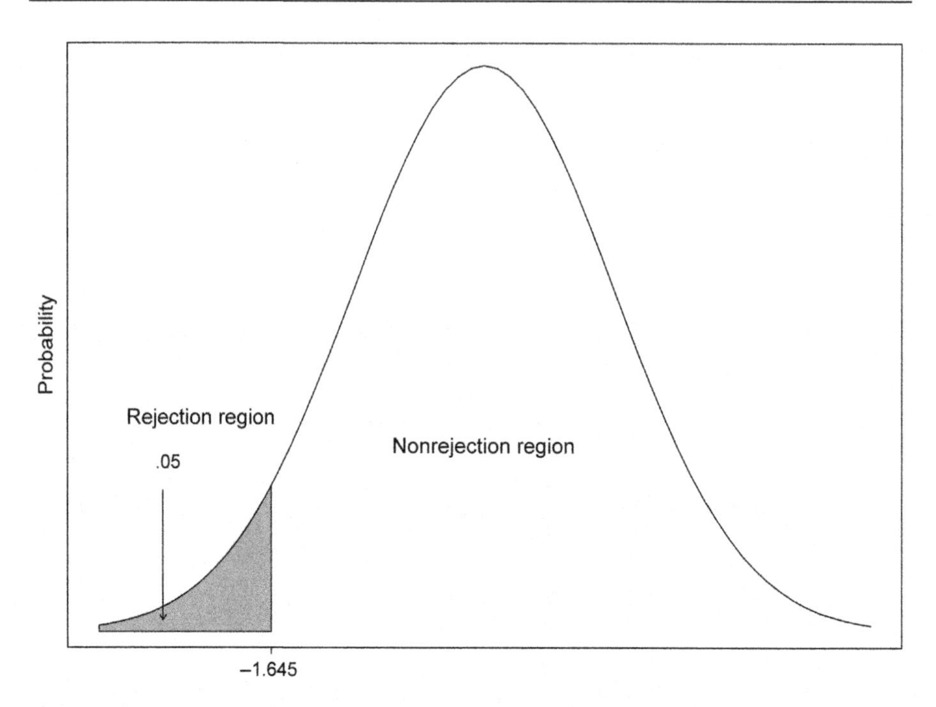

Figure 1.3 The rejection and nonrejection regions for $H_1: \mu < 0.17$.

Step 5: Make a decision and interpret the results

The last step in the general procedure for hypothesis testing is to make a decision to reject the null hypothesis (H_0) or not. Interpretation of the decision is needed and it is important to draw a conclusion regarding the study under investigation and reach the objective of the study.

1.4 Procedures for performing hypothesis testing

There are three procedures for making a decision in statistical hypothesis testing, which are the critical value procedure, *P*-value procedure, and confidence interval procedure. We will present the critical value procedure in this chapter, while the *P*-value and confidence interval procedures will be given in Chapter 6, The observed significance level (*P*-value) procedure, and Chapter 7, Interval estimation for one population, respectively.

1. *Critical value procedure*

The critical value procedure is used to guide researchers and help in making a decision in statistical hypothesis testing. This procedure is also known as a traditional procedure. We can use the critical value procedure to make a decision regarding the null hypothesis using the test statistic value and the critical value related to the distribution used for solving the problem. The decision is made by comparing the two values (test statistic value and critical value). If the absolute test statistic value is greater than the critical value, the null hypothesis is rejected; otherwise the null hypothesis is not rejected. Or using the probability distribution curve, if the test statistic value falls in the critical region, the null hypothesis is rejected; otherwise the null hypothesis is not rejected.

We will employ the critical value procedure for making a decision regarding the one-sample test for the mean value, proportion value, and variance value, while the other two procedures, the P-value and confidence interval procedures, will be given in Chapter 6, The observed significance level (P-value) procedure, and Chapter 7, Interval estimation for one population, respectively, and then compare the results of various procedures.

1.5 Types of errors

Two types of errors are associated with hypothesis testing procedures in making a decision regarding the null hypothesis. A decision is usually made based on the sample information, and this decision could be correct or incorrect. The two types of errors are type I and type II errors.

The first type of error is called a *type I error*. This error occurs when we incorrectly decide to reject a true null hypothesis. A type I error is called the significance level (α).

The second type of error is called a *type II error*. This error occurs when we fail to reject a false null hypothesis. The symbol for the probability of a type II error is β.

Note

- Type I and type II errors are inversely proportional; decreasing type I error (α) would increase type II error β, and increasing type I error (α) would decrease type II error (β).

Example 1.7: Specify type I and type II errors: Specify type I and type II errors for the claim which states that the mean concentration of cadmium (Cd) is 0.17 mg/L.

We need to identify the null and alternative hypotheses for this claim. The two hypotheses are given below:

$$H_0:\mu = 0.17$$
$$H_1:\mu \neq 0.17$$

Type I error

A type I error would occur when we incorrectly decide to reject a true null hypothesis. For the mean concentration of cadmium, a type I error would occur if the mean concentration of cadmium of surface water is 0.17 mg/L ($\mu = 0.17$), but the information provided by the selected sample tells us to reject the claim that the mean concentration of cadmium of surface water is 0.17 mg/L and conclude that the mean concentration of cadmium of surface water is not equal to 0.17 mg/L ($\mu \neq 0.17$).

Type II error

A type II error would occur when we fail to reject a false null hypothesis. For the mean concentration of cadmium of surface water, a type II error would occur if the mean concentration of cadmium of surface water is not equal to 0.17 mg/L ($\mu \neq 0.17$), but the information provided by the selected sample tells us not to reject the null hypothesis and conclude that the mean concentration of cadmium of surface water is equal to 0.17 mg/L ($\mu = 0.17$).

Correct decision

A correct decision can be achieved in two cases: in the first case, if we do not reject the null hypothesis when it is true. The correct decision for this example is we do not reject the null hypothesis if the mean concentration of cadmium of surface water is equal to 0.17 mg/L ($\mu = 0.17$).

In the second case, if we reject the null hypothesis when it is false. The correct decision for this example is we reject the null hypothesis if the mean concentration of cadmium of surface water is not equal to 0.17 mg/L ($\mu \neq 0.17$).

Further reading

Alkarkhi, A. F. M., & ALqaraghuli, W. A. A. (2020). *Applied statistics for environmental science with R* (1st ed.). Elsevier.

Alkarkhi, A. F. M., & Chin, L. H. (2012). *Elementary statistics for technologist* (1st ed.). Malaysia: Universiti Sains Malaysia.

Bluman, A. G. (1998). *Elementary statistics: A step by step approach* (3rd ed.). Boston: WCB McGraw-Hill.

Weiss, N. A. (2012). *Introductory statistics* (9th ed.). Pearson.

Z-test for one-sample mean

2

Abstract

This chapter presents hypothesis testing based on normal distribution including the concept of normal distribution, standard normal distribution, and related terms. The area under normal distribution curve with the concept of rejection and nonrejection regions is delivered with illustrations by examples for each region using a normal distribution curve. Moreover, the general procedure for hypothesis testing using Z-test for a large sample is given and explained clearly. Examples from the field of environmental science are selected and used to illustrate the steps of hypothesis testing using Z-test and making a decision regarding the study with sufficient explanation for each step.

Keywords: Normal distribution; standard normal distribution; critical and critical regions; critical value; area under the curve; Z-test

Learning outcomes

After completing this chapter, readers will be able to:

- Understand the importance of normal distribution
- Explain the concept of standard normal distribution
- Compute the area under the standard normal curve
- Determine the probabilities for normal distribution
- Describe the common steps for performing hypothesis testing
- Compute the Z-test statistic value for a single population mean
- Know the rejection and nonrejection regions for Z-test
- Know one-tailed and two-tailed tests for Z-test
- Describe the steps for providing correct interpretation of the results
- Write useful conclusions

2.1 Introduction

Z-test for one-sample mean and other statistical tests employs the concept of normal distribution to perform hypothesis testing. Thus the concept and properties of normal distribution should be delivered before giving the procedure of hypothesis testing for one sample Z-test.

Applications of Hypothesis Testing for Environmental Science. DOI: https://doi.org/10.1016/B978-0-12-824301-5.00007-1

The concept and properties of normal distribution and standard normal distribution are covered in this chapter. Moreover, the procedure for calculating the area under the standard normal curve is covered and the general procedure for performing a Z-test for a one-sample mean also is given.

2.2 What is normal distribution?

There are many distributions in statistics used to describe the behavior of different phenomena. One of the most important distributions is called normal distribution, or the bell curve due to its shape (bell-shaped). A Gaussian distribution is another name for normal distribution. The normal distribution curve is presented in Fig. 2.1.

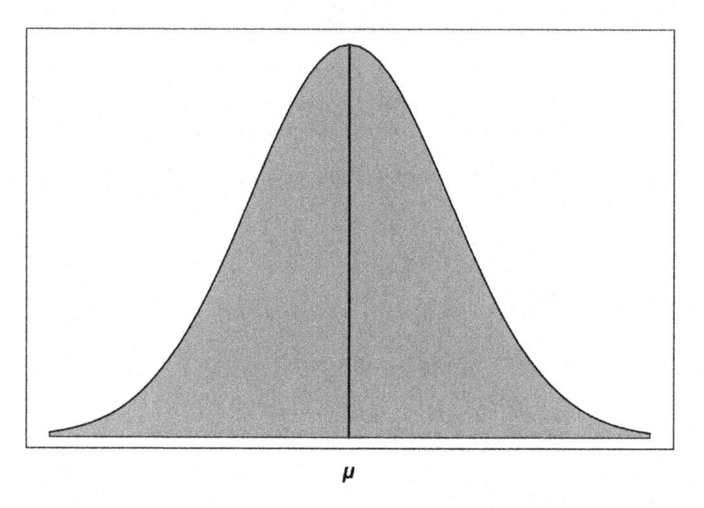

Figure 2.1 The normal distribution curve.

Consider Y represents a random variable that follows a normal distribution with mean (μ) and variance (σ^2), then the normal probability distribution function is presented in Eq. (2.1).

$$f(Y) = \frac{1}{\sqrt{2\pi\sigma^2}} e^{-(Y-\mu)^2} / 2\sigma^2 \quad -\infty < Y < \infty \tag{2.1}$$

where,

Y represents a random variable that is distributed normally $N(\mu, \sigma^2)$ or symbolized as $Y \sim N(\mu, \sigma^2)$,

$\pi = 3.14$, and

$e \cong 2.71828$.

The properties of normal distribution curve are as follows:

1. The normal distribution curve is bell-shaped.
2. The shape of the normal distribution curve is symmetric about the mean.
3. The mean is at the center and divides the area into two equal parts.
4. The total area under the normal distribution curve is equal to 1.
5. Most of the area under the curve falls between three standard deviations of the mean. The percentage of the area under one, two, and three standard deviations is given below.
   - 68% of the points fall between one standard deviation of the mean;
   - 95% of the points fall between two standard deviations of the mean;
   - 99.7% of the points fall between three standard deviations of the mean.

2.3 What is standard normal distribution?

Consider that Y represents a random variable that follows a normal distribution with mean (0) and variance (1), then the standard normal distribution function is presented in Eq. (2.2).

$$f(Y) = \frac{1}{\sqrt{2\pi}} e^{-Y^2/2} \tag{2.2}$$

Y represents a random variable that is distributed normally $N(0, 1)$ or symbolized as $Y \sim N(0, 1)$. The standard normal distribution has the same properties of normal distribution, the standard normal distribution curve is presented in Fig. 2.2.

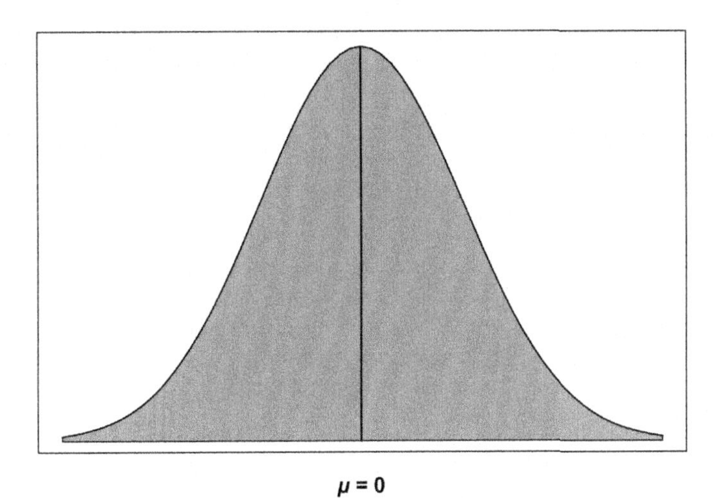

Figure 2.2 The standard normal curve with $\mu = 0$ and $\sigma^2 = 1$.

We can convert any normal random variable to the standard normal distribution employing the formula given in Eq. (2.3).

$$Z = \frac{Y - \mu}{\sigma} \tag{2.3}$$

The formula in Eq. (2.3) is called the standard score.

2.4 Finding the area under the normal curve

The area under the standard normal curve can be computed using the standard score. The Z score provides information on the position of a value compared to the average value using the distance to express the position by how many standard deviations the value falls above or below the average. The steps for finding the area under the standard normal curve require describing the standard normal table and then follow a few steps to extract the area from the table.

We use Table A in the Appendix to extract the area under the curve for Z values from 0 to 3. The value of Z should be divided into two parts, the first part of the Z value in Table A is represented in the left-hand (first) column, the values of Z in the table start from 0 to 3 (nearest tenth), while the first (upper) row provides the second part of the Z value (second decimal place). We can use the same table to find the area under the curve for negative Z values because of the symmetrical property of normal distribution.

Example 2.1: Compute the area to the left of a Z value: Compute the area under the standard normal curve to the left of a positive Z value of 2.13.

Finding the area to the left of a positive Z value of 2.13 requires using Table A in the Appendix for the standard normal. The value "2.13" should be divided into two pieces, the first piece is "2.1" and the second is "0.03." The exact area to the left of 2.13 can be computed employing the steps below.

- The first step is to search for the position of "2.1" in the first vertical column of Table 2.1 labeled Z to specify the first piece "2.1" of the number 2.13 (highlighted row) [Table 2.1 is a portion of the standard normal table in the Appendix (Table A)].
- The second step in finding the area is to move on the row of "2.1" to the column labeled "0.03" (highlighted column), the point of intersection represents the required value which is **0.9834** as shown in Table 2.1 (bold value).

The exact area to the left of 2.13 is shown in Fig. 2.3 (shaded area).

Table 2.1 The area to the left of a Z value of 2.13.

Z	Second decimal in Z									
	0.0	0.01	0.02	0.03	0.04	0.05	0.06	0.07	0.08	0.09
0	0.5000	0.5040	0.5080	0.5120	0.5160	0.5199	0.5239	0.5279	0.5319	0.5359
0.1	0.5398	0.5438	0.5478	0.5517	0.5557	0.5596	0.5636	0.5675	0.5714	0.5753
0.2	0.5793	0.5832	0.5871	0.591	0.5948	0.5987	0.6026	0.6064	0.6103	0.6141
0.3	0.6179	0.6217	0.6255	0.6293	0.6331	0.6368	0.6406	0.6443	0.648	0.6517
0.4	0.6554	0.6591	0.6628	0.6664	0.6700	0.6736	0.6772	0.6808	0.6844	0.6879
0.5	0.6915	0.695	0.6985	0.7019	0.7054	0.7088	0.7123	0.7157	0.719	0.7224
0.6	0.7257	0.7291	0.7324	0.7357	0.7389	0.7422	0.7454	0.7486	0.7517	0.7549
0.7	0.758	0.7611	0.7642	0.7673	0.7704	0.7734	0.7764	0.7794	0.7823	0.7852
0.8	0.7881	0.791	0.7939	0.7967	0.7995	0.8023	0.8051	0.8078	0.8106	0.8133
0.9	0.8159	0.8186	0.8212	0.8238	0.8264	0.8289	0.8315	0.834	0.8365	0.8389
1	0.8413	0.8438	0.8461	0.8485	0.8508	0.8531	0.8554	0.8577	0.8599	0.8621
1.1	0.8643	0.8665	0.8686	0.8708	0.8729	0.8749	0.8770	0.8790	0.8810	0.8830
1.2	0.8849	0.8869	0.8888	0.8907	0.8925	0.8944	0.8962	0.898	0.8997	0.9015
1.3	0.9032	0.9049	0.9066	0.9082	0.9099	0.9115	0.9131	0.9147	0.9162	0.9177
1.4	0.9192	0.9207	0.9222	0.9236	0.9251	0.9265	0.9279	0.9292	0.9306	0.9319
1.5	0.9332	0.9345	0.9357	0.937	0.9382	0.9394	0.9406	0.9418	0.9429	0.9441
1.6	0.9452	0.9463	0.9474	0.9484	0.9495	0.9505	0.9515	0.9525	0.9535	0.9545
1.7	0.9554	0.9564	0.9573	0.9582	0.9591	0.9599	0.9608	0.9616	0.9625	0.9633
1.8	0.9641	0.9649	0.9656	0.9664	0.9671	0.9678	0.9686	0.9693	0.9699	0.9706
1.9	0.9713	0.9719	0.9726	0.9732	0.9738	0.9744	0.975	0.9756	0.9761	0.9767
2	0.9772	0.9778	0.9783	0.9788	0.9793	0.9798	0.9803	0.9808	0.9812	0.9817
2.1	0.9821	0.9826	0.983	**0.9834**	0.9838	0.9842	0.9846	0.985	0.9854	0.9857
2.2	0.9861	0.9864	0.9868	0.9871	0.9875	0.9878	0.9881	0.9884	0.9887	0.989

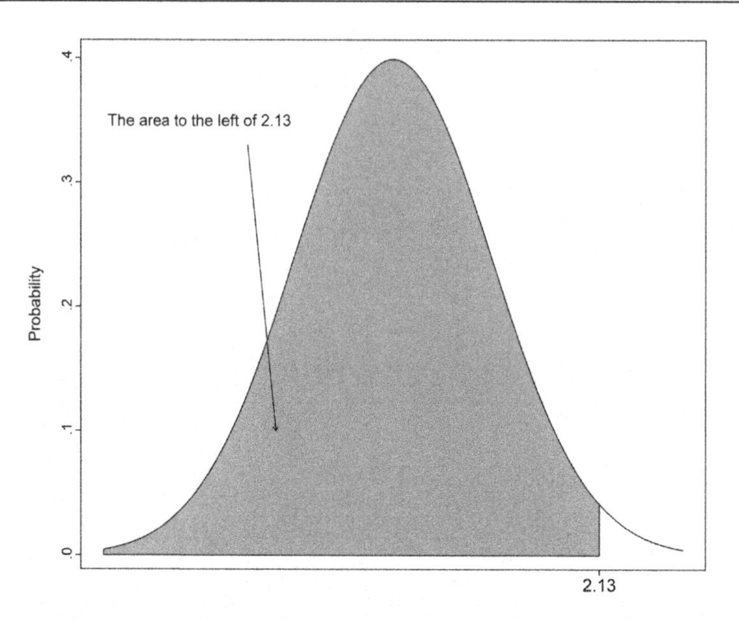

Figure 2.3 Showing the area to the left of 2.13.

Example 2.2: Compute the area to the right of a Z value: Compute the area under the standard normal curve to the right of a positive Z value of 2.03.

Finding the area to the right of a positive Z value of 2.03 requires using Table A in the Appendix for the standard normal. The value "2.03" should be divided into two pieces, the first piece is "2.0" and the second is "0.03." The exact area to the right of 2.03 can be computed employing the steps below.

- The first step is to search for the position of "2.0" in the first column of Table 2.2 labeled Z to specify the first piece "2.0" of the number 2.03 (highlighted row) [Table 2.2 is a portion of the standard normal table in the Appendix (Table A)].
- The second step in finding the area is to move on the row of "2.0" to the column labeled "0.03" (highlighted column), the point of intersection represents the area to the left of 2.03 which is "**0.9788**" as shown in Table 2.1 (bold value).

Table 2.2 The area to the right of a Z value of 2.03

Z	Second decimal in Z									
	0.0	0.01	0.02	0.03	0.04	0.05	0.06	0.07	0.08	0.09
0	0.5000	0.5040	0.5080	0.5120	0.5160	0.5199	0.5239	0.5279	0.5319	0.5359
0.1	0.5398	0.5438	0.5478	0.5517	0.5557	0.5596	0.5636	0.5675	0.5714	0.5753
0.2	0.5793	0.5832	0.5871	0.591	0.5948	0.5987	0.6026	0.6064	0.6103	0.6141
0.3	0.6179	0.6217	0.6255	0.6293	0.6331	0.6368	0.6406	0.6443	0.648	0.6517
0.4	0.6554	0.6591	0.6628	0.6664	0.6700	0.6736	0.6772	0.6808	0.6844	0.6879
0.5	0.6915	0.695	0.6985	0.7019	0.7054	0.7088	0.7123	0.7157	0.719	0.7224
0.6	0.7257	0.7291	0.7324	0.7357	0.7389	0.7422	0.7454	0.7486	0.7517	0.7549
0.7	0.758	0.7611	0.7642	0.7673	0.7704	0.7734	0.7764	0.7794	0.7823	0.7852
0.8	0.7881	0.791	0.7939	0.7967	0.7995	0.8023	0.8051	0.8078	0.8106	0.8133
0.9	0.8159	0.8186	0.8212	0.8238	0.8264	0.8289	0.8315	0.834	0.8365	0.8389
1	0.8413	0.8438	0.8461	0.8485	0.8508	0.8531	0.8554	0.8577	0.8599	0.8621
1.1	0.8643	0.8665	0.8686	0.8708	0.8729	0.8749	0.8770	0.8790	0.8810	0.8830
1.2	0.8849	0.8869	0.8888	0.8907	0.8925	0.8944	0.8962	0.898	0.8997	0.9015
1.3	0.9032	0.9049	0.9066	0.9082	0.9099	0.9115	0.9131	0.9147	0.9162	0.9177
1.4	0.9192	0.9207	0.9222	0.9236	0.9251	0.9265	0.9279	0.9292	0.9306	0.9319
1.5	0.9332	0.9345	0.9357	0.937	0.9382	0.9394	0.9406	0.9418	0.9429	0.9441
1.6	0.9452	0.9463	0.9474	0.9484	0.9495	0.9505	0.9515	0.9525	0.9535	0.9545
1.7	0.9554	0.9564	0.9573	0.9582	0.9591	0.9599	0.9608	0.9616	0.9625	0.9633
1.8	0.9641	0.9649	0.9656	0.9664	0.9671	0.9678	0.9686	0.9693	0.9699	0.9706
1.9	0.9713	0.9719	0.9726	0.9732	0.9738	0.9744	0.975	0.9756	0.9761	0.9767
2	0.9772	0.9778	0.9783	**0.9788**	0.9793	0.9798	0.9803	0.9808	0.9812	0.9817
2.1	0.9821	0.9826	0.983	0.9834	0.9838	0.9842	0.9846	0.985	0.9854	0.9857
2.2	0.9861	0.9864	0.9868	0.9871	0.9875	0.9878	0.9881	0.9884	0.9887	0.989

The area to the right of a Z value can be computed employing the fact that "the total area under the curve is 1," thus the area to the right of a Z value is equal to 1 minus the area to the left of a Z value. The area to the right of 2.03 is 0.0212 (1 − 0.9788 = 0.0212). The area to the right of 2.03 is 0.0212 as shown in Fig. 2.4 (shaded area).

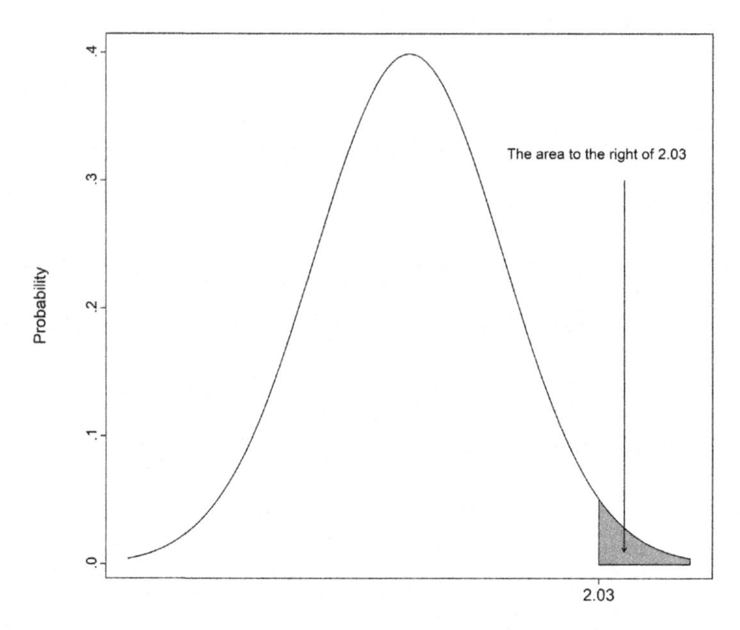

Figure 2.4 Showing the area to the right of 2.03.

Example 2.3: Compute the area between a negative Z value and a positive Z value: Compute the area under the standard normal curve between a Z value of −1.22 and a Z value of 1.53.

The area between two Z values can be represented by the difference between the area to the left of a larger value and the area to the left of a smaller value, thus the area between −1.22 and 1.53 can be calculated using Table A in the Appendix for the standard normal. Each value should be divided into two pieces as before and then we employ the steps below.

- For the first value 1.53, we follow similar steps to identify the area, search for the position of "1.5" in the first column of Table 2.3 labeled Z to specify the first piece "1.5" of the number 1.53 (highlighted row), and then move to the column labeled "0.03" (highlighted column). The exact area to the left of 1.53 is "**0.9370**."
- The second step in finding the area between two values is to compute the area to the left of −1.22. The area to the left of −1.22 is "**0.1112**." The value of 0.1112 can be computed using the symmetry property of standard normal distribution to find the area to the left of 1.22. The area to the left of 1.22 is "**0.8888**" (Table 2.3), and then employing the concept of the total area under normal curve is 1, thus the area to the right of 1.22 is "**0.1112**" (1 − 0.8888), the area to the right of 1.22 is equal to the area to the left of − 1.22 (symmetry property). The exact area between the two values of Z is equal to the difference between the two values (0.9370 − 0.1112 = 0.8250) as shown in Fig. 2.5.

The area between − 1.22 and 1.53 is represented by the shaded area in Fig. 2.5.

Table 2.3 The area between two Z values, −1.22 and 1.53.

Z	Second decimal in Z									
	0	0.01	0.02	0.03	0.04	0.05	0.06	0.07	0.08	0.09
0	0.5000	0.5040	0.5080	0.5120	0.5160	0.5199	0.5239	0.5279	0.5319	0.5359
0.1	0.5398	0.5438	0.5478	0.5517	0.5557	0.5596	0.5636	0.5675	0.5714	0.5753
0.2	0.5793	0.5832	0.5871	0.591	0.5948	0.5987	0.6026	0.6064	0.6103	0.6141
0.3	0.6179	0.6217	0.6255	0.6293	0.6331	0.6368	0.6406	0.6443	0.648	0.6517
0.4	0.6554	0.6591	0.6628	0.6664	0.6700	0.6736	0.6772	0.6808	0.6844	0.6879
0.5	0.6915	0.695	0.6985	0.7019	0.7054	0.7088	0.7123	0.7157	0.719	0.7224
0.6	0.7257	0.7291	0.7324	0.7357	0.7389	0.7422	0.7454	0.7486	0.7517	0.7549
0.7	0.758	0.7611	0.7642	0.7673	0.7704	0.7734	0.7764	0.7794	0.7823	0.7852
0.8	0.7881	0.791	0.7939	0.7967	0.7995	0.8023	0.8051	0.8078	0.8106	0.8133
0.9	0.8159	0.8186	0.8212	0.8238	0.8264	0.8289	0.8315	0.834	0.8365	0.8389
1	0.8413	0.8438	0.8461	0.8485	0.8508	0.8531	0.8554	0.8577	0.8599	0.8621
1.1	0.8643	0.8665	0.8686	0.8708	0.8729	0.8749	0.8770	0.8790	0.8810	0.8830
1.2	0.8849	0.8869	**0.8888**	0.8907	0.8925	0.8944	0.8962	0.898	0.8997	0.9015
1.3	0.9032	0.9049	0.9066	0.9082	0.9099	0.9115	0.9131	0.9147	0.9162	0.9177
1.4	0.9192	0.9207	0.9222	0.9236	0.9251	0.9265	0.9279	0.9292	0.9306	0.9319
1.5	0.9332	0.9345	0.9357	**0.9370**	0.9382	0.9394	0.9406	0.9418	0.9429	0.9441
1.6	0.9452	0.9463	0.9474	0.9484	0.9495	0.9505	0.9515	0.9525	0.9535	0.9545
1.7	0.9554	0.9564	0.9573	0.9582	0.9591	0.9599	0.9608	0.9616	0.9625	0.9633

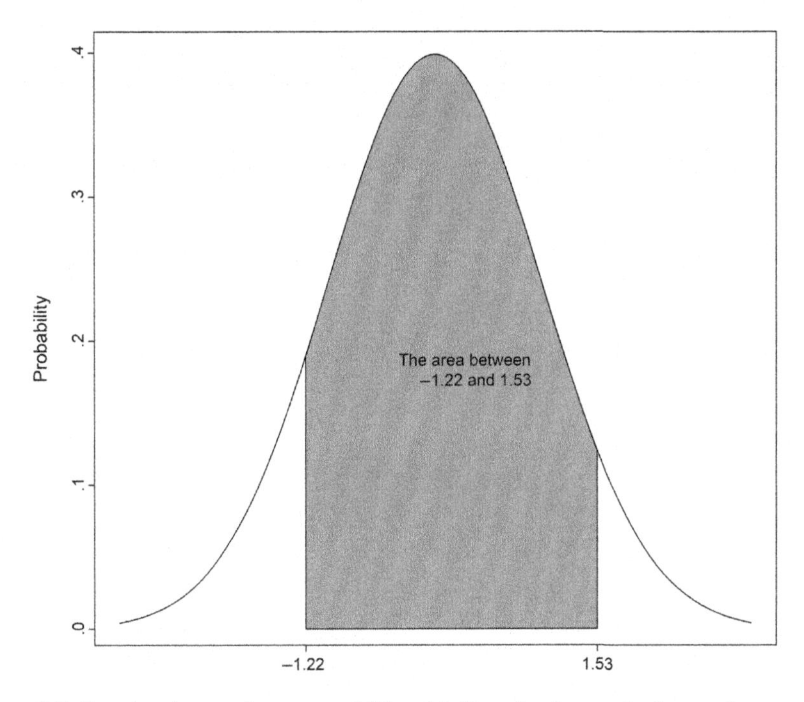

Figure 2.5 Showing the area between -1.22 and 1.53 under the standard normal curve.

Example 2.4: Compute the area on the two tails: Compute the area under the standard normal curve to the left of -1.22 and to the right of 1.53.

The area on the tails of the standard normal curve can be computed employing the procedure used in the previous examples. Thus the area under the standard normal curve to the left of -1.22 and to the right of 1.53 can be computed as shown below.

- The first step is to compute the area under the curve to the left of -1.22, which is equal to 0.1112. Table 2.3 for standard normal shows that the area to the left of 1.22 is 0.8888 (employing the concept of the total area under normal curve is 1), thus the area equals 1 minus the area to the left of 1.22, the area to the left of -1.22 is 0.1112 ($1-0.8888=0.1112$) (the area to the right of 1.22 is equal to the area to the left of -1.22), and the area to the right of 1.53 is 0.0630 ($1-0.9370=0.0630$) as shown in Table 2.4.
- Second, the total area is the sum of both regions ($0.1112+0.0630=0.1742$).

Table 2.4 The area to the left of -1.22 and to the right of 1.53.

Z	Second decimal in Z									
	0	0.01	0.02	0.03	0.04	0.05	0.06	0.07	0.08	0.09
0	0.5000	0.5040	0.5080	0.5120	0.5160	0.5199	0.5239	0.5279	0.5319	0.5359
0.1	0.5398	0.5438	0.5478	0.5517	0.5557	0.5596	0.5636	0.5675	0.5714	0.5753
0.2	0.5793	0.5832	0.5871	0.591	0.5948	0.5987	0.6026	0.6064	0.6103	0.6141
0.3	0.6179	0.6217	0.6255	0.6293	0.6331	0.6368	0.6406	0.6443	0.648	0.6517
0.4	0.6554	0.6591	0.6628	0.6664	0.6700	0.6736	0.6772	0.6808	0.6844	0.6879
0.5	0.6915	0.695	0.6985	0.7019	0.7054	0.7088	0.7123	0.7157	0.719	0.7224
0.6	0.7257	0.7291	0.7324	0.7357	0.7389	0.7422	0.7454	0.7486	0.7517	0.7549
0.7	0.758	0.7611	0.7642	0.7673	0.7704	0.7734	0.7764	0.7794	0.7823	0.7852
0.8	0.7881	0.791	0.7939	0.7967	0.7995	0.8023	0.8051	0.8078	0.8106	0.8133
0.9	0.8159	0.8186	0.8212	0.8238	0.8264	0.8289	0.8315	0.834	0.8365	0.8389
1	0.8413	0.8438	0.8461	0.8485	0.8508	0.8531	0.8554	0.8577	0.8599	0.8621
1.1	0.8643	0.8665	0.8686	0.8708	0.8729	0.8749	0.8770	0.8790	0.8810	0.8830
1.2	0.8849	0.8869	**0.8888**	0.8907	0.8925	0.8944	0.8962	0.898	0.8997	0.9015
1.3	0.9032	0.9049	0.9066	0.9082	0.9099	0.9115	0.9131	0.9147	0.9162	0.9177
1.4	0.9192	0.9207	0.9222	0.9236	0.9251	0.9265	0.9279	0.9292	0.9306	0.9319
1.5	0.9332	0.9345	0.9357	**0.9370**	0.9382	0.9394	0.9406	0.9418	0.9429	0.9441
1.6	0.9452	0.9463	0.9474	0.9484	0.9495	0.9505	0.9515	0.9525	0.9535	0.9545
1.7	0.9554	0.9564	0.9573	0.9582	0.9591	0.9599	0.9608	0.9616	0.9625	0.9633

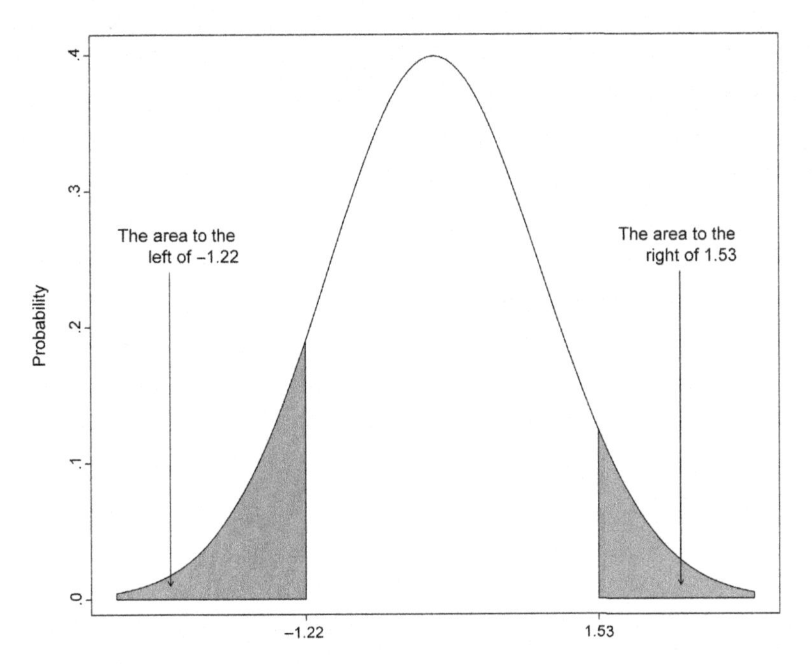

Figure 2.6 Showing the area to the right of 1.53 and to the left of -1.22 under the standard normal curve.

The area to the left of -1.22 and to the right of 1.53 is represented by the shaded area as shown in Fig. 2.6.

2.5 Hypothesis testing for one sample mean (Z-test)

Consider a large sample of size n ($n \geq 30$) that is selected from a normally distributed population and Y represents a random variable of interest. A claim regarding the mean value of the variable of interest can be tested employing Z-test for one sample mean to make a decision regarding the mean value. The mathematical formula for computing the test statistic value for one sample Z-test is presented in Eq. (2.4).

$$Z = \frac{\overline{Y} - \mu_0}{\sigma/\sqrt{n}} \tag{2.4}$$

where,
$\overline{Y}$ represents the sample mean,
μ_0 represents the claimed value (hypothesized mean),
σ represents the population standard deviation, and
n represents the sample size used.

The computed test statistic value obtained from the sample data (2.4) is usually used to make a decision to reject or not to reject the null hypothesis regarding the mean value of the variable of interest. The procedure of making a decision is to compare the test statistic value with the theoretical value (critical value) of the normal distribution or using a normal distribution curve.

Example 2.5: The concentration of cadmium of surface water: A professor at an environmental section wanted to verify the claim that the mean concentration of cadmium (Cd) of surface water in Juru River is 1.4 (mg/L). He selected 35 samples and tested for the cadmium concentration. The collected data showed that the mean concentration of cadmium is 1.6 and the standard deviation of the population is 0.4. A significance level of $\alpha = 0.01$ is chosen to test the claim. Assume that the population is normally distributed.

The general procedure for conducting hypothesis testing can be used to make the decision regarding the mean concentration of cadmium of surface water in Juru River.

Step 1: Specify the null and alternative hypotheses

The mean concentration of cadmium of surface water in Juru River (μ) is 1.4, this claim should be under the null hypothesis because the claim represents equality (=). If the mean concentration of cadmium of surface water in Juru River is not equal to 1.4, then two directions should be considered, the first direction could be the mean concentration of cadmium is greater than 1.4 and the second direction could be the mean concentration of cadmium is less than 1.4. The two directions (greater than and less than) can be represented mathematically as $\neq$. Thus we can write the two hypotheses (null and alternative) as presented in Eq. (2.5).

$$H_0: \mu = 1.4 \quad \text{vs} \quad H_1: \mu \neq 1.4 \tag{2.5}$$

We should make a decision regarding the null hypothesis whether the mean concentration of cadmium is exactly equal to 1.4, or the mean concentration of cadmium of surface water in Juru River differs (more or less) from 1.4.

Step 2: Select the significance level (α) for the study

The significance level of 0.01 ($\alpha = 0.01$) is selected to test the hypothesis. We divide the value of the significance level (0.01) by 2 to represent the two-tailed (more or less) test of the alternative hypothesis, thus $\frac{\alpha}{2} = \frac{0.01}{2} = 0.005$, which represents the rejection region in each tail of the standard normal curve (the left and right tails). The Z critical value for two-tailed test with $\alpha = 0.01$ is 2.58 as appeared in the standard normal table (Table A in the Appendix). Thus the Z critical values for both sides are ± 2.58, we use the two Z critical values to make a decision whether to reject or fail to reject the null hypothesis.

Step 3: Use the sample information to calculate the test statistic value

The entries for the Z-test statistic formula as presented in Eq. (2.4) are already provided. The claimed value for the mean concentration of cadmium (μ_0) is 1.4, the mean concentration of cadmium calculated from the sample data is 1.6, the population standard deviation for the concentration of cadmium is 0.4, and the sample size (n) is 35.

We apply the formula presented in Eq. (2.4) to test the hypothesis regarding the mean concentration of cadmium of surface water in Juru River.

$$Z = \frac{\overline{Y} - \mu_0}{\sigma/\sqrt{n}} = \frac{1.6 - 1.4}{0.4/\sqrt{35}}$$

$$Z = 2.95804 = 2.96$$

The test statistic value for the mean concentration of cadmium of surface water in Juru River is found to be 2.96.

Step 4: Identify the critical and noncritical regions for the study

We can easily identify the rejection and nonrejection regions for two-tailed test employing the Z critical values found in step 2. The rejection and nonrejection regions for the mean concentration of cadmium of surface water in Juru River are presented in Fig. 2.7 for the standard normal curve (shaded area).

Step 5: Make a decision and interpret the results

One can observe that the null hypothesis should be rejected because the test statistic value of the Z-test calculated from the sample data (2.96) is greater than the Z critical value for the two-tailed test with $\alpha = 0.01$ (2.58). Moreover, one can use a standard normal curve to decide, the same decision (reject the null hypothesis) can be reached using the normal curve and it can be seen that the test statistic value (2.96) falls in the rejection region on the right tail, as shown in Fig. 2.7. The null hypothesis is rejected in favor of the alternative hypothesis and it is believed that the mean concentration of cadmium of surface water in Juru River is not equal to 1.4.

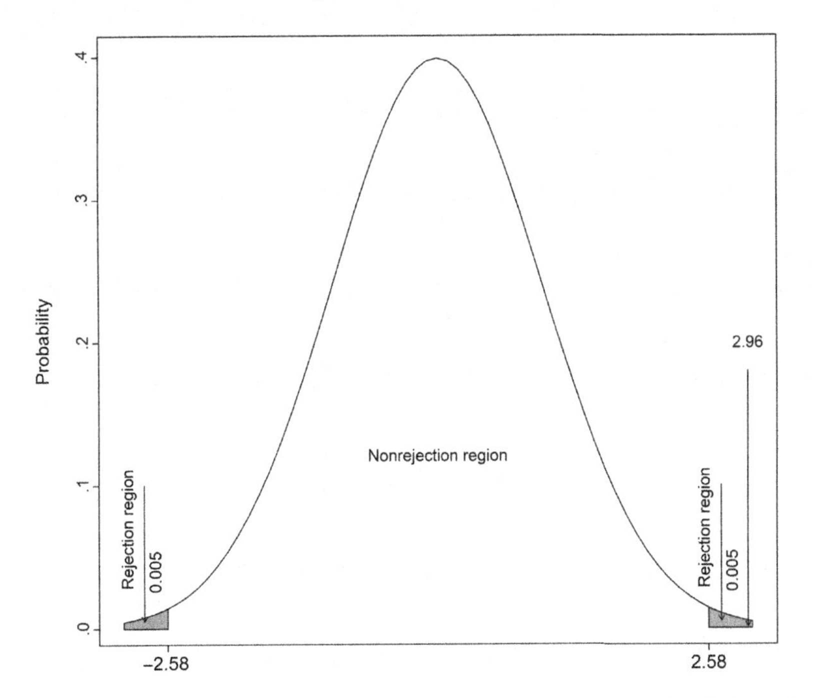

Figure 2.7 The rejection and nonrejection regions for the concentration of cadmium of surface water in Juru River.

We can conclude at a 1% significance level, that there is sufficient evidence provided by the collected data to believe that the mean concentration of cadmium of surface water in Juru River differs from 1.4.

Example 2.6: The concentration of lead in sediment: A researcher at a research institution wanted to verify the claim that the mean concentration of lead (Pb) in sediment of Jejawi River is less than 0.23 (mg/L). He selected 33 samples and tested for lead concentration. The collected data showed that the mean concentration of lead is 0.235 and the standard deviation of the population is 0.106. A significance level of $\alpha = 0.01$ is chosen to test the claim. Assume that the population is normally distributed.

The general procedure for conducting hypothesis testing can be used to make the decision regarding the mean concentration of lead in sediment of Jejawi River.

Step 1: Specify the null and alternative hypotheses

The mean concentration of lead in sediment of Jejawi River (μ) is less than 0.23, this claim should be under the alternative hypothesis because the claim represents only one direction which is less than 0.23. If the mean concentration of lead in sediment is not less than 0.23, then two directions should be considered, the first direction could be the mean concentration of lead is greater than 0.23 and the second direction could be the mean concentration of lead is equal to 0.23. The two directions (greater than and equal) can be represented mathematically as $\geq$. Thus we can write the two hypotheses (null and alternative) as presented in Eq. (2.6).

$$H_0: \mu \geq 0.23 \quad \text{vs} \quad H_1: \mu < 0.23 \tag{2.6}$$

We should make a decision regarding the null hypothesis whether the mean concentration of lead in sediment is greater than or equal to 0.23, or the mean concentration of lead in sediment is less than 0.23.

Step 2: Select the significance level (α) for the study

The significance level of 0.01 ($\alpha = 0.01$) is selected to test the hypothesis. Because the test is a one-tailed test as presented by the alternative hypothesis (less than), we should represent the rejection region on the left tail of the standard normal curve. The Z critical value for a one-tailed test with $\alpha = 0.01$ is 2.326 as appeared in the standard normal table (Table A in the Appendix). Thus, the Z critical value for the left-tailed test is -2.326, and we use the Z critical value to decide whether to reject or fail to reject the null hypothesis.

Step 3: Use the sample information to calculate the test statistic value

The entries for the Z-test statistic formula as presented in Eq. (2.4) are already provided. The claimed value for the mean concentration of lead (μ_0) is 0.23, the mean concentration of lead calculated from the sample data is 0.235, the population standard deviation for the concentration of lead is 0.106, and the sample size (n) is 33.

We apply the formula presented in Eq. (2.4) to test the hypothesis regarding the mean concentration of lead in sediment of Jejawi River.

$$Z = \frac{\overline{Y} - \mu_0}{\sigma/\sqrt{n}} = \frac{0.235 - 0.23}{0.106/\sqrt{33}}$$

$$Z = 0.2709699 = 0.27$$

The test statistic value for the mean concentration of lead in sediment of Jejawi River is found to be 0.27.

Step 4: Identify the critical and noncritical regions for the study

We can easily identify the rejection and nonrejection regions for a one-tailed test employing the Z critical value found in step 2. The rejection (on the left tail) and nonrejection regions for the mean concentration of lead in sediment of Jejawi River are presented in Fig. 2.8 for the standard normal curve (shaded area).

Step 5: Make a decision and interpret the results

One can observe that we fail to reject the null hypothesis because the test statistic value of the Z-test calculated from the sample data 0.27 is greater than the Z critical value for a one-tailed test with $\alpha = 0.01$ (-2.326). Moreover, one can use a standard normal curve to decide, the same decision (fail to reject the null hypothesis) can be reached using the normal curve and see that the test statistic value (0.27) falls in the nonrejection region as shown in Fig. 2.8. The null hypothesis is not rejected, and we should believe that the mean concentration of lead in sediment of Jejawi River is more than or equal to 0.23.

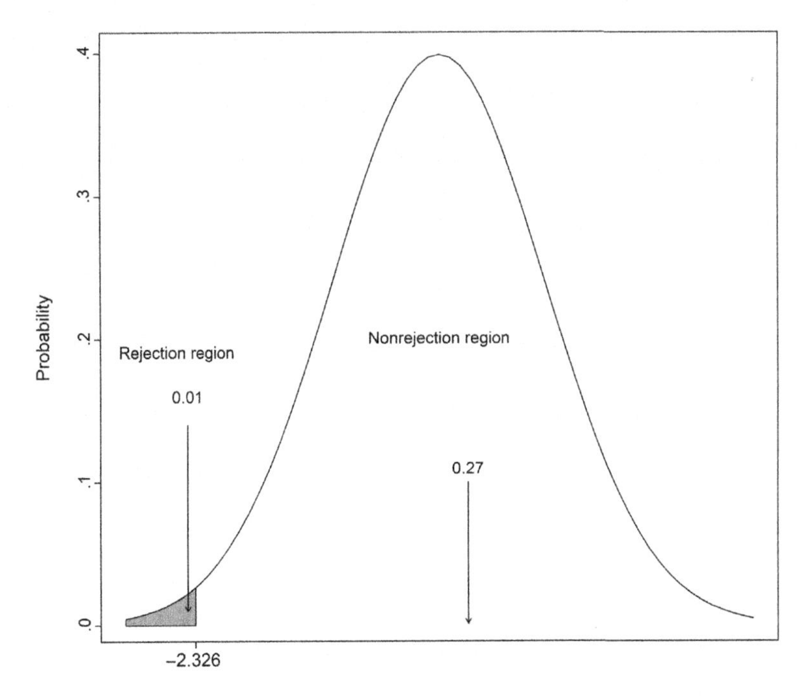

Figure 2.8 The rejection and nonrejection regions for the concentration of lead in sediment of Jejawi River.

We can conclude at a 1% significance level, that there is sufficient evidence provided by the collected data to believe that the mean concentration of lead in sediment of Jejawi River is more than or equal to 0.23.

Example 2.7: The concentration of sodium in particulate matter in the air: The concentration of sodium (Na) as an inorganic element in particulate matter (PM_{10}) in the air environment of an equatorial urban coastal location during the summer season was investigated. A researcher wants to verify that the concentration of sodium in the air during the summer season is less than or equal to 6.20. He selected 34 samples and the concentration of sodium was measured. The average concentration of sodium provided by the sample data is 7.46 ($\mu g\ m^{-3}$) and the population standard deviation is 2.18. A significance level of $\alpha = 0.01$ is chosen to test the claim. Assume that the population is normally distributed.

The general procedure for conducting hypothesis testing can be used to make the decision regarding the mean concentration of sodium in particulate matter in the air environment of an equatorial urban coastal location during the summer season.

Step 1: Specify the null and alternative hypotheses

The mean concentration of sodium in the air environment of an equatorial urban coastal location during the summer season (μ) is less than or equal to 6.20, this claim should be under the null hypothesis because the claim represents two directions and contains the equality sign. If the mean concentration of sodium in the air environment of an equatorial urban coastal location during the summer season is not less than or equal to 6.20, then the mean concentration of sodium is greater than 6.20, this direction (greater than) can be represented mathematically as $>$ and places it under the alternative hypothesis. Thus we can write the two hypotheses (null and alternative) as presented in Eq. (2.7).

$$H_0: \mu \leq 6.20 \quad \text{vs} \quad H_1: \mu > 6.20 \tag{2.7}$$

We should make a decision regarding the null hypothesis whether the mean concentration of sodium in the air is less than or equal to 6.20, or the mean concentration of sodium in the air is greater than 6.20.

Step 2: Select the significance level (α) for the study

The significance level of 0.01 ($\alpha = 0.01$) is selected to test the hypothesis. Because the test is a one-tailed test as presented by the alternative hypothesis (greater than), we should represent the rejection region on the right tail of the standard normal curve. The Z critical value for a one-tailed test with $\alpha = 0.01$ is 2.326 as appeared in the standard normal table (Table A in the Appendix). Thus the Z critical value for the right-tailed test is 2.326, and we use the Z critical value to decide whether to reject or fail to reject the null hypothesis.

Step 3: Use the sample information to calculate the test statistic value

The entries for the Z-test statistic formula as presented in Eq. (2.4) are already provided. The claimed value for the mean concentration of sodium (μ_0) is 6.20, the mean concentration of sodium calculated from the sample data is 7.46, the population standard deviation for the concentration of sodium is 2.18, and the sample size (n) is 34.

We apply the formula presented in Eq. (2.4) to test the hypothesis regarding the mean concentration of sodium in the air environment of an equatorial urban coastal location during the summer season.

$$Z = \frac{\overline{Y} - \mu_0}{\sigma/\sqrt{n}} = \frac{7.46 - 6.20}{2.18/\sqrt{34}}$$

$$Z = 3.370183 = 3.37$$

The test statistic value for the mean concentration of sodium in the air environment is found to be 3.37.

Step 4: Identify the critical and noncritical regions for the study

We can easily identify the rejection and nonrejection regions for a one-tailed test employing the Z critical value found in step 2. The rejection (on the right tail) and nonrejection regions for the mean concentration of sodium in the air environment are presented in Fig. 2.9 for the standard normal curve (shaded area).

Step 5: Make a decision and interpret the results

One can observe that we should reject the null hypothesis because the test statistic value of the Z-test calculated from the sample data (3.37) is greater than the Z critical value for a one-tailed test with $\alpha = 0.01$ (2.326). Moreover, one can use a standard normal curve to decide, the same decision (reject the null hypothesis) can be reached using the normal curve and we can see that the test statistic value (3.37) falls in the rejection region as shown in Fig. 2.9. The null hypothesis is rejected and we should believe that the mean concentration of sodium in the air environment is more than the claimed value of 6.20.

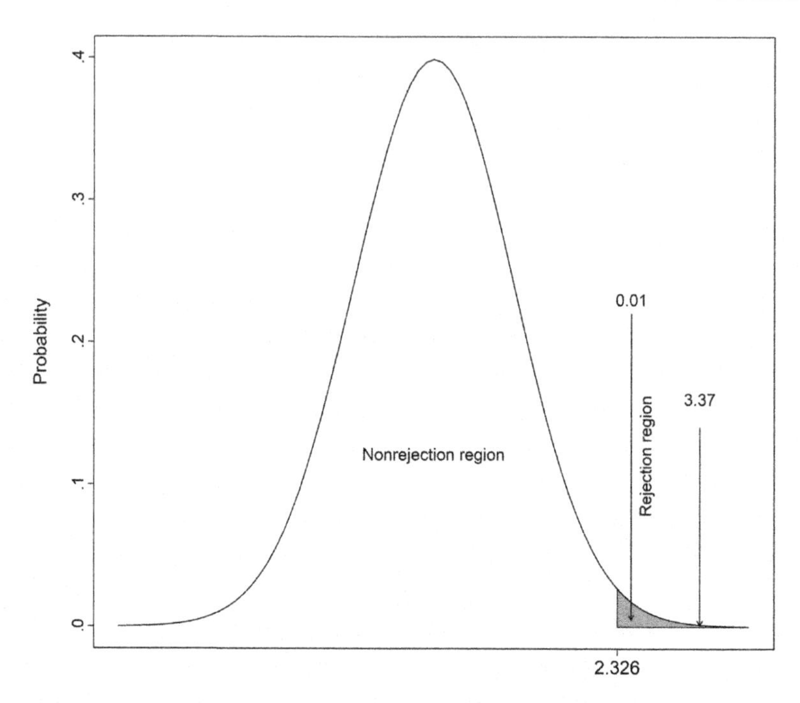

Figure 2.9 The rejection and nonrejection regions for the concentration of sodium (Na) in the air environment.

We can conclude at a 1% significance level, that there is sufficient evidence provided by the collected data to believe that the mean concentration of sodium in the air environment is more than 6.20.

Further reading

Alkarkhi, A. F. M., Ahmad, A., & Easa, A. M. (2009a). Assessment of surface water quality of selected estuaries of Malaysia: multivariate statistical techniques. *Environmentalist, 29*, 255–262.

Alkarkhi, A. F. M., & Alqaraghuli, W. A. A. (2020). *Applied statistics for environmental science with R* (1st ed.). Elsevier.

Alkarkhi, A. F. M., & Chin, L. H. (2012). *Elementary statistics for technologist* (1st ed.). Malaysia: Universiti Sains Malaysia.

Alkarkhi, A. F. M., Ismail, N., Ahmed, A., & Easa, A. M. (2009b). Analysis of heavy metal concentrations in sediments of selected estuaries of Malaysia—a statistical assessment. *Environ Monit Assess, 153*, 179–185.

Bluman, A. G. (1998). *Elementary statistics: a step by step approach* (3rd ed.). Boston: WCB, McGraw-Hill.

Weiss, N. A. (2012). *Introductory statistics* (9th ed.). Pearson.

Yusup, Y., & Alkarkhi, A. F. M. (2011). Cluster analysis of inorganic elements in particulate matter in the air environment of an equatorial urban coastal location. *Chemistry and Ecology, 27*(3), 273–286.

t-test for one-sample mean

3

Abstract

This chapter presents hypothesis testing based on t distribution including the concept of t distribution and related terms. The area under the t distribution curve with the concept of rejection and nonrejection regions are delivered with illustration by examples for each region using a t distribution curve. Moreover, the general procedure for hypothesis testing using a t-test for a small sample is given and explained clearly. Examples from the field of environmental science are selected and used to illustrate the steps of hypothesis testing using the t-test and making a decision regarding the study with sufficient explanations for each step.

Keywords: t Distribution; rejection and nonrejection regions; critical value; area under the t curve; t-test

Learning outcomes

After completing this chapter, readers will be able to:

- Understand the importance of t distribution
- Explain how to obtain the t critical values
- Compute the area under a t distribution curve
- Describe the common steps for testing a hypothesis about the average value when a small sample size is used
- Compute the t-test statistic value for a single population mean
- Identify the rejection and nonrejection regions related to a t-test
- Know the procedure for interpreting the results and match it to the field of study
- Describe one-sided and two-sided tests related to a t-test
- Write smart conclusions regarding the problem under study

3.1 Introduction

Hypothesis testing for one population mean is discussed using the Z-test for large samples or the variance of the population is given. Z-test cannot be employed when the sample size is small and the variance is not given. Thus the t-test is the correct technique that should be employed to test a hypothesis for one population mean when the sample size is small. The t-test depends on a new distribution called t

Applications of Hypothesis Testing for Environmental Science. DOI: https://doi.org/10.1016/B978-0-12-824301-5.00011-3

distribution; the concept and properties of t distribution are covered before giving the t-test for one population mean.

3.2 What is t distribution?

t Distribution, also known as Student's t distribution, is a continuous probability distribution that is close to normal distribution in the shape and some properties. t Distribution is usually used to test population means when the sample size is small. t Distribution depends on the concept of degrees of freedom. The properties of t distribution are:

1. t Distribution has a bell shape and is symmetric about the mean which is equal to 0;
2. The variance of t distribution is always greater than 1;
3. t Distribution becomes closer to standard normal distribution when the sample size increases;
4. The variable of t distribution takes values within the range from $-\infty$ to ∞ (infinity);
5. The number of degrees of freedom (*d.f*) affects the shape of the t distribution as is shown clearly in Fig. 3.1 for t distribution curves for various degrees of freedom.

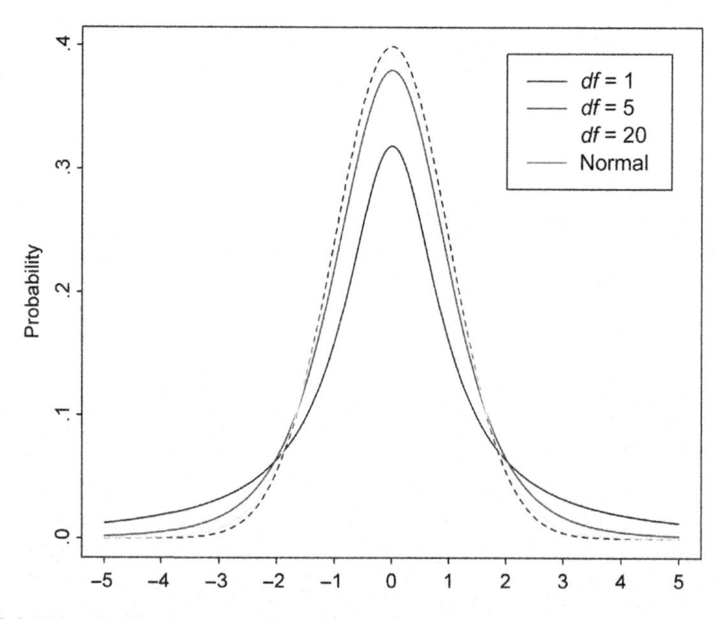

Figure 3.1 The t distribution curve for $d.f = 1$, $d.f = 5$, $d.f = 20$, and normal.

3.3 Finding the t critical values

Critical values are usually used to make a decision regarding a hypothesis. Getting t critical values requires providing information about the degrees of freedom (*d.f*) and the level of significance (α) then using the t table for critical values. The t critical values are provided in Table B in the Appendix. The first column on the left represents the degrees of freedom (*d.f*) from 1 to ∞, while the first upper row represents the level of significance (α).

Example 3.1: Finding the critical value for a right-tailed t-test: Use a significance level of 0.01 ($\alpha = 0.01$) and degrees of freedom of 11 ($d.f = 11$) to obtain the t critical value for the right-tailed test.

We can obtain the t critical value for a right-tailed test as long as two values are provided, the two values are the degrees of freedom "$d.f = 11$" and significance level "$\alpha = 0.01$."

- The first step in extracting the t critical value is to specify the location of "$d.f = 11$" in the first column of Table 3.1, labeled *d.f* (highlighted row 11). Table 3.1 is a portion of Table B in the Appendix.
- The second step is to specify the location of $\alpha = 0.01$ in the first upper row and then move on the column (highlighted column 0.01) to the row labeled $d.f = 11$; the value that represents the point of intersection between $d.f = 11$ and $\alpha = 0.01$ is the t critical value. One can observe that the t critical value is **2.718** as shown in Table 3.1 (bold value).

Table 3.1 The t critical value for a right-tailed test.

α d.f	0.25	0.10	0.05	0.025	0.01	0.005	0.001
1	1.000	3.078	6.314	12.706	31.821	63.657	318.310
2	0.816	1.886	2.920	4.303	6.965	9.925	22.326
3	0.765	1.638	2.353	3.182	4.541	5.841	10.213
4	0.741	1.533	2.132	2.776	3.747	4.604	7.173
5	0.727	1.476	2.015	2.571	3.365	4.032	5.893
6	0.718	1.440	1.943	2.447	3.143	3.707	5.208
7	0.711	1.415	1.895	2.365	2.998	3.499	4.785
8	0.706	1.397	1.860	2.306	2.896	3.355	4.501
9	0.703	1.383	1.833	2.262	2.821	3.250	4.297
10	0.700	1.372	1.812	2.228	2.764	3.169	4.144
11	0.697	1.363	1.796	2.201	**2.718**	3.106	4.025
12	0.695	1.356	1.782	2.179	2.681	3.055	3.930
13	0.694	1.350	1.771	2.160	2.650	3.012	3.852
14	0.692	1.345	1.761	2.145	2.624	2.977	3.787
15	0.691	1.341	1.753	2.131	2.602	2.947	3.733
16	0.690	1.337	1.746	2.120	2.583	2.921	3.686
17	0.689	1.333	1.740	2.110	2.567	2.898	3.646

The t critical value for right-tailed $t_{(\alpha,d.f)} = t_{(0.01,11)} = 2.718$ is shown in Fig. 3.2.

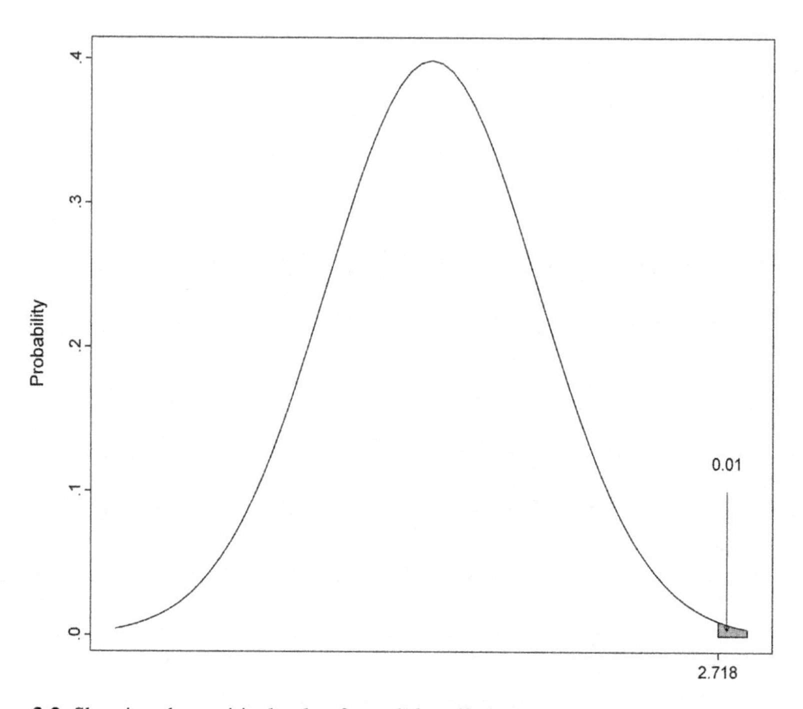

Figure 3.2 Showing the t critical value for a right-tailed test.

Example 3.2: Finding the critical value for a left-tailed t-test: Use a significance level of 0.01 ($\alpha = 0.01$) and degrees of freedom of 11 ($d.f = 11$) to obtain the t critical value for the left-tailed test. We can obtain the t critical value for the left-tailed test as long as the two values are already provided, the degrees of freedom "$d.f = 11$" and significance level "$\alpha = 0.01$."

- The first step in extracting the t critical value is to specify the location of "$d.f = 11$" in the first column of Table 3.2, labeled *d.f* (highlighted row 11). Table 3.2 is a portion of Table B in the Appendix.
- The second step is to specify the location of $\alpha = 0.01$ in the first upper row and then move on the column (highlighted column 0.01) to the row labeled $d.f = 11$; the value that represents the point of intersection between $d.f = 11$ and $\alpha = 0.01$ is the t critical value. One can observe that the t critical value is **2.718** as shown in Table 3.2 (bold value).

Table 3.2 The t critical value for a left-tailed test.

α d.f	0.25	0.10	0.05	0.025	0.01	0.005	0.001
1	1.000	3.078	6.314	12.706	31.821	63.657	318.310
2	0.816	1.886	2.920	4.303	6.965	9.925	22.326
3	0.765	1.638	2.353	3.182	4.541	5.841	10.213
4	0.741	1.533	2.132	2.776	3.747	4.604	7.173
5	0.727	1.476	2.015	2.571	3.365	4.032	5.893
6	0.718	1.440	1.943	2.447	3.143	3.707	5.208
7	0.711	1.415	1.895	2.365	2.998	3.499	4.785
8	0.706	1.397	1.860	2.306	2.896	3.355	4.501
9	0.703	1.383	1.833	2.262	2.821	3.250	4.297
10	0.700	1.372	1.812	2.228	2.764	3.169	4.144
11	0.697	1.363	1.796	2.201	**2.718**	3.106	4.025
12	0.695	1.356	1.782	2.179	2.681	3.055	3.930
13	0.694	1.350	1.771	2.160	2.650	3.012	3.852
14	0.692	1.345	1.761	2.145	2.624	2.977	3.787
15	0.691	1.341	1.753	2.131	2.602	2.947	3.733
16	0.690	1.337	1.746	2.120	2.583	2.921	3.686
17	0.689	1.333	1.740	2.110	2.567	2.898	3.646

The test in this case is a left-tailed problem; thus, the t critical value for the left-tailed test is $t_{(\alpha,d.f)} = t_{(0.01,11)} = -2.718$.

The t critical value for left-tailed $t_{(\alpha,d.f)} = t_{(0.01,11)} = -2.718$ is shown in Fig. 3.3.

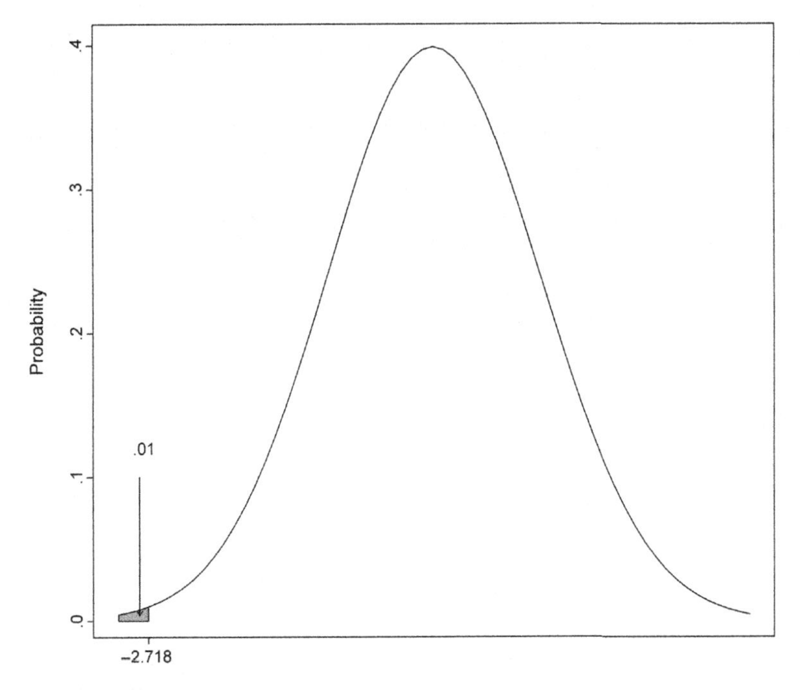

Figure 3.3 Showing the t critical value for a left-tailed test.

Example 3.3: Finding the critical value for a two-tailed t-test: Use a significance level of 0.01 ($\alpha = 0.01$) and degrees of freedom of 11 ($d.f = 11$) to obtain the t critical value for the two-tailed test.

We can obtain the t critical value for the two-tailed test as long as the two values are provided, the degrees of freedom "$d.f = 11$" and significance level "$\alpha = 0.01$."

- The first step in extracting the t critical value is to specify the location of "$d.f = 11$" in the first column of Table 3.3 labeled *d.f* (highlighted row 11). Table 3.3 is a portion of Table B in the Appendix.
- The second step is to specify the location of $\frac{\alpha}{2} = 0.005$ (the significance level is divided by 2 because the test is a two-sided test) in the first upper row and then move on the column (highlighted column 0.005) to the row labeled $d.f = 11$; the value that represents the point of intersection between $d.f = 11$ and $\alpha = 0.005$ is the t critical value. One can observe that the t critical value is **3.106** as shown in Table 3.3 (bold value).

Table 3.3 The t critical value for a two-tailed test.

α d.f	**0.25**	**0.10**	**0.05**	**0.025**	**0.01**	**0.005**	**0.001**
1	1.000	3.078	6.314	12.706	31.821	63.657	318.310
2	0.816	1.886	2.920	4.303	6.965	9.925	22.326
3	0.765	1.638	2.353	3.182	4.541	5.841	10.213
4	0.741	1.533	2.132	2.776	3.747	4.604	7.173
5	0.727	1.476	2.015	2.571	3.365	4.032	5.893
6	0.718	1.440	1.943	2.447	3.143	3.707	5.208
7	0.711	1.415	1.895	2.365	2.998	3.499	4.785
8	0.706	1.397	1.860	2.306	2.896	3.355	4.501
9	0.703	1.383	1.833	2.262	2.821	3.250	4.297
10	0.700	1.372	1.812	2.228	2.764	3.169	4.144
11	0.697	1.363	1.796	2.201	2.718	**3.106**	4.025
12	0.695	1.356	1.782	2.179	2.681	3.055	3.930
13	0.694	1.350	1.771	2.160	2.650	3.012	3.852
14	0.692	1.345	1.761	2.145	2.624	2.977	3.787
15	0.691	1.341	1.753	**2.131**	2.602	2.947	3.733
16	0.690	1.337	1.746	2.120	2.583	2.921	3.686
17	0.689	1.333	1.740	2.110	2.567	2.898	3.646

The test in this case is a two-tailed problem; the t critical values for a two-tailed test are $t_{(\alpha,d.f)} = t_{(0.005,11)} = \mp 3.106$.

The t critical values for two-tailed $t_{(\alpha,d.f)} = t_{(0.005,11)} = \mp 3.106$ are shown in Fig. 3.4.

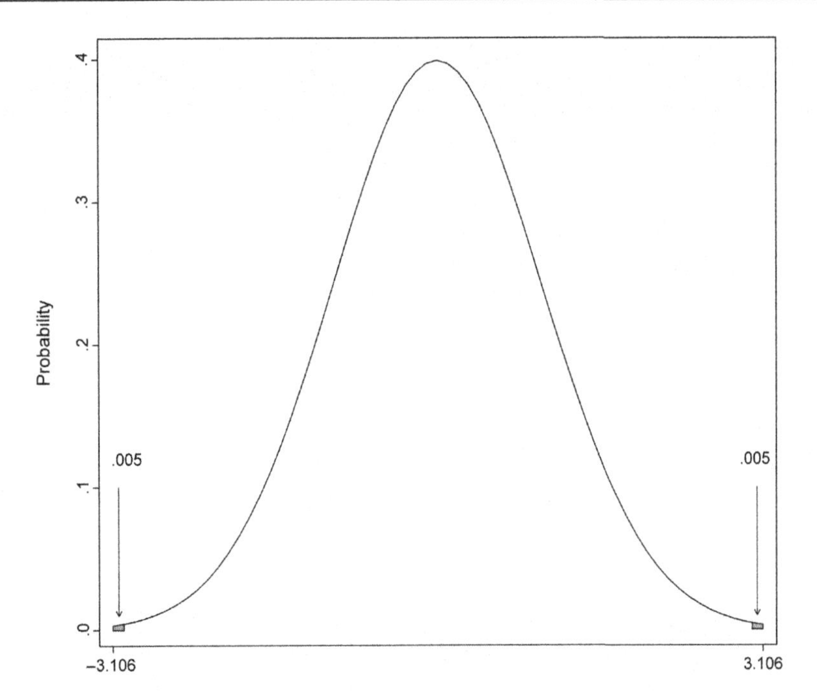

Figure 3.4 Showing the t critical value for a two-tailed test.

3.4 Hypothesis testing for a one-sample mean (t-test)

Consider a sample of size n that is selected from a normally distributed population and Y represents a random variable of interest. A claim regarding the mean value of the variable of interest when the variance of the population (σ^2) is unknown and the sample size is small ($n < 30$) can be tested employing a t-test for one sample mean to make a decision regarding the mean value. The mathematical formula for computing the test statistic value for one sample mean when the variance of the population (σ^2) is unknown and the sample size is small ($n < 30$) employing the t-test is presented in Eq. (3.1).

$$t = \frac{\overline{Y} - \mu_0}{S/\sqrt{n}} \tag{3.1}$$

Follow t distribution with ($n - 1$) degrees of freedom, where $\overline{Y}$ represents the sample mean, μ_0 represents the claimed value (hypothesized mean), S represents the sample standard deviation, and n represents the sample size used.

The computed test statistic value obtained from the sample data (3.1) is used to make a decision to reject or fail to reject the null hypothesis regarding the mean value of the variable of interest. The procedure to make a decision is to compare the test statistic value with the theoretical value (critical value) of the t distribution or using t distribution curve.

Example 3.4: The concentration of trace metal iron in the air: A researcher at an environmental section wishes to verify the claim that the mean concentration of trace metal iron (Fe) on coarse particulate matter (PM_{10}) in the air is 0.36 ($\mu g/m^3$) during the summer season of an equatorial urban coastal location. Twelve samples were selected, and the concentration of iron was measured. The collected data showed that the mean concentration of iron is 0.44 and the standard deviation is 0.34. A significance level of $\alpha = 0.01$ is chosen to test the claim. Assume that the population is normally distributed.

The general procedure for conducting hypothesis testing can be used to make the decision regarding the mean concentration of iron in the air. Because the sample size is small $n = 12$, a t-test should be employed to test the hypothesis regarding the mean concentration of iron in the air.

Step 1: Specify the null and alternative hypotheses

The mean concentration of iron in the air (μ) is 0.36, this claim should be under the null hypothesis because the claim represents equality (=). If the mean concentration of iron in the air is not equal to 0.36, then two cases should be considered, the first case could be the mean concentration of iron in the air is greater than 0.36 and the second case could be the mean concentration of iron in the air is less than 0.36. The two cases (greater than and less than) can be represented mathematically as $\neq$. Thus we can write the two hypotheses (null and alternative) as presented in Eq. (3.2).

$$H_0{:}\mu = 0.36 \text{ vs } H_1{:}\mu \neq 0.36 \tag{3.2}$$

We should make a decision regarding the null hypothesis whether the mean concentration of iron in the air is exactly equal to 0.36, or the mean concentration of iron in the air differs (more or less) from 0.36.

Step 2: Select the significance level (α) for the study

The significance level of 0.01 ($\alpha = 0.01$) is selected to test the hypothesis. We divide the value of the significance level (0.01) by 2 to represent the two-tailed test (more or less) of the alternative hypothesis, thus $\frac{\alpha}{2} = \frac{0.01}{2} = 0.005$ which represents the rejection region in each tail of the t distribution curve (the left and right tails). The t critical value for the two-tailed test with $\alpha = 0.01$ and $d.f = 11$ is 3.106 as appeared in the t table (Table B in the Appendix). Thus the t critical values for both sides are ∓ 3.106, we use the two t critical values to make a decision whether to reject or fail to reject the null hypothesis.

Step 3: Use the sample information to calculate the test statistic value

The entries for the t-test statistic formula as presented in Eq. (3.1) are already provided. The claimed value for the mean concentration of iron (μ_0) is 0.36, the mean concentration of iron calculated from the sample data is 0.44, the standard deviation for the concentration of iron is 0.34 and the sample size (n) is 12.

We apply the formula presented in Eq. (3.1) to test the hypothesis regarding the mean concentration of iron in the air.

$$t = \frac{\overline{Y} - \mu_0}{S/\sqrt{n}} = \frac{0.44 - 0.36}{0.34/\sqrt{12}}$$

$$t = 0.8150827 = 0.82$$

The test statistic value for the mean concentration of iron in the air is found to be 0.82.

Step 4: Identify the critical and noncritical regions for the study

We can easily identify the rejection and nonrejection regions for a two-tailed test employing the t critical values found in step 2. The rejection and nonrejection regions for the mean concentration of iron in the air are presented in Fig. 3.5 for the t distribution curve (shaded area).

Step 5: Make a decision and interpret the results

One can observe that the null hypothesis should not be rejected because the test statistic value of the t-test calculated from the sample data is 0.82, which is less than the t critical value for the two-tailed test (3.106). Moreover, one can use a t distribution curve to decide, the same decision (fail to reject the null hypothesis) can be reached using the t distribution curve and it can be seen that the test statistic value falls in the nonrejection region as shown in Fig. 3.5. The null hypothesis is not rejected and we believe that the mean concentration of iron in the air is equal to 0.36.

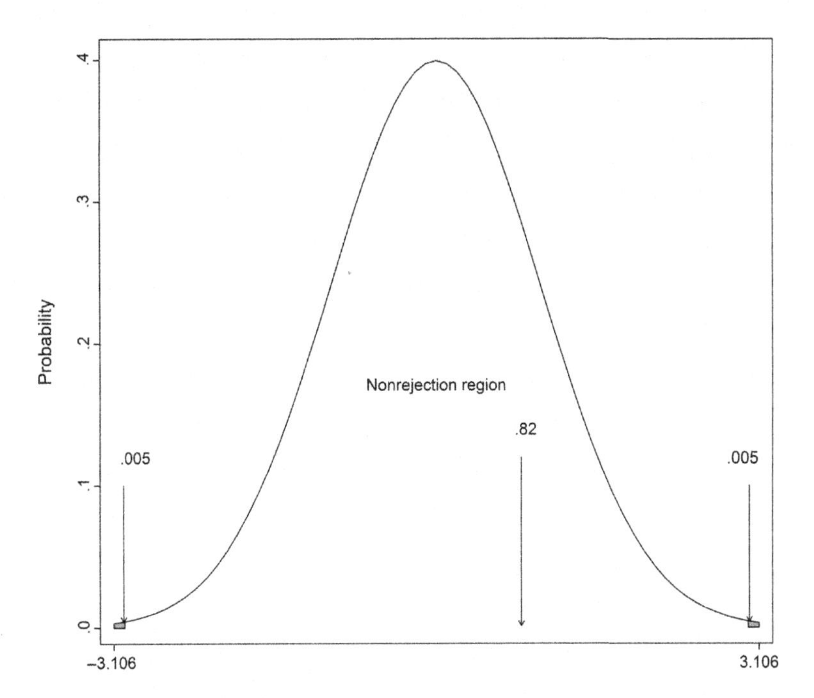

Figure 3.5 The rejection and nonrejection regions for iron in the air.

We can conclude at a 1% significance level that there is sufficient evidence provided by the collected data to believe that the mean concentration of iron in the air is 0.36.

Example 3.5: The concentration of chemical oxygen demand of surface water: An environmentalist at a research center wishes to verify the claim that the mean concentration of the chemical oxygen demand (COD) in surface water of Juru River is less than 650 (mg/L). Fourteen samples were selected and tested for chemical oxygen demand concentration. The collected data showed that the mean concentration of chemical oxygen demand is 470 and the standard deviation is 245. A significance level of $\alpha = 0.01$ is chosen to test the claim. Assume that the population is normally distributed.

The general procedure for conducting hypothesis testing can be used to make the decision regarding the mean concentration of chemical oxygen demand in surface water of Juru River.

Step 1: Specify the null and alternative hypotheses

The mean concentration of chemical oxygen demand in surface water of Juru River (μ) is less than 650, this claim should be under the alternative hypothesis because the claim represents only one side which is less than ($<$) 650. If the mean concentration of chemical oxygen demand in surface water of Juru River is not less than 650, then two cases should be considered, the first case could be the mean concentration of chemical oxygen demand being greater than 650 and the second case could be the mean concentration of chemical oxygen demand being equal to 650. The two cases (greater than and equal to) can be represented mathematically as $\geq$. Thus we can write the two hypotheses (null and alternative) as presented in Eq. (3.3).

$$H_0{:}\mu \geq 650 \text{ vs } H_1{:}\mu < 650 \tag{3.3}$$

We should make a decision regarding the null hypothesis whether the mean concentration of chemical oxygen demand is greater than or equal to 650, or the mean concentration of chemical oxygen demand is less than 650.

Step 2: Select the significance level (α) for the study

The significance level of 0.01 ($\alpha = 0.01$) is selected to test the hypothesis. Because the test is a one-tailed test as presented by the alternative hypothesis (less than), we should represent the rejection region on the left tail of the t distribution curve. The t critical value for a one-tailed test with $\alpha = 0.01$ and $d.f = 13$ is 2.65 as appeared in the t table for critical values (Table B in the Appendix). Thus, the t critical value for a left-tailed test is -2.65, we use the t critical value to decide whether to reject or fail to reject the null hypothesis.

Step 3: Use the sample information to calculate the test statistic value

The entries for the t-test statistic formula as presented in Eq. (3.1) are already provided. The claimed value for the mean concentration of chemical oxygen demand (μ_0) is 650, the mean concentration of chemical oxygen demand calculated from the sample data is 470, the standard deviation for the concentration of chemical oxygen demand is 245, and the sample size (n) is 14.

We apply the formula presented in Eq. (3.1) to test the hypothesis regarding the mean concentration of chemical oxygen demand in surface water of Juru River.

$$t = \frac{\overline{Y} - \mu_0}{S/\sqrt{n}} = \frac{470 - 650}{245/\sqrt{14}}$$

$$t = -2.748973 = -2.75$$

The test statistic value for the mean concentration of chemical oxygen demand in surface water of Juru River is found to be −2.75.

Step 4: Identify the critical and noncritical regions for the study

We can easily identify the rejection and nonrejection regions for a left-tailed test employing the t critical value found in step 2. The rejection and nonrejection regions for the mean concentration of chemical oxygen demand in surface water of Juru River are presented in Fig. 3.6 for the t distribution curve (shaded area).

Step 5: Make a decision and interpret the results

One can observe that the null hypothesis should be rejected because the test statistic value of the t-test calculated from the sample data is −2.75, which is smaller than the t critical value for the left-tailed test (−2.65). Moreover, one can use a t distribution curve to decide, the same decision (reject the null hypothesis) can be reached using the t distribution curve and it can be seen that the test statistic value falls in the rejection region as shown in Fig. 3.6. The null hypothesis is rejected and we believe that the mean concentration of chemical oxygen demand in surface water of Juru River is less than 650.

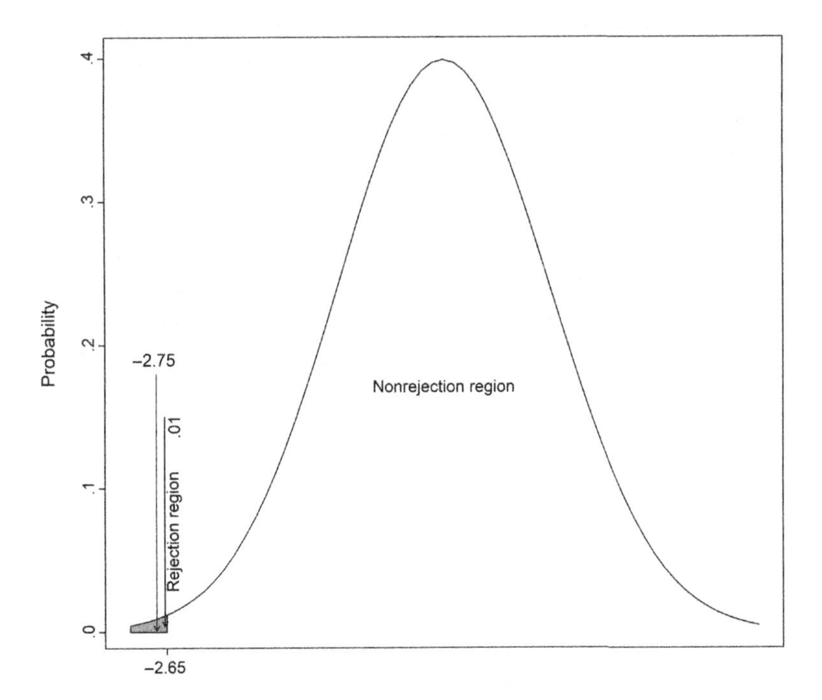

Figure 3.6 The rejection and nonrejection regions for the mean concentration of chemical oxygen demand of surface water of Juru River.

We can conclude at a 1% significance level that there is sufficient evidence provided by the collected data to believe that the mean concentration of chemical oxygen demand of surface water in Juru River is less than 650.

Example 3.6: The concentration of total phosphate in surface water: The concentration of total phosphate of surface water of Juru River was studied. An environmentalist wishes to verify that the concentration of total phosphate of surface water of Juru River is less than or equal to 0.85 (mg/L). Thirteen samples were selected and the concentration of total phosphate of surface water of Juru River was measured. The average concentration of total phosphate provided by the sample data is 0.97 and the standard deviation is 0.38. A significance level of $\alpha = 0.01$ is chosen to test the claim. Assume that the population is normally distributed.

The general procedure for conducting hypothesis testing can be used to make the decision regarding the mean concentration of total phosphate in surface water of Juru River.

Step 1: Specify the null and alternative hypotheses

The mean concentration of total phosphate in surface water of Juru River (μ) is less than or equal to 0.85, this claim should be under the null hypothesis because the claim represents two directions and contains the equality sign ($\leq$). If the mean concentration of total phosphate in surface water of Juru River is not less than or equal to 0.85, then the mean concentration of total phosphate is greater than 0.85, this direction (greater than) can be represented mathematically as $>$ and places it under the alternative hypothesis. Thus we can write the two hypotheses (null and alternative) as presented in Eq. (3.4).

$$H_0: \mu \leq 0.85 \text{ vs } H_1: \mu > 0.85 \tag{3.4}$$

We should make a decision regarding the null hypothesis whether the mean concentration of total phosphate is less than or equal 0.85, or the mean concentration of total phosphate is greater than 0.85.

Step 2: Select the significance level (α) for the study

The significance level of 0.01 ($\alpha = 0.01$) is selected to test the hypothesis. Because the test is a one-tailed test as presented by the alternative hypothesis (greater than), we should represent the rejection region on the right tail of the t distribution curve. The t critical value for a one-tailed test with $\alpha = 0.01$ and $d.f = 12$ is 2.681 as appeared in the t distribution table for critical values (Table B in the Appendix). Thus the t critical value for a right-tailed test is 2.681, we use the t critical value to decide whether to reject or fail to reject the null hypothesis.

Step 3: Use the sample information to calculate the test statistic value

The entries for the t-test statistic formula as presented in Eq. (3.1) are already provided. The claimed value for the mean concentration of total phosphate in surface water of Juru River (μ_0) is 0.85, the mean concentration of total phosphate calculated from the sample data is 0.97, the standard deviation for the total phosphate is 0.38, and the sample size (n) is 13.

We apply the formula presented in Eq. (3.1) to test the hypothesis regarding the mean concentration of total phosphate in surface water of Juru River.

$$t = \frac{\overline{Y} - \mu_0}{S/\sqrt{n}} = \frac{0.97 - 0.85}{0.38/\sqrt{13}}$$

$$t = 1.138595 = 1.14$$

The test statistic value for the mean concentration of total phosphate in surface water of Juru River is found to be 1.14.

Step 4: Identify the critical and noncritical regions for the study

We can easily identify the rejection and nonrejection regions for a right-tailed test employing the t critical value found in step 2. The rejection and nonrejection regions for the mean concentration of total phosphate in surface water of Juru River are presented in Fig. 3.7 for the t distribution curve (shaded area).

Step 5: Make a decision and interpret the results

One can observe that the null hypothesis should not be rejected because the test statistic value of the t-test calculated from the sample data is 1.14, which is less than the t critical value for a one-tailed test (2.681). Moreover, one can use the t distribution curve to decide, and the same decision (fail to reject the null hypothesis) can be reached using the t distribution curve and it can be seen that the test statistic value falls in the nonrejection region as shown in Fig. 3.7. The null hypothesis is not rejected and we believe that the mean concentration of total phosphate in surface water of Juru River is less than or equal to 0.85.

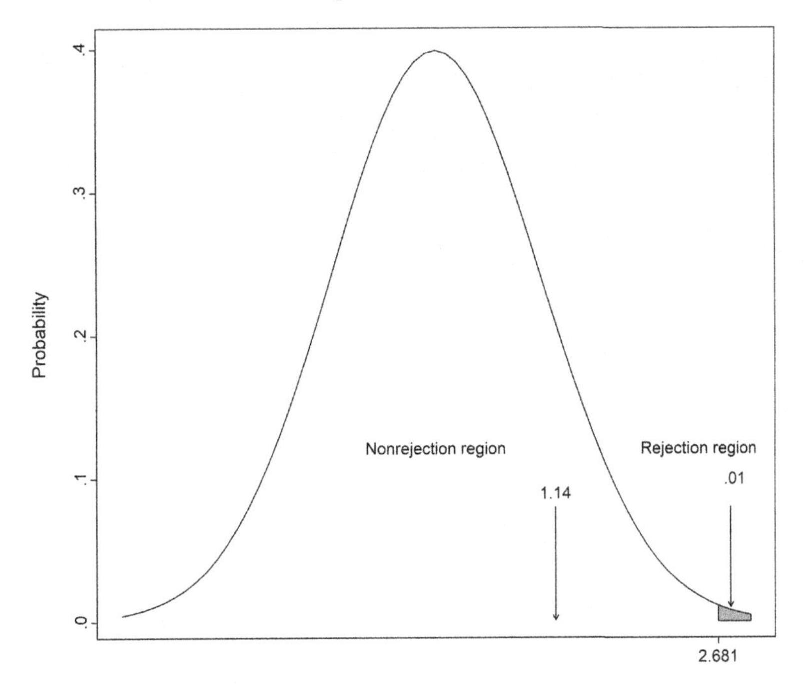

Figure 3.7 The rejection and nonrejection regions for the total phosphate in the water of Juru River.

We can conclude at a 1% significance level that there is sufficient evidence provided by the collected data to believe that the mean concentration of total phosphate in surface water of Juru River is less than or equal to 0.85.

Further reading

Alkarkhi, A. F. M., Ahmad, A., & Easa, A. M. (2009). Assessment of surface water quality of selected estuaries of Malaysia: multivariate statistical techniques. *Environmentalist, 29*, 255–262.

Alkarkhi, A. F. M., & Chin, L. H. (2012). *Elementary statistics for technologist* (1st ed.). Malaysia: Universiti Sains Malaysia.

Bluman, A. G. (1998). Elementary statistics: A Step by Step *Approach* (3rd ed.). Boston: WCB, McGraw-Hill.

Weiss, N. A. (2012). *Introductory statistics* (9th ed.). Pearson.

Yusup, Y., & Alkarkhi, A. F. M. (2011). Cluster analysis of inorganic elements in particulate matter in the air environment of an equatorial urban coastal location. *Chemistry and Ecology*, *27*(3), 273–286.

Z-test for one sample proportion

4

Abstract

This chapter addresses hypothesis testing based on binomial distribution including the concept of Bernoulli and binomial distributions and related terms. Moreover, the concepts of proportion, success, and failure are described in a simple way with examples. The area under the standard normal distribution curve and the concept of rejection and nonrejection regions are delivered with illustration by examples for each region using a normal distribution curve. Moreover, the general procedure for hypothesis testing using Z-test for proportions is given and explained clearly. Examples from the field of environmental science are selected and used to illustrate the steps of hypothesis testing using Z-test and making a decision regarding the study with sufficient explanation for each step.

Keywords: Binomial distribution; normal distribution; rejection and nonrejection regions; critical value; area under the curve; Z-test

Learning outcomes

After completing this chapter, readers will be able to:

- Understand the concept and importance of Bernoulli experiment;
- Understand the concept and importance of binomial experiment;
- Know the properties of binomial distribution;
- Know the meaning and concept of proportion;
- Apply a Z-test for solving problems related to proportions;
- Understand the common steps for performing hypothesis testing for one sample proportion;
- Compute a test statistic value for a single population proportion;
- Know the procedure for interpreting the results and match it to the field of study;
- Write smart conclusions.

4.1 Introduction

Problems regarding the mean value (following normal distribution) have been discussed with Z-tests and t-tests including large and small sample sizes. This chapter covers the concept of Bernoulli experiment, binomial experiment, binomial distribution, and related topics such as success and failure with the application of

Applications of Hypothesis Testing for Environmental Science. DOI: https://doi.org/10.1016/B978-0-12-824301-5.00001-0

binomial distribution for testing various hypotheses regarding variables that follow binomial distribution.

4.2 What is Bernoulli distribution?

A distribution that describes the results of an experiment with only two outcomes is called Bernoulli distribution and the experiment is called a Bernoulli experiment. The two results of Bernoulli experiments are called success and failure; success happens with probability p and the complement of success is the failure which happens with probability $q = (1 - p)$, for example, present or absent, dead or alive, normal or defective. For instance, testing one sample of water wells for pesticide residues whether the water wells that supply drinking water showed positive or negative results regarding the pesticide.

The probability distribution for a Bernoulli variable is called Bernoulli distribution having the mathematical formula given in Eq. (4.1).

$$P(Y) = p^Y q^{1-Y}, Y = 0, 1 \tag{4.1}$$

where, 0 represents the failure $q = (1 - p)$ and 1 represents the success p.

Or it can be written as

$$P(Y) = \begin{cases} 1 - p & Y = 0 \\ p & Y = 1 \end{cases}$$

4.3 What is Binomial distribution?

A single experiment with two outcomes can be described by Bernoulli distribution. Suppose there are n trials (experiments) with only two outcomes each, this experiment is called a binomial experiment, and the distribution that describes a binomial experiment is called binomial distribution.

Each binomial experiment should satisfy the following properties:

1. There are n trials (experiments), each with two outcomes: success and failure.
2. The trials are independent.
3. The probability of success remains constant for all experiments.

Suppose there is a binomial experiment consisting of n experiments, then the probability distribution for binomial variable Y is called binomial distribution having the mathematical formula presented in Eq. (4.2).

$$P(Y) = C_Y^n p^Y q^{n-Y} \quad Y = 1, 2, \ldots, n \tag{4.2}$$

$Y \sim b(Y, n, p)$ represents a variable Y that follows binomial distribution with probability p where, Y represents the number of successes obtained from a binomial experiment. n represents the total number of experiments in a binomial experiment. p represents the probability of success for each experiment. q represents the probability of failure for the experiment, ($q = 1 - p$). $P(Y)$ represents the probability of obtaining exactly Y successes.

The mean and variance of binomial distribution are:

$$\mu = np$$

$$\sigma^2 = npq$$

A graphical presentation for binomial distribution is given in Fig. 4.1.

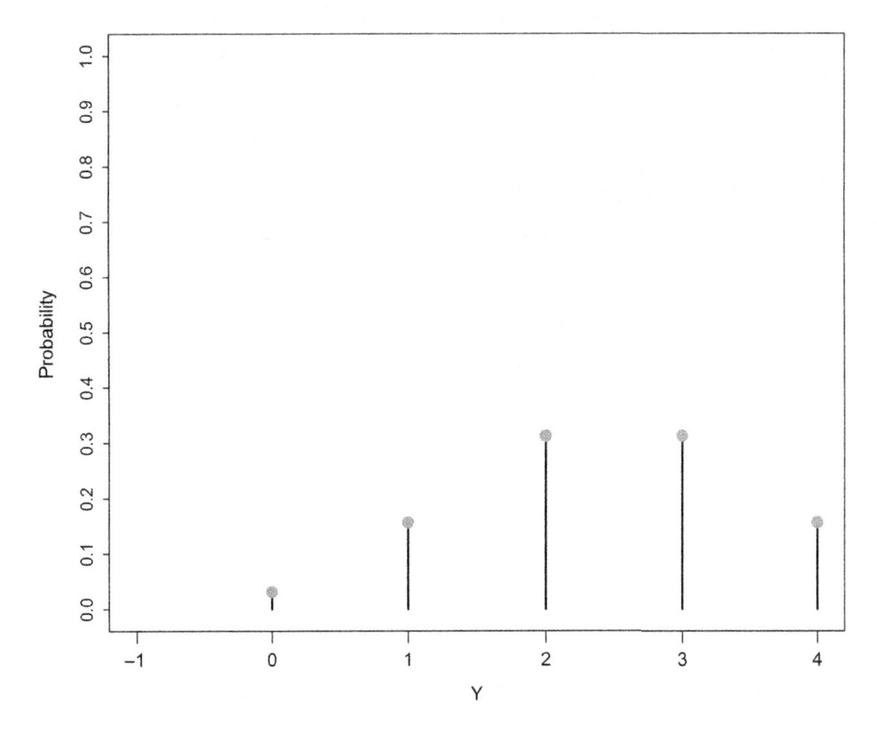

Figure 4.1 Graphical presentation for binomial distribution.

Example 4.1: Testing mineral water quality: A factory for producing mineral water has a quality control department to check the final product. The quality control department usually selects 20 bottles randomly from each batch to test the quality, whether the product meets the criterion set by the factory or not. If at most one bottle out of 20 does not meet the criterion, the decision is to approve the whole batch. The sample data showed that 99% of the batches meet the criterion. What is the probability that a batch will meet the criterion?

One should check the properties of the binomial experiment.

1. There are n trials (experiments), each with two outcomes, success and failure.
   There are 20 bottles (limited number) to represent the sample and the test of each bottle will result either in meeting the criterion or not, and the first condition is satisfied.
2. The trials are independent.
   The test of each bottle is not affected by the result of other bottles whether they meet the criterion or not, this indicates that the testing is independent. The second condition is satisfied.
3. The probability of success remains constant for all experiments.

The probability of success is fixed for each test, which is 99%. The third condition is satisfied.

One can say that this experiment is a binomial experiment because all the conditions are met. Thus, the binomial mathematical formula presented in Eq. (4.2) can be used to calculate the probabilities.

The probability of at most one out of 20 bottles not meeting the criterion of the factory is equivalent to obtaining at least 19 bottles that meet the criterion; this means that 19 or 20 bottles meet the criterion.

One can observe that the probability of success (approve) is $P = 0.99$, probability of failure (reject) is $q = (1 - 0.99) = 0.01$, the total number of bottles to be tested is $n = 20$, and the number of bottles that meet the criterion of the factory is at least 19.

Suppose Y represents the number of bottles that meet the criterion of the factory. Finding the probability of at least 19 bottles that meet the criterion of the factory which is equivalent to $Y \geq 19$. The probability of obtaining at least 19 bottles meeting the criterion can be calculated using the binomial distribution formula presented in Eq. (4.2).

$$P(Y) = C_{Y}^{n} p^{Y} q^{n-Y}$$

$$P(Y \geq 19) = C_{19}^{20}(0.99)^{19}(0.01)^{20-19} + C_{20}^{20}(0.99)^{20}(0.01)^{20-20}$$

$$P(Y \geq 19) = 0.9831407 = 0.983$$

The probability that the batch will meet the criterion set by the factory is 0.98.

Example 4.2: Testing for bacteria: A professor published a study regarding the presence of a specific bacteria in some landfills. The study showed that 0.03 of the landfills have this type of bacteria. A researcher wants to verify the claim, he selected 10 landfills and tested for this type of bacteria. What is the probability that only one landfill contains this type of bacteria?

Let us start checking the properties of binomial experiment.

1. There are n trials (experiments), each with two outcomes, success and failure.
   There are 10 landfills (limited number) to represent the sample and the test of each landfill will show either the landfill contains the specific bacteria or not, and the first condition is satisfied.
2. The trials are independent.
   The test of each landfill is not affected by the result of other landfills whether a landfill contains the bacteria or not, this indicates that the testing is independent. The second condition is satisfied.
3. The probability of success remains constant for all experiments.

The probability of success (presence of a bacteria) is fixed for each test, which is 0.03. The third condition is satisfied.

One can say that this experiment is a binomial experiment because all the conditions are met. Thus, the binomial mathematical formula presented in Eq. (4.2) can be used to calculate the probabilities.

The probability that only 1 out of 10 landfills contains a specific bacteria is equivalent to obtaining $P(Y = 1)$.

One can observe that the probability of success (presence of a bacteria) is $P = 0.03$, probability of failure (absence of a bacteria) is $q = (1 - 0.03) = 0.97$, the total number of landfills to be tested is $n = 10$, and the number of landfills that contain the specific bacteria is 1.

Suppose Y represents the number of landfills that contain the specific bacteria. Finding the probability of 1 landfill containing the specific bacteria is equivalent to $Y = 1$. The probability of obtaining 1 landfill containing specific bacteria can be calculated using the binomial distribution formula presented in Eq. (4.2).

$$P(Y) = C_Y^n p^Y q^{n-Y}$$

$$P(Y = 1) = C_1^{10}(0.03)^1(0.97)^{10-1}$$

$$P(Y = 1) = 0.2280693 = 0.228$$

The probability that a landfill will contain specific bacteria is 0.228.

Example 4.3: The concentration of Zn in PM_{10}: A research group claimed that in an average 4% of the places in a certain country, the concentration of Zn in PM_{10} in the air is more than 1.6 (μgm^{-3}). If a random sample of 11 places is selected and tested for Zn concentration in PM_{10}, what is the probability that two places will have a concentration of Zn in PM_{10} more than 1.6.

Let us start checking the properties of the binomial experiment.

1. There are n trials (experiments), each with two outcomes, success and failure.
   There are 11 places (limited number) to represent the sample and the testing of each place will show either the concentration of Zn in PM_{10} in the air is more than 1.6 or not, and the first condition is satisfied.
2. The trials are independent.
   The testing of each place is not affected by the results of other places, whether the concentration of Zn in PM_{10} in the air is more than 1.6 or not, this indicates that the test is independent. The second condition is satisfied.
3. The probability of success remains constant for all experiments.

The probability of success (concentration of Zn in PM_{10} in the air is more than 1.6) is fixed for each test, which is 4%. The third condition is satisfied.

One can say that this experiment is a binomial experiment because all the conditions are met. Thus, binomial mathematical formula (4.2) can be used to calculate the probabilities.

The probability that in two out of 11 places, the concentration of Zn in PM_{10} in the air is more than 1.6 is equivalent to obtaining $P(Y=2)$.

One can observe that the probability of success (more than 1.6) is $P=0.04$, probability of failure (not more than 1.6) is $q=(1-0.04)=0.96$, the total number of places to be tested is $n=11$, and the concentration of Zn in PM_{10} in the air is more than 1.6 (2).

Suppose Y represents the number of places that the concentration of Zn in PM_{10} in the air is more than 1.6. Finding the probability of two places, the concentration of Zn in PM_{10} in the air is more than 1.6 and is equivalent to $Y=2$. The probability of obtaining two places can be calculated using the binomial distribution formula presented in Eq. (4.2).

$$P(Y) = C_Y^n p^Y q^{n-Y}$$

$$P(Y=2) = C_2^{11}(0.04)^2(0.96)^{11-2}$$

$$P(Y=2) = 0.06094299 = 0.061$$

The probability that two places have a concentration of Zn in PM_{10} in the air greater than 1.6 is 0.061.

4.4 Hypothesis testing for one sample proportion (Z-test)

Hypothesis testing for proportion needs to know the concept of proportion, how to calculate the proportion, and other related concepts and terms. As an example for the proportion, we might wish to know the proportion of landfills that show a specific bacteria among 20 landfills. Suppose that only four out of 20 landfills have this type of bacteria, thus, $\frac{4}{20} = \frac{1}{5} = 0.20$ or 20%. The value of 20% represents the proportion of landfills that have a specific bacteria.

Proportion ($\hat{p}$) is a value calculated from the sample data to show the percentage of a part that carry a certain attribute to the whole. The formula for computing the proportion is given in Eq. (4.3).

$$\hat{p} = \frac{Y}{n} \tag{4.3}$$

where Y is the number of observations in the selected sample that carry the certain attribute, and n is the sample size.

The concept of proportion belongs to the family of experiments of two outcomes which follow binomial distribution. If the sample size is large and the probability of success P is small, such that $np \geq 5$ and $nq \geq 5$, then binomial distribution can be approximated by the normal distribution.

Consider a large sample that is selected from a normally distributed population and Y represents a random variable of interest. A claim regarding the proportion value of the variable of interest can be tested employing Z-test for one sample proportion to make a decision regarding the hypothesis of the proportion value. The mathematical formula for computing the test statistic value for one sample proportion employing Z-test is presented in Eq. (4.4).

$$Z = \frac{\hat{p} - p}{\sqrt{pq/n}} \tag{4.4}$$

where Z is the test statistic, p is the claimed proportion (hypothesized proportion), and $\hat{p}$ is the sample proportion which is calculated as:

$$\hat{p} = \frac{Y}{n}$$

where Y is the number of individuals (units) in the sample that possess the characteristic of interest, and $q = \frac{n-Y}{n}$ or $1 - p$, and n is the sample size.

Example 4.4: Mineral water: A research group at a university wishes to verify the claim announced by a specific factory for mineral water which says that 98% of people are satisfied with the quality of the mineral water. The research group selected 700 people randomly and recorded their responses; they found that 26 were not satisfied. Use the sample information to make a decision whether to support the claim or not. A significance level of $\alpha = 0.01$ is chosen to test the claim. Assume that the population is normally distributed.

The general procedure for conducting hypothesis testing can be used to make the decision regarding the proportion of people who feel satisfied with the mineral water produced by the factory.

Step 1: Specify the null and alternative hypotheses

The proportion announced by the factory for the people who are satisfied with the mineral water (p) is 0.98, which represents the population proportion. This claim should be under the null hypothesis because the claim represents equality (=). If the proportion of people who are satisfied is not equal to 0.98, then two cases should be considered, the first case is whether the proportion could be greater than 0.98, and the second case is whether the proportion could be less than 0.98. The two cases (greater than and less than) can be represented mathematically as $\neq$. Thus, we can write the two hypotheses (null and alternative) as presented in Eq. (4.5).

$$H_0{:}p = 0.98 \text{ vs } H_1{:}p \neq 0.98 \tag{4.5}$$

We should make a decision regarding the null hypothesis whether the proportion of satisfied people is exactly equal to 0.98, or the proportion differs (more or less) from 0.98.

Step 2: Select the significance level (α) for the study

The significance level of 0.01 ($\alpha = 0.01$) is selected to test the hypothesis. We divide the value of significance level (0.01) by 2 to represent the two-tailed test (more or less) of the alternative hypothesis, thus, $\frac{\alpha}{2} = \frac{0.01}{2} = 0.005$ which represents the rejection region in each tail of the standard normal curve (the left and right tails). The Z critical value for the two-tailed test with $\alpha = 0.01$ is 2.58, as appeared in the standard normal table (Table A in the Appendix). Thus, the Z critical values for both sides are ± 2.58, we use the two Z critical values to make a decision whether to reject or fail to reject the null hypothesis.

Step 3: Use the sample information to calculate the test statistic value

The number of people who are not satisfied with the quality is 26, and the number of people who are satisfied with the quality is $700 - 26 = 674$. Thus, the proportion of satisfied people in the sample is:

$$\hat{p} = \frac{Y}{n} = \frac{674}{700} = 0.963 = 0.96$$

96% of the people in the sample were satisfied with the quality of the mineral water and 4% were dissatisfied.

The proportion claimed by the factory regarding the satisfaction of people with the factory mineral water is $p = 0.98$ and the proportion of dissatisfaction is:

$$q = \frac{n - Y}{n} = 1 - p = 1 - 0.98 = 0.02$$

The entries for the Z-test statistic formula as presented in Eq. (4.4) are already provided. The announced proportion of people who are satisfied with the quality (p) is 0.98, the proportion of people who are dissatisfied with the quality (q) is 0.02, the proportion of people who are satisfied with the quality calculated from the sample data is 0.96, and the sample size (n) is 700.

We apply the formula presented in Eq. (4.4) to test the hypothesis about the proportion claimed by the factory regarding the satisfaction of people with the mineral water produced by the factory.

$$Z = \frac{\hat{p} - p}{\sqrt{pq/n}} = \frac{0.96 - 0.98}{\sqrt{(0.02)(0.98)/700}}$$

$$Z = -3.779645 = -3.78$$

The computed test statistic value for the proportion of people who are satisfied is found to be -3.78.

Step 4: Identify the rejection and nonrejection regions

The Z critical values found in step 2 can be used to locate the rejection and nonrejection regions as shown in Fig. 4.2 for the standard normal curve. It can be seen that the rejection region is located on both tails because the alternative hypothesis is unequal.

Step 5: Make a decision and interpret the results

One can observe that the null hypothesis should be rejected because the test statistic value of Z-test calculated from the sample data is -3.78, which is smaller than the Z critical value for the two-tailed test with $\alpha = 0.01$ (-2.58). Moreover, one can use a standard normal curve to decide, the same decision (reject the null hypothesis) can be reached using the normal curve and see that the test statistic value falls in the rejection region on the left tail, as shown in Fig. 4.2. The null hypothesis is rejected in favor of the alternative hypothesis and we believe that the proportion of people who are satisfied with the factory differs from 0.98.

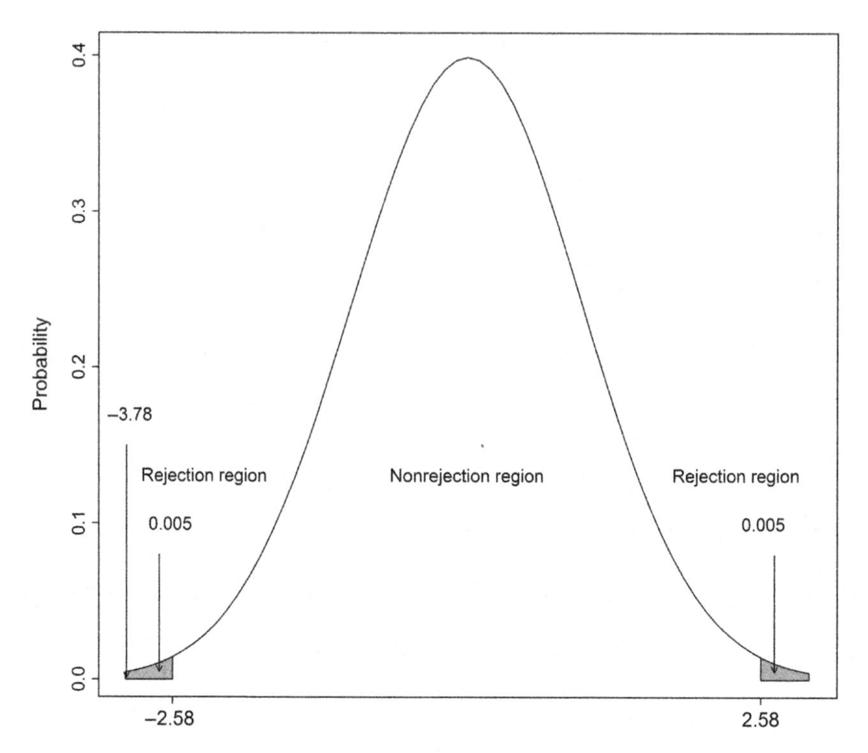

Figure 4.2 The rejection and nonrejection regions for the mineral water.

We can conclude at a 1% significance level that there is sufficient evidence provided by the collected data to believe that the proportion of people who are satisfied with the factory mineral water differs from the announced proportion of 0.98.\

Example 4.5: Lab skills: A department of the environment in a university announced that less than or equal to 0.03 of the students have low lab skills. A professor wishes to evaluate her students' skills in the lab work. She selected a group of 200 students and the results were checked. It was found that seven students had the wrong result. Use the sample information to make a decision as to whether to support the claim or not. A significance level of $\alpha = 0.01$ is chosen to test the claim. Assume that the population is normally distributed.

The general procedure for conducting hypothesis testing can be used to make the decision regarding the proportion of students who have low lab skills.

Step 1: Specify the null and alternative hypotheses

The proportion of students who have low lab skills (p) is less than or equal to 0.03 as announced by the department which represents the population proportion. This claim should be under the null hypothesis because the claim includes equality (=). If the proportion of student with low skills is not less than or equal to 0.03, then the proportion is greater than 0.03, this direction (greater than) can be represented mathematically as > and placed under the alternative hypothesis. Thus, we can write the two hypotheses (null and alternative) as presented in Eq. (4.6).

$$H_0: p \leq 0.03 \text{ vs } H_1: p > 0.03 \tag{4.6}$$

We should make a decision regarding the null hypothesis as to whether the proportion of students who have low lab skills is less than or equal to 0.03, or the proportion is greater than 0.03.

Step 2: Select the significance level (α) for the study

The significance level of 0.01 ($\alpha = 0.01$) is selected to test the hypothesis. Because the test is a one-tailed test as presented by the alternative hypothesis (greater than), we should represent the rejection region on the right tail of the standard normal curve. The Z critical value for a one-tailed test with $\alpha = 0.01$ is 2.326, as appeared in the standard normal table (Table A in the Appendix). Thus, the Z critical value for the right-tailed test is 2.326, and we use the Z critical value to decide whether to reject or fail to reject the null hypothesis.

Step 3: Use the sample information to calculate the test statistic value

The number of students that have low skills is seven, and the students that have high skills is 200 − 7 = 193. Thus, the proportion of students with low lab skills is:

$$\hat{p} = \frac{Y}{n} = \frac{7}{200} = 0.035 = 0.04$$

4% of the students have low lab skills and 0.97 have high lab skills.

The proportion claimed by the department regarding the students with low lab skills is $p = 0.03$ and the proportion of students with high lab skills is:

$$q = \frac{n - Y}{n} = 1 - p = 1 - 0.03 = 0.97$$

The entries for the Z-test statistic formula as presented in Eq. (4.4) are already provided, The announced proportion of students with low lab skills (p) is 0.03, the proportion of students with high lab skills is (q) 0.97, the proportion of students with low lab skills calculated from the sample data is 0.02, and the sample size (n) is 200.

We apply the formula presented in Eq. (4.4) to test the hypothesis regarding the proportion claimed by the department regarding the lab skill for students.

$$Z = \frac{\hat{p} - p}{\sqrt{pq/n}} = \frac{0.04 - 0.03}{\sqrt{(0.03)(0.97)/200}}$$

$$Z = 0.8290267 = 0.83$$

The computed test statistic value for the students who have low lab skills is found to be 0.83.

Step 4: Identify the rejection and nonrejection regions

The Z critical value found in step 2 can be used to locate the rejection and nonrejection regions as shown in Fig. 4.3 for the standard normal curve. It can be seen that the rejection region is located on the right tail in Fig. 4.3.

Step 5: Make a decision and interpret the results

One can observe that the null hypothesis should not be rejected because the test statistic value of the Z-test calculated from the sample data is 0.83, which is smaller than the Z critical value for the right-tailed test with $\alpha = 0.01$ (2.33). Moreover, one can use a standard normal curve to decide, and the same decision (fail to reject the null hypothesis) can be reached using the normal curve and it can be seen that the test statistic value falls in the nonrejection region as shown in Fig. 4.3. The null hypothesis is not rejected and we believe that the proportion of students with low lab skills is less than or equal to 0.03.

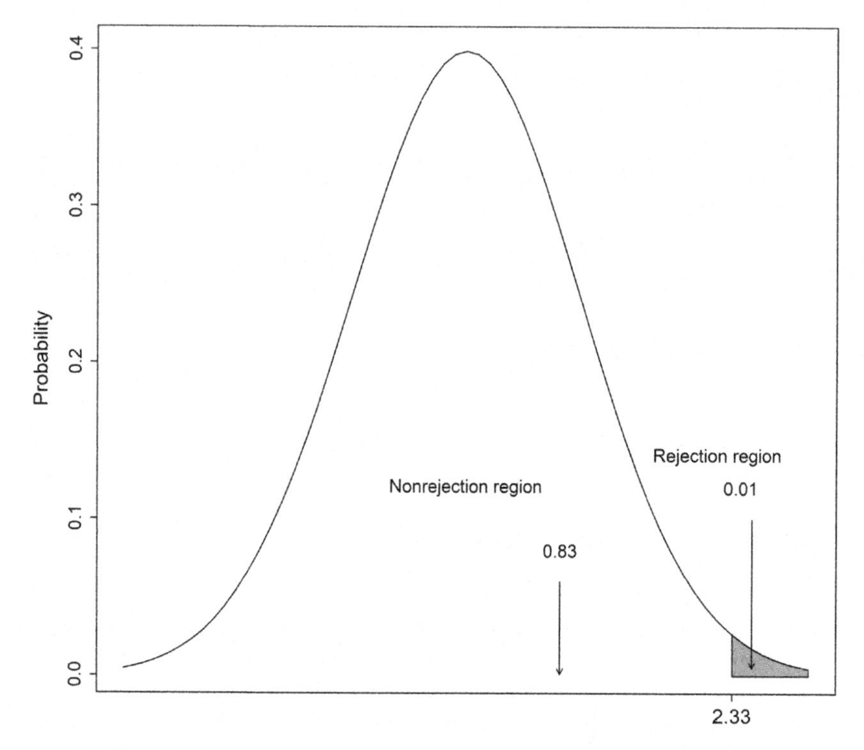

Figure 4.3 The rejection and nonrejection regions for the students with low lab skills.

We can conclude at a 1% significance level that there is sufficient evidence provided by the collected data to believe that the proportion of students with low lab skills is less than or equal to 0.03.

Example 4.6: Beyond permissible limits: A study showed that at least 10% of water samples selected from a specific company exceed the permissible limit of heavy metals in water. A research team wishes to investigate the truth regarding the announced proportion. The team selected 130 samples and tested for the heavy metals. The sample data showed three samples with heavy metals beyond the permissible limits. Use the sample information to make a decision whether to support the claim or not. A significance level of $\alpha = 0.01$ is chosen to test the claim. Assume that the population is normally distributed.

The general procedure for conducting hypothesis testing can be used to make the decision regarding the proportion of water samples that exceed the permissible limit of heavy metals.

Step 1: Specify the null and alternative hypotheses

The proportion of water samples that exceed the permissible limit of heavy metals as announced by the study is at least (more than or equal) (p)0.10, which represents the population proportion. This claim should be under the null hypothesis because the claim includes equality (=). If the proportion of water samples that exceed the permissible limit of heavy metals is not more than or equal to 0.10, then the proportion is less than 0.10, this direction (less than) can be represented mathematically as $<$, and place it under the alternative hypothesis. Thus, we can write the two hypotheses (null and alternative) as presented in Eq. (4.7).

$$H_0: p \geq 0.10 \text{ vs } H_1: p < 0.10 \tag{4.7}$$

We should make a decision regarding the null hypothesis as to whether the proportion of water samples that exceed the permissible limit of heavy metals is more than or equal to 0.10, or the proportion is less than 0.10.

Step 2: Select the significance level (α) for the study

The significance level of 0.01 ($\alpha = 0.01$) is selected to test the hypothesis. Because the test is a one-tailed test as presented by the alternative hypothesis (less than), we should represent the rejection region on the left tail of the standard normal curve. The Z critical value for a one-tailed test with $\alpha = 0.01$ is 2.326, as appeared in the standard normal table (Table A in the Appendix). Thus, the Z critical value for the left-tailed test is -2.326, and we use the Z critical value to decide whether to reject or fail to reject the null hypothesis.

Step 3: Use the sample information to calculate the test statistic value

The number of water samples that exceed the permissible limit of heavy metals is three, and the number of samples within the specification limits is $130 - 3 = 127$. Thus, the proportion of water samples that exceed the permissible limit of heavy metals is:

$$\hat{p} = \frac{Y}{n} = \frac{3}{130} = 0.023 = 0.02$$

2% of the samples exceed the permissible limits and 98% are within the permissible limits.

The proportion claimed by the study regarding the samples exceeding the permissible limits is $p = 0.10$ and the proportion of water samples that are within permissible limit of heavy metals is:

$$q = \frac{n - Y}{n} = 1 - p = 1 - 0.02 = 0.98$$

The entries for the Z-test statistic formula as presented in Eq. (4.4) are already provided, The proportion of samples that exceeded the permissible limits (p) is 0.10, the proportion of samples within the specification limits (q) is 0.90, the proportion of samples that exceeded the permissible limits calculated from the sample data is 0.02, and the sample size (n) is 130.

We apply the formula presented in Eq. (4.4) to test the hypothesis regarding the proportion of water samples that exceed the permissible limit of heavy metals.

$$Z = \frac{\hat{p} - p}{\sqrt{pq/n}} = \frac{0.02 - 0.10}{\sqrt{(0.10)(0.90)/130}}$$

$$Z = -3.040468 = -3.04$$

The computed test statistic value for the water samples that exceed the permissible limit of heavy metals is found to be -3.04.

Step 4: Identify the rejection and nonrejection regions

The Z critical value found in step 2 can be used to locate the rejection and nonrejection regions as shown in Fig. 4.4 for the standard normal curve. It can be seen that the rejection region is located on the left tail in Fig. 4.4.

Step 5: Make a decision and interpret the results

One can observe that the null hypothesis should be rejected because the test statistic value of the Z-test calculated from the sample data is -3.04, which is smaller than the Z critical value for the left-tailed test with $\alpha = 0.01$ (-2.326). Moreover, one can use the standard normal curve to decide, the same decision (reject the null hypothesis) can be reached using the normal curve and it can be seen that the test statistic value falls in the rejection region on the left tail as shown in Fig. 4.4. The null hypothesis is rejected and we believe that the proportion of samples that exceed the permissible limit is less than 0.10.

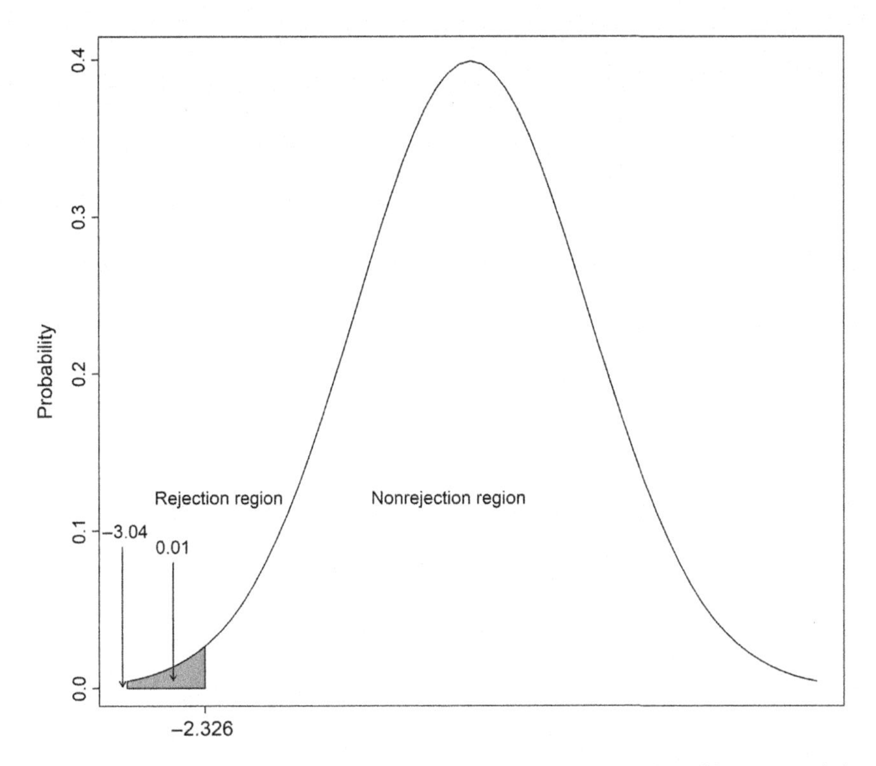

Figure 4.4 The rejection and nonrejection regions for the concentration of heavy metals in water samples.

We can conclude at a 1% significance level that there is sufficient evidence provided by the collected data to believe that the proportion of samples that exceed the permissible limit of heavy metals in water samples is less than 0.10.

Further reading

Alkarkhi, A. F. M., & Chin, L. H. (2012). *Elementary statistics for technologist* (1st ed.). Malaysia: Universiti Sains Malaysia.

Bluman, A. G. (1998). *Elementary statistics: A step by step approach* (3rd ed.). Boston: WCB, McGraw-Hill.

Weiss, N. A. (2012). *Introductory statistics* (9th ed.). Pearson.

Chi-square test for one sample variance

Abstract

This chapter addresses hypothesis testing based on chi-square distribution including the concept of chi-square distributions and related terms. Moreover, the area under the chi-square distribution curve and the concept of rejection and nonrejection regions are delivered with illustration by examples for each region using a chi-square distribution curve. Furthermore, the general procedure for hypothesis testing using a chi-square test for population variance is given and explained clearly. Examples from the field of environmental science are selected and used to illustrate the steps of hypothesis testing using the chi-square test and making a decision regarding the study with sufficient explanation for each step.

Keywords: Chi-square distribution; rejection and nonrejection regions; critical value; area under the curve; chi-square test

Learning outcomes

After completing this chapter, readers will be able to:

- Understand the concept and importance of chi-square distribution
- Explain how to obtain chi-square critical values
- Compute the area under the chi-square distribution curve
- Describe the common steps for testing a hypothesis about one population variance
- Compute a test statistic value for a single population variance
- Identify the rejection and nonrejection regions related to chi-square test
- Know the procedure for interpreting the result and match it to the field of study
- Describe one-sided and two-sided tests for a chi-square test
- Write smart conclusions regarding the problem under study

5.1 Introduction

Hypothesis testing uses several distributions to test different hypotheses regarding various parameters such as the normal distribution which is used to test hypotheses regarding the mean value and hypotheses regarding proportions, and t distribution which is used for testing hypotheses regarding the mean value when the sample size is small. However, these distributions are not valid to test hypotheses regarding a population variance, thus we need to produce another distribution that tests hypotheses regarding the population standard deviation or variance and draw smart

Applications of Hypothesis Testing for Environmental Science. DOI: https://doi.org/10.1016/B978-0-12-824301-5.00003-4

conclusions about the dispersion of the population. Hypotheses about the variance or standard deviation can be tested using a chi-square distribution.

5.2 What is chi-square distribution?

A chi-square distribution (χ^2) (pronounced "chi" as "kai") is a statistical distribution that is used in many situations to test various hypotheses such as goodness-of-fit test, independence test, and hypothesis testing for population variance or standard deviation. This distribution belongs to the same family as t distribution which depends on the concept of degrees of freedom.

Suppose $Y > 0$ is a random variable that follows a chi-square distribution with n degrees of freedom ($d.f$); $Y \sim \chi^2(n)$. A chi-square curve for various degrees of freedom (1, 4, 7, 10, 17) is presented in Fig. 5.1.

A chi-square distribution has the following properties:

1. A chi-square variable is always nonnegative, $Y > 0$, and the shape of the curve tends to be skewed to the right.
2. The mean and variance for a chi-square distribution are $\mu = n$ and $\sigma^2 = 2n$, respectively.
3. The number of degrees of freedom influences the shape of the distribution, the chi-square curve approaches a normal distribution with a large number of degrees of freedom.

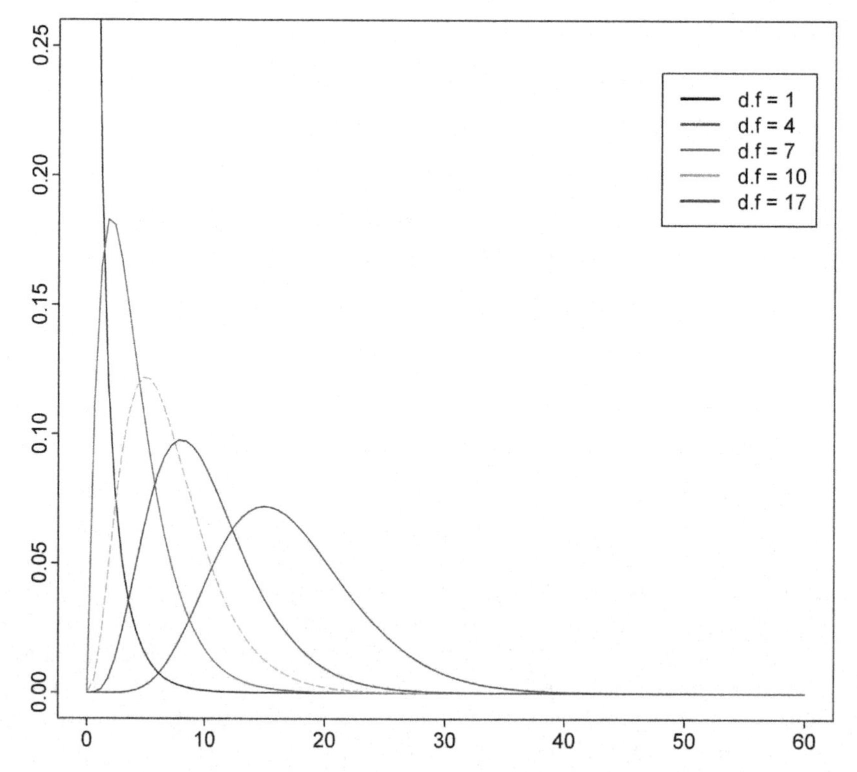

Figure 5.1 Showing various chi-square curves with 1, 4, 7, 10, and 17 degrees of freedom.

5.3 Finding the chi-square values (area under the chi-square curve)

Critical values for chi-square distribution are usually used to help and guide researchers making a decision about a hypothesis of interest. We can obtain chi-square critical values for various degrees of freedom (*d.f*) and the level of significance (α) from Table C in the Appendix. Thus we need to prepare two values to use a chi-square table and obtain the required critical value; the two values are the degrees of freedom and the significance level. The first column on the left of Table C represents the degrees of freedom (*d.f*) from 1 to ∞, while the first upper row represents the level of significance (α).

Example 5.1: Finding the critical value for the left-tailed chi-square test: Use a significance level of 0.01 ($\alpha = 0.01$) and degrees of freedom of 11 ($d.f = 11$) to obtain the chi-square critical value for the left-tailed chi-square test.

We can obtain the chi-square critical value for the left-tailed test as long as two values are provided, the two values are the degrees of freedom "$d.f = 11$" and significance level "$\alpha = 0.01$."

- The first step in extracting the chi-square critical value is to specify the position of "$d.f = 11$" in the first column of Table C, labeled *d.f.* Table 5.1 is a portion of Table C in the Appendix.
- The second step is to specify the position of $1 - \alpha = 0.99$ (because the area under the chi-square curve is to the right of χ^2 and the required value is to the left side, thus the required area equals to $1 - \alpha$) in the first upper row (highlighted column) and then move on the column to the row labeled $d.f = 11$ (highlighted row), the value that represents the point of intersection between $d.f = 11$ and $1 - \alpha = 0.99$ is the χ^2 critical value. One can observe that the χ_L^2 critical value is 3.053 as shown in Table 5.1 (bold value).

This problem is a left-tailed test; thus the critical value for the left-tailed test is $\chi_L^2 = \chi^2_{(1-\alpha,d.f)} = \chi^2_{(1-0.01,11)} = \chi^2_{(0.99,11)} = 3.053$.

Fig. 5.2 shows the area to the left of $\chi^2_{(0.99,11)} = 3.053$ (shaded area under the curve).

Table 5.1 Chi-square value for the left-tailed test.

α d.f	0.995	0.990	0.975	0.950	0.900	0.750	0.500	0.250	0.100
1	0.000	0.000	0.001	0.004	0.016	0.102	0.455	1.323	2.706
2	0.010	0.020	0.051	0.103	0.211	0.575	1.386	2.773	4.605
3	0.072	0.115	0.216	0.352	0.584	1.213	2.366	4.108	6.251
4	0.207	0.297	0.484	0.711	1.064	1.923	3.357	5.385	7.779
5	0.412	0.554	0.831	1.145	1.610	2.675	4.351	6.626	9.236
6	0.676	0.872	1.237	1.635	2.204	3.455	5.348	7.841	10.645
7	0.989	1.239	1.690	2.167	2.833	4.255	6.346	9.037	12.017
8	1.344	1.647	2.180	2.733	3.490	5.071	7.344	10.219	13.362
9	1.735	2.088	2.700	3.325	4.168	5.899	8.343	11.389	14.684
10	2.156	2.558	3.247	3.940	4.865	6.737	9.342	12.549	15.987
11	2.603	**3.053**	3.816	4.575	5.578	7.584	10.341	13.701	17.275
12	3.074	3.571	4.404	5.226	6.304	8.438	11.340	14.845	18.549
13	3.565	4.107	5.009	5.892	7.042	9.299	12.340	15.984	19.812
14	4.075	4.660	5.629	6.571	7.790	10.165	13.339	17.117	21.064

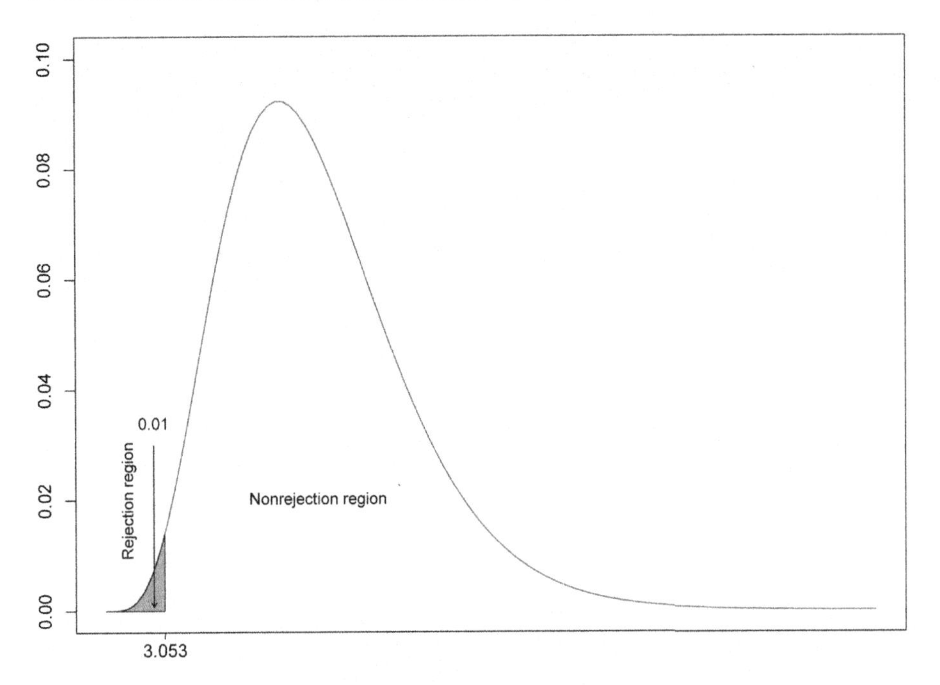

Figure 5.2 Showing the area to the left of the chi-square test.

Example 5.2: Finding the critical value for the right-tailed chi-square test: Use a significance level of 0.01 ($\alpha = 0.01$) and degrees of freedom of 11 ($d.f = 11$) to obtain the chi-square critical value for the right-tailed chi-square test.

We can obtain the chi-square critical value for the right-tailed test as long as the degrees of freedom "$d.f = 11$" and significance level "$\alpha = 0.01$" are provided.

- The first step in extracting the chi-square critical value is to specify the position of "$d.f = 11$" in the first column of Table C labeled *d.f*. Table 5.2 is a portion of Table C in the Appendix.
- The second step is to specify the position of $\alpha = 0.01$ in the first upper row (highlighted column) and then move on the column to the row labeled $d.f = 11$ (highlighted row) the value that represents the point of intersection between $d.f = 11$ and $\alpha = 0.01$ is the χ^2 critical value. One can observe that the χ^2_R critical value is 24.725 as shown in Table 5.2 (bold value).

This problem is a right-tailed test; thus the critical value for the right-tailed test is $\chi^2_R = \chi^2_{(\alpha,d.f)} = \chi^2_{(0.01,11)} = \chi^2_{(0.01,11)} = 24.725$.

Fig. 5.3 shows the area to the right of $\chi^2_{(0.01,11)} = 24.725$ (shaded area under the curve).

Table 5.2 Chi-square value for the right-tailed test.

α d.f	0.900	0.750	0.500	0.250	0.100	0.050	0.025	0.010	0.005
1	0.016	0.102	0.455	1.323	2.706	3.841	5.024	6.635	7.879
2	0.211	0.575	1.386	2.773	4.605	5.991	7.378	9.210	10.597
3	0.584	1.213	2.366	4.108	6.251	7.815	9.348	11.345	12.838
4	1.064	1.923	3.357	5.385	7.779	9.488	11.143	13.277	14.860
5	1.610	2.675	4.351	6.626	9.236	11.071	12.833	15.086	16.750
6	2.204	3.455	5.348	7.841	10.645	12.592	14.449	16.812	18.548
7	2.833	4.255	6.346	9.037	12.017	14.067	16.013	18.475	20.278
8	3.490	5.071	7.344	10.219	13.362	15.507	17.535	20.090	21.955
9	4.168	5.899	8.343	11.389	14.684	16.919	19.023	21.666	23.589
10	4.865	6.737	9.342	12.549	15.987	18.307	20.483	23.209	25.188
11	5.578	7.584	10.341	13.701	17.275	19.675	21.920	**24.725**	26.757
12	6.304	8.438	11.340	14.845	18.549	21.026	23.337	26.217	28.300
13	7.042	9.299	12.340	15.984	19.812	22.362	24.736	27.688	29.819
14	7.790	10.165	13.339	17.117	21.064	23.685	26.119	29.141	31.319

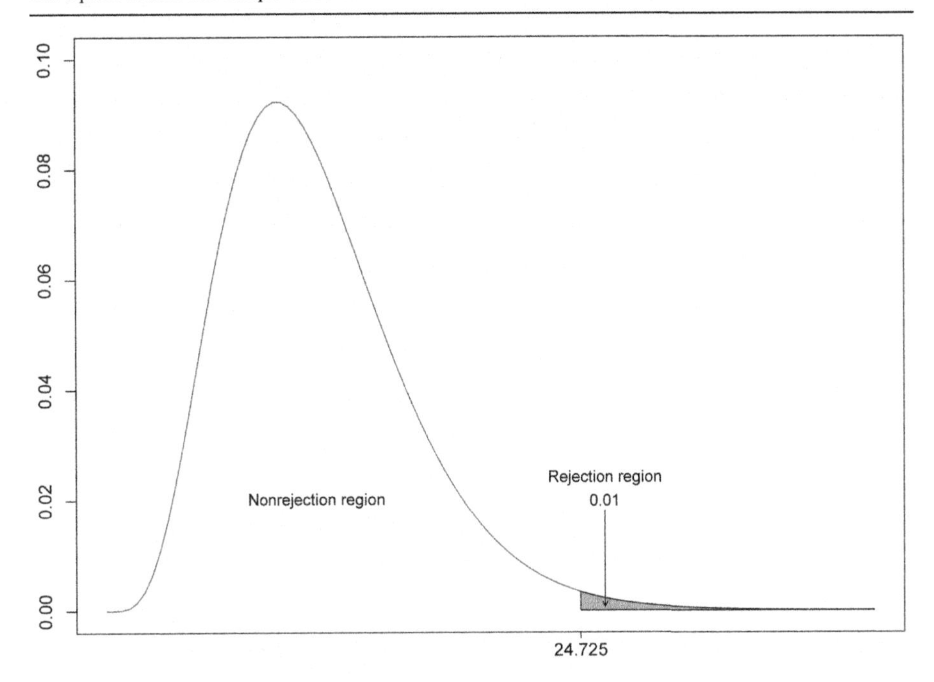

Figure 5.3 Showing the area to the right of the chi-square test.

Example 5.3: Finding the critical value for the two-tailed chi-square test: Use a significance level of 0.01 ($\alpha = 0.01$) and degrees of freedom of 11 ($d.f = 11$) to obtain the chi-square critical value for the two-tailed chi-square test.

We can obtain the chi-square critical value for the two-tailed test as long as the degrees of freedom "$d.f = 11$" and significance level "$\alpha = 0.01$" are provided.

- The first step in extracting the chi-square critical value is to specify the position of "$d.f = 11$" in the first column of Table C labeled *d.f.* Table 5.3 is a portion of Table C in the Appendix.
- The second step is to specify the location of $\frac{\alpha}{2} = 0.005$ (the significance level is divided by 2 because the test is two-sided test) in the first upper row.
- Two chi-square critical values should be extracted from Table 5.3 as long as the test is two-tailed test. The two values can be extracted as shown below.

The left-tailed value: The chi-square critical value for a left-tailed test with $d.f = 11$ (highlighted row) and $1 - \frac{\alpha}{2} = 0.995$ (highlighted column) is 2.603, the value that represents the point of intersection between $d.f = 11$ and $1 - \frac{\alpha}{2} = 0.995$ is 2.603, $\chi_L^2 = \chi^2_{(1-\frac{\alpha}{2},d.f)} = \chi^2_{(0.995,11)} = 2.603$ as shown in Table 5.3 (bold value). The area to the left (2.603) is presented in Fig. 5.4.

The right-tailed value: The chi-square critical value for a right-tailed test with $d.f = 11$ and $\frac{\alpha}{2} = 0.005$ is 26.757, the value that represents the point of intersection between $d.f = 11$ and $\frac{\alpha}{2} = 0.005$ is 26.757, $\chi_R^2 = \chi^2_{(\frac{\alpha}{2},d.f)} = \chi^2_{(0.005,11)} = 26.757$ as shown in Table 5.3 (bold value). The area to the right (26.757) is presented in Fig. 5.4.

Fig. 5.4 shows the areas to the left and right for $\alpha = 0.01$, $d.f = 11$ (shaded area under the curve).

Table 5.3 Chi-square values for the two-tailed test.

α d.f	**0.995**	**0.990**	**0.975**	**0.500**	**0.250**	**0.100**	**0.050**	**0.025**	**0.010**	**0.005**
1	0.000	0.000	0.001	0.455	1.323	2.706	3.841	5.024	6.635	7.879
2	0.010	0.020	0.051	1.386	2.773	4.605	5.991	7.378	9.210	10.597
3	0.072	0.115	0.216	2.366	4.108	6.251	7.815	9.348	11.345	12.838
4	0.207	0.297	0.484	3.357	5.385	7.779	9.488	11.143	13.277	14.860
5	0.412	0.554	0.831	4.351	6.626	9.236	11.071	12.833	15.086	16.750
6	0.676	0.872	1.237	5.348	7.841	10.645	12.592	14.449	16.812	18.548
7	0.989	1.239	1.690	6.346	9.037	12.017	14.067	16.013	18.475	20.278
8	1.344	1.647	2.180	7.344	10.219	13.362	15.507	17.535	20.090	21.955
9	1.735	2.088	2.700	8.343	11.389	14.684	16.919	19.023	21.666	23.589
10	2.156	2.558	3.247	9.342	12.549	15.987	18.307	20.483	23.209	25.188
11	**2.603**	3.053	3.816	10.341	13.701	17.275	19.675	21.920	24.725	**26.757**
12	3.074	3.571	4.404	11.340	14.845	18.549	21.026	23.337	26.217	28.300
13	3.565	4.107	5.009	12.340	15.984	19.812	22.362	24.736	27.688	29.819
14	4.075	4.660	5.629	13.339	17.117	21.064	23.685	26.119	29.141	31.319
15	4.601	5.229	6.262	14.339	18.245	22.307	24.996	27.488	30.578	32.801

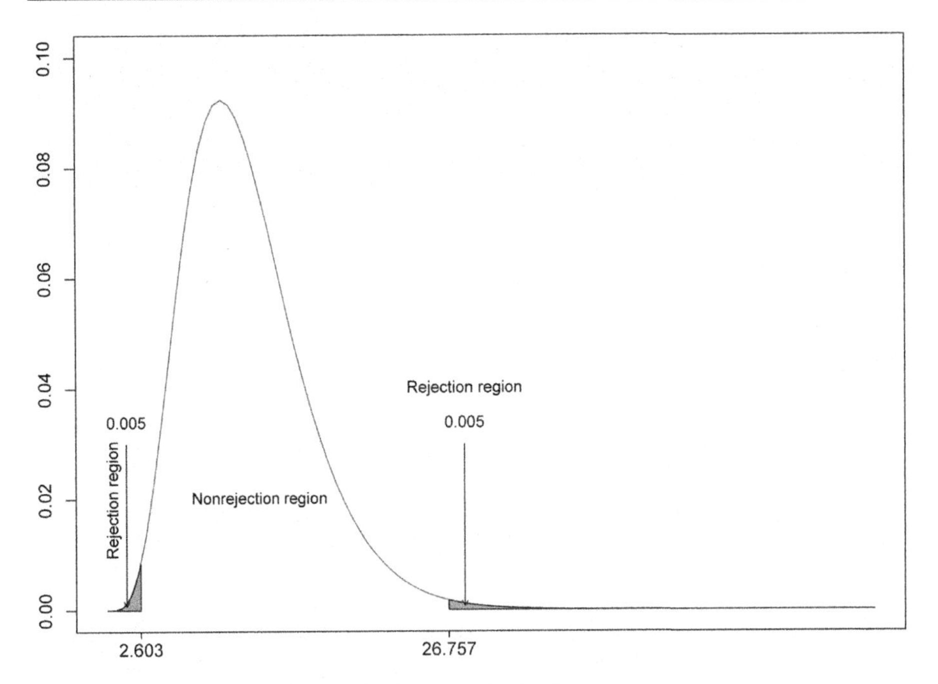

Figure 5.4 Showing the χ^2 value for the two-tailed chi-square test.

5.4 Hypothesis testing for one-sample variance or standard deviation

Hypothesis testing about the variance or standard deviation can be carried out using a new test called a χ^2 test.

Consider a random sample of size n that is selected from a normally distributed population, a claim regarding the variance (standard deviation) value of the variable of interest can be tested employing the χ^2 test for one sample variance (standard deviation) to make a decision regarding the variance (standard deviation) value. The mathematical formula for computing the test statistic value for one sample variance (standard deviation) employing the χ^2 test is presented in Eq. (5.1).

$$\chi^2 = \frac{(n-1)s^2}{\sigma_0^2} \tag{5.1}$$

Follow chi-square distribution with $(n-1)$ degrees of freedom, where
n: the sample size,
s^2: the sample variance, and
σ_0^2: the population variance.

Example 5.4: The concentration of total suspended solid of surface water: A researcher at an environmental section wishes to verify the claim that the variance of total suspended solids concentration (TSS) of Beris dam surface water is 1.25 (mg/L). Twelve samples were selected and the total suspended solids concentration was measured. The collected data showed that the standard deviation of total suspended solids concentration is 1.80. A significance level of $\alpha = 0.01$ is chosen to test the claim. Assume that the population is normally distributed.

The general procedure for conducting hypothesis testing can be used to make the decision regarding the variance of total suspended solids concentration in the surface water of Beris dam.

Step 1: Specify the null and alternative hypotheses

The population variance of total suspended solids concentration (σ^2) is 1.25; this claim should be under the null hypothesis because the claim represents equality (=). If the variance of total suspended solids concentration of surface water is not equal to 1.25, then two cases should be considered; in the first case, the variance of total suspended solids concentration is greater than 1.25, and in the second case, the variance is less than 1.25. The two cases (greater than and less than) can be represented mathematically as $\neq$. Thus we can write the two hypotheses (null and alternative) as presented in Eq. (5.2).

$$H_0{:}\sigma^2 = 1.25 \text{ vs } H_1{:}\sigma^2 = 1.25 \tag{5.2}$$

We should make a decision regarding the null hypothesis as to whether the variance of total suspended solids concentration exactly equals 1.25, or the variance of total suspended solids concentration differs (more or less) from 1.25.

Step 2: Select the significance level (α) for the study

The significance level of 0.01 ($\alpha = 0.01$) is selected to test the hypothesis regarding the variance of total suspended solids concentration. We divide the value of significance level (0.01) by 2 to represent the two-tailed test (more or less) of the alternative hypothesis, thus $\frac{\alpha}{2} = \frac{0.01}{2} = 0.005$ which represents the rejection region in each tail of the χ^2 distribution curve (the left and right tails). The χ^2 critical value for a two-tailed test with $\alpha = 0.01$ and $d.f = 12 - 1 = 11$ can be computed as follows:

The left-tailed value: The chi-square critical value for the left-tailed test with $d.f = 11$ and $1 - \frac{\alpha}{2} = 0.995$ is 2.603, the value that represents the point of intersection between $d.f = 11$ and $1 - \frac{\alpha}{2} = 0.995$ is 2.603, $\chi_L^2 = \chi^2_{(1-\frac{\alpha}{2},d.f)} = \chi^2_{(0.995,11)} = 2.603$ as shown in Table C in the Appendix. The area to the left (2.603) is presented in Fig. 5.5.

The right-tailed value: The chi-square value for a right-tailed test with $d.f = 11$ and $\frac{\alpha}{2} = 0.005$ is 26.757, the value that represents the point of intersection between $d.f = 11$ and $\frac{\alpha}{2} = 0.005$ is 26.757, $\chi_R^2 = \chi^2_{(\frac{\alpha}{2},d.f)} = \chi^2_{(0.005,11)} = 26.757$ as shown in Table C in the Appendix and presented in Fig. 5.5.

Step 3: Use the sample information to calculate the test statistic value

The entries for the χ^2 test statistic formula, as presented in Eq. (5.1), are already provided. The claimed value for the variance of total suspended solids concentration (σ^2) is 1.25, the standard deviation calculated from the sample data is (S) 1.80, and the sample size (n) is 12.

We apply the formula presented in Eq. (5.1) to test the hypothesis regarding the variance of total suspended solids concentration of surface water for Beris dam.

$$\chi^2 = \frac{(n-1)s^2}{\sigma_0^2} = \frac{(12-1)(1.80)^2}{1.25}$$
$$= 28.512$$

The test statistic values for the variance of total suspended solids concentration of surface water for Beris dam is found to be 28.512.

Step 4: Identify the critical and noncritical regions for the study

We can easily identify the rejection and nonrejection regions for a two-tailed test employing the χ^2 critical value found in step 2. The rejection and nonrejection regions for the variance of total suspended solids concentration of surface water for Beris dam are presented in Fig. 5.5 for the χ^2 distribution curve (shaded area).

Step 5: Make a decision and interpret the results

One can observe that the null hypothesis should be rejected because the test statistic value of the χ^2 test calculated from the sample data is 28.512, which is greater than the χ^2 critical value for the right-tailed test (26.757). Moreover, one can use the χ^2 distribution curve to decide, and the same decision (reject the null hypothesis) can be reached using the χ^2 distribution curve and to see that the test statistic value falls in the rejection region (right-tailed) as shown in Fig. 5.5. The null hypothesis is rejected and we believe that the variance of total suspended solids concentration of surface water for Beris dam is not equal to 1.25.

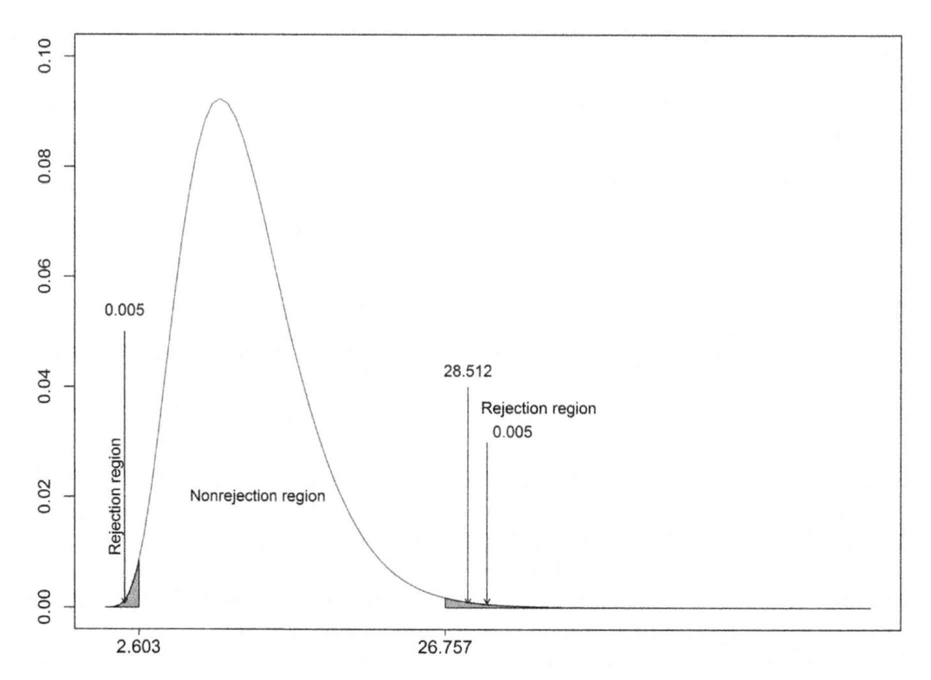

Figure 5.5 Rejection and nonrejection regions for the variance of total suspended solids concentration of surface water.

We can conclude at a 1% significance level that there is sufficient evidence provided by the collected data to believe that the variance of total suspended solids concentration of surface water differs from 1.25.

Example 5.5: The concentration of lead in sediment: A professor at a research center wishes to verify the claim that the variance of lead (Pb) concentration in sediment of Juru River is less than or equal 0.20 (mg/L). Ten samples were selected and tested for lead concentration. The collected data showed that the standard deviation of lead concentration is 0.23. A significance level of $\alpha = 0.01$ is chosen to test the claim. Assume that the population is normally distributed.

The general procedure for conducting hypothesis testing can be used to make the decision regarding the variance of lead concentration in sediment of Juru River.

Step 1: Specify the null and alternative hypotheses

The population variance of lead concentration in sediment (σ^2) is less than or equal to 0.20, this claim should be under the null hypothesis because the claim represents equality (=). If the variance of lead concentration in sediment is not less than or equal to 0.20, then one case should be considered, the variance of lead concentration is greater than 0.20 and can be represented mathematically as $>$.

Thus we can write the two hypotheses (null and alternative) as presented in Eq. (5.3).

$$H_0:\sigma^2 \leq 0.20 \text{ vs } H_1:\sigma^2 > 0.20 \tag{5.3}$$

We should make a decision regarding the null hypothesis whether the variance of lead concentration in sediment is less than or equal to 0.20, or the variance of lead concentration is more than 0.20.

Step 2: Select the significance level (α) for the study

The significance level of 0.01 ($\alpha = 0.01$) is selected to test the hypothesis. Because the test is a one-tailed test as presented by the alternative hypothesis (greater than), we should represent the rejection region on the right tail of the χ^2 distribution curve. The χ^2 critical value for a one-tailed test with $\alpha = 0.01$ and $d.f = 10 - 1 = 9$ is 21.666, as appeared in the χ^2 distribution table for critical values (Table C in the Appendix). Thus the χ^2 critical value for the right-tailed test is 21.666, and we use the χ^2 critical value to decide whether to reject or fail to reject the null hypothesis.

Step 3: Use the sample information to calculate the test statistic value

The entries for the χ^2 test statistic formula as presented in Eq. (5.1) are already provided. The claimed value for the variance of lead concentration (σ^2) is 0.20, the standard deviation for lead concentration calculated from the sample data (S) is 0.23, and the sample size (n) is 10.

We apply the formula presented in Eq. (5.1) to test the hypothesis regarding the variance of lead concentration in sediment.

$$\chi^2 = \frac{(n-1)s^2}{\sigma_0^2} = \frac{(10-1)(0.23)^2}{0.20}$$
$$= 2.3805 = 2.3805 = 2.381$$

The test statistic value for the variance of lead concentration in sediment is found to be 2.381.

Step 4: Identify the critical and noncritical regions for the study

We can easily identify the rejection and nonrejection regions for the right-tailed test employing the χ^2 critical value found in step 2. The rejection and nonrejection regions for the variance of lead concentration in sediment are presented in Fig. 5.6 for the χ^2 distribution curve (shaded area).

Step 5: Make a decision and interpret the results

One can observe that the null hypothesis should not be rejected because the test statistic value of the χ^2 test calculated from the sample data is 2.381, which is smaller than the χ^2 critical value for the right-tailed test (21.666). Moreover, one can use the χ^2 distribution curve to decide, and the same decision (fail to reject the null hypothesis) can be reached using the χ^2 distribution curve and to see that the test statistic value falls in the nonrejection region as shown in Fig. 5.6. The null hypothesis is not rejected and we believe that the variance of lead concentration is less than or equal to 0.20.

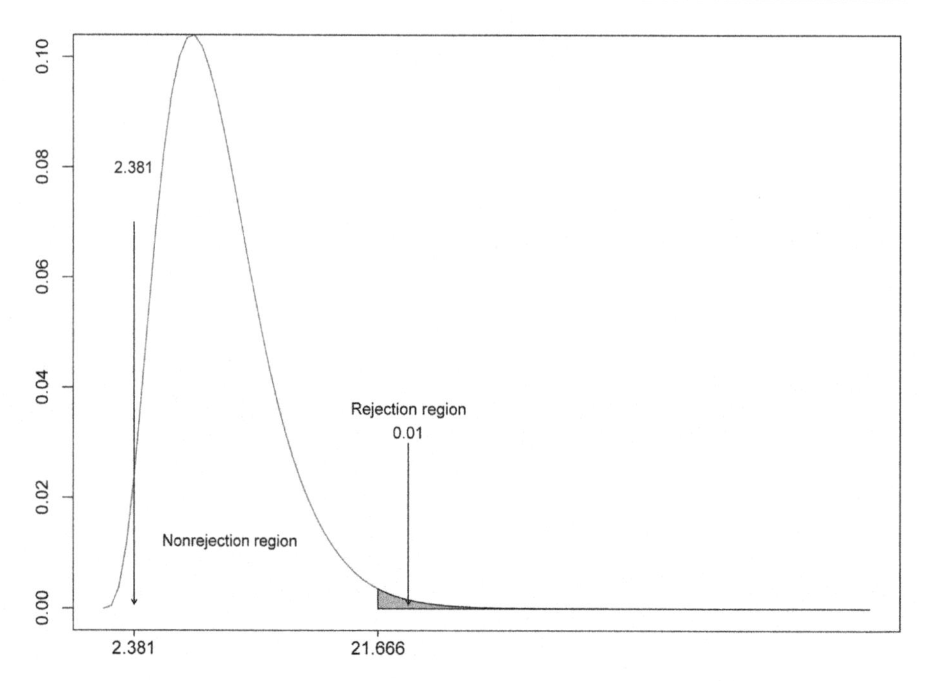

Figure 5.6 Rejection and nonrejection regions for the variance of lead concentration in sediment.

We can conclude at a 1% significance level that there is sufficient evidence provided by the collected data to believe that the variance of lead concentration in sediment is less than or equal to 0.20.

Example 5.6: The level of particulate matter in a palm oil mill: A research group wishes to verify that the standard deviation of the particulate matter (PM_1) level in the air of a palm oil mill is less 23.50 ($\mu g/m\hat{3}$). Twelve samples were selected and the PM_1 level was measured. The sample data showed that the standard deviation of PM_1 level is 26.89. A significance level of $\alpha = 0.01$ is chosen to test the claim. Assume that the population is normally distributed.

The general procedure for conducting hypothesis testing can be used to make the decision regarding the standard deviation of the PM_1 level in the air of the palm oil mill.

Step 1: Specify the null and alternative hypotheses

The population standard deviation of the PM_1 level (σ^2) is less than 23.50, this claim should be under the alternative hypothesis because the claim represents only one side which is less than 23.50. If the standard deviation of the PM_1 level in the air of the palm oil mill is not less than 23.50, then two cases should be considered; in the first case, the standard deviation of the PM_1 level is greater than 23.50, and in the second case it would be equal to 23.50. The two cases (greater than and equal to) can be represented mathematically as $\geq$. Thus we can write the two hypotheses (null and alternative) as presented in Eq. (5.4).

$$H_0{:}\sigma \geq 23.50 \text{ vs } H_1{:}\sigma < 23.50 \tag{5.4}$$

We should make a decision regarding the null hypothesis as to whether the standard deviation of the PM_1 level is greater than or equal to 23.50, or the variance of the PM_1 level in the air of the palm oil mill is less than 23.50.

Step 2: Select the significance level (α) for the study

The significance level of 0.01 ($\alpha = 0.01$) is selected to test the hypothesis. Because the test is a one-tailed test as presented by the alternative hypothesis (less than), we should represent the rejection region on the left tail of the χ^2 distribution curve. The χ^2 critical value for a one-tailed test with $1 - \alpha = 1 - 0.01 = 0.99$ and $d.f = 11$ is 3.053, as appeared in the χ^2 distribution table for critical values (Table C in the Appendix). Thus the χ^2 critical value for the left-tailed test is 3.053, we use the χ^2 critical value to decide whether to reject or fail to reject the null hypothesis.

Step 3: Use the sample information to calculate the test statistic value

The entries for the χ^2 test statistic formula as presented in Eq. (5.1) are already provided. The claimed value for the standard deviation of the PM_1 level (σ) is 23.50, the standard deviation for the PM_1 level calculated from the sample data (S) is 26.89, and the sample size (n) is 12.

We apply the formula presented in Eq. (5.1) to test the hypothesis regarding the standard deviation of the PM_1 level in the air of the palm oil mill.

$$\chi^2 = \frac{(n-1)s^2}{\sigma_0^2} = \frac{(12-1)(26.89)^2}{(23.50)^2}$$
$$= 14.40252 = 14.40$$

The test statistic value for the standard deviation of the PM_1 level in the air of the palm oil mill is found to be 14.40.

Step 4: Identify the critical and noncritical regions for the study

We can easily identify the rejection and nonrejection regions for the left-tailed test employing the χ^2 critical value found in step 2. The rejection and nonrejection regions for the standard deviation of the PM_1 level in the air of the palm oil mill are presented in Fig. 5.7 for the χ^2 distribution curve (shaded area).

Step 5: Make a decision and interpret the results

One can observe that the null hypothesis should not be rejected because the test statistic value of the χ^2 test calculated from the sample data is 14.40, which is greater than the χ^2 critical value for the left-tailed test (3.053). Moreover, one can use the χ^2 distribution curve to decide, the same decision (fail to reject the null hypothesis) can be reached using the χ^2 distribution curve and to see that the test statistic value falls in the nonrejection region as shown in Fig. 5.7. The null hypothesis is not rejected and we believe that the standard deviation of the PM_1 level is more than or equal to 23.50.

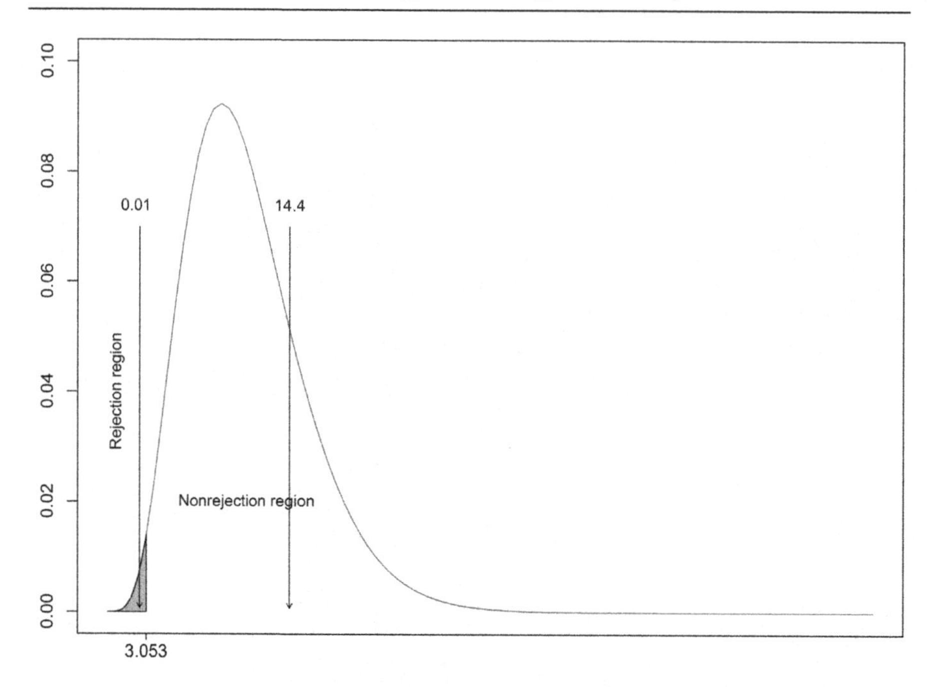

Figure 5.7 Rejection and nonrejection regions for the variance of the PM_1 level in the air of the palm oil mill.

We can conclude at a 1% significance level, there is sufficient evidence provided by the collected data to believe that the variance of the level of PM_1 is more than or equal to 23.50.

Further reading

Alkarkhi, A. F. M., Ahmad, A., & Easa, A. M. (2009). Assessment of surface water quality of selected estuaries of Malaysia: multivariate statistical techniques. *The Environmentalist*, *29*, 255–262.

Alkarkhi, A. F. M., & Chin, L. H. (2012). *Elementary statistics for technologist* (1st ed.). Malaysia: Universiti Sains Malaysia.

Alkarkhi, A. F. M., Ismail, N., Ahmed, A., & Easa, A. M. (2009). Analysis of heavy metal concentrations in sediments of selected estuaries of Malaysia—a statistical assessment. *Environmental Monitoring and Assessment*, *153*, 179–185.

Bluman, A. G. (1998). *Elementary statistics: A step by step approach* (3rd ed.). Boston: WCB: McGraw-Hill.

NIST/SEMATECH. (2013). Engineering statistics handbook. In, vol. 2020: NIST/SEMATECH. https://www.itl.nist.gov/div898/handbook/eda/section3/eda3674.htm.

Rahman, N. N. N. A., Chard, N. C., AlKarkhi, A. F. M., Rafatullah, M., & Kadir, M. O. A. (2017). Analysis of particulate matters in air of palm oil mills - A statistical assessment. *Environmental Engineering and Management Journal*, *16*(11), 2537–2543.

Weiss, N. A. (2012). *Introductory statistics* (9th ed.). Pearson.

The observed significance level (*P*-value) procedure

6

Abstract

This chapter addresses the concept and meaning of *P*-value with the general procedure on how to use it in making a decision for various tests including one-tailed and two-tailed tests. The calculations of *P*-value for normal distribution, t distribution, and chi-square distribution are presented clearly with detailed explanation on how to get each value from critical value tables and also technology. Various examples are presented and solved using the *P*-value procedure in making a decision and the results compared with the traditional procedure. The chapter includes 21 examples to demonstrate the calculation and use of *P*-value for environmental science problems covering Z-tests, t-tests, and chi-square tests.

Keywords: *P*-value; traditional procedure; Z-test; t-test; chi-square test; mean; proportion; variance

Learning outcomes

After completing this chapter, readers will be able to:

- Understand the idea and the meaning of *P*-value;
- Know how to compute and use the *P*-value;
- Describe the common steps for computing the *P*-value regarding one-sided and two-sided tests;
- Know how to compute the *P*-value for Z-tests, t-tests, and chi-square tests;
- Understand the steps for interpreting the *P*-value and match it to the problem under study;
- Draw smart inferences concerning the problem under investigation.

6.1 Introduction

We have studied the critical value procedure (traditional procedure) as the first procedure used to test various hypotheses regarding population parameters. The second procedure used for performing hypothesis testing is called the observed significance level, also known as the *P*-value procedure.

This chapter covers the idea of the observed significance level which is better known as the *P*-value, the procedure used to perform various hypothesis testing, and interpretation of the results are delivered in a simple and step-by-step way including hypothesis testing based on t distribution, normal distribution, and chi-square distribution.

Applications of Hypothesis Testing for Environmental Science. DOI: https://doi.org/10.1016/B978-0-12-824301-5.00010-1

6.2 What is the observed significance level?

The observed significance level (*P*-value) is considered to be the second procedure for making a decision regarding the hypothesis under investigation.

Let us give the concept and definition of *P*-value and then illustrate the procedure of using the *P*-value for hypothesis testing employing examples taken from the environmental field.

Suppose the null hypothesis is correct, the probability of obtaining a more extreme value of the test statistic than the one actually observed is called the observed significance level (*P*-value). Because *P*-value is a probability value, thus its value falls between 0 and 1.

We can perform hypothesis testing using the *P*-value procedure using similar steps to those used for the critical value (traditional) procedure. The five steps for conducting hypothesis testing employing the *P*-value procedure are presented below.

Step 1: Specify the null and alternative hypotheses

Step 2: Select the significance level (α) for the study

Step 3: Use the sample information to calculate the test statistic value

Step 4: Calculate the *P*-value and identify the critical and noncritical regions for the study

Step 5: Make a decision using *P*-value and interpret the results

A small *P*-value leads to rejecting the null hypothesis while a large *P*-value leads to not rejecting the null hypothesis.

Reject the null hypothesis if the $P-\text{value} \leq \alpha$ and fail to reject the null hypothesis if the $P\text{-value} > \alpha$.

We discuss the general procedure step by step, supported by examples where necessary.

6.3 Computing the *P*-value for a Z-test

A Z-test for testing the mean and proportion values for different types of hypotheses using the critical value procedure was studied. The steps for computing the *P*-value for a Z-test including one-sided and two-sided tests will be illustrated step by step, with examples using the general procedure for hypothesis testing.

Example 6.1: Compute the *P*-value to the left of a Z value: Compute the *P*-value to the left of a negative Z value ($Z = -1.25$) (left-tailed test). Use a significance level of 0.01 ($\alpha = 0.01$).

Computing the *P*-value for a left-tailed Z-test can be achieved employing the general procedure for testing a hypothesis.

Step 1: Specify the null and alternative hypotheses

The two hypotheses (the null and alternative) for a left-tailed test can be written as presented in Eq. (6.1):

$$H_0{:}\mu \geq c \text{ vs } H_1{:}\mu < c \tag{6.1}$$

The hypothesis in Eq. (6.1) represents a one-tailed test because the alternative hypothesis is $H_1{:}\mu < c$ (left-tailed), where c is a given value.

Step 2: Select the significance level (α) for the study

The level of significance is chosen to be 0.01.

Step 3: Use the sample information to calculate the test statistic value

The test statistic value of Z is given to be $Z = -1.25$, otherwise we have to calculate it using a formula.

Step 4: Calculate the *P*-value and identify the critical and noncritical regions for the study

The *P*-value for the Z-test can be computed easily by employing the standard normal table (Table A in the Appendix) to calculate the probability of Z (-1.25) or less, which represents the required *P*-value. The area to the left of -1.25 is 0.1056 ($1 - 0.8944 = 01056$) which represents the *P*-value. The shaded area in Fig. 6.1 represents the required *P*-value.

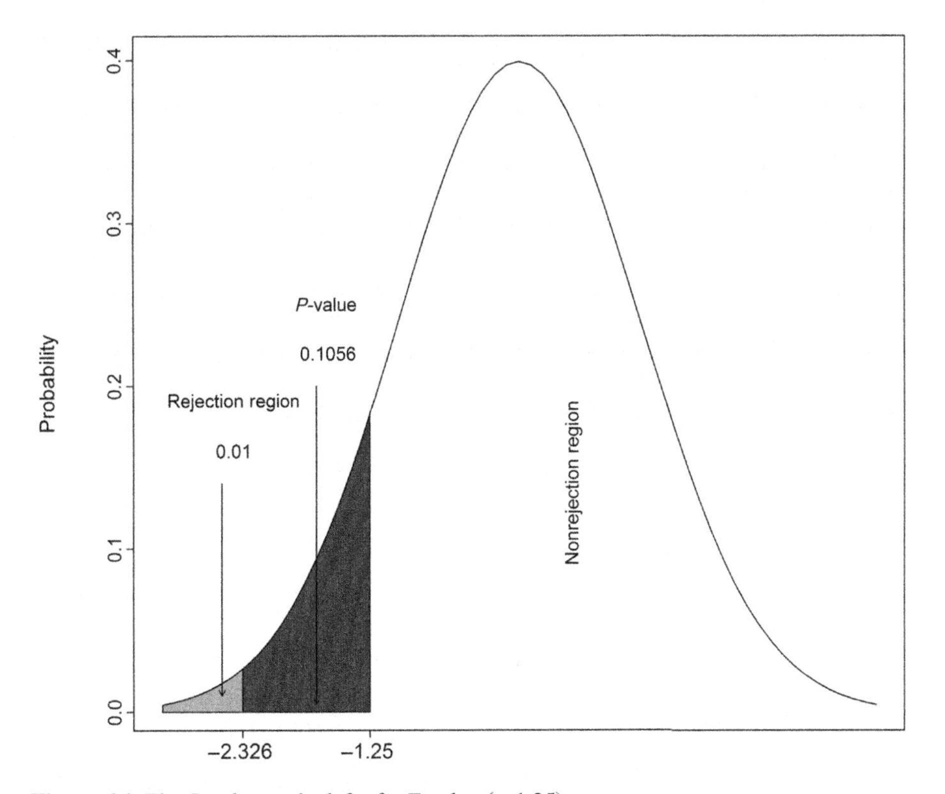

Figure 6.1 The *P*-value to the left of a Z value (-1.25).

Step 5: Make a decision using *P*-value and interpret the results

The null hypothesis should not be rejected because the *P*-value is large and greater than the level of significance (0.01), (0.1056 > 0.01). We can say that the sample data supplied sufficient proof to make a decision not to reject the null hypothesis.

Example 6.2: Compute the area to the right of a Z value: Compute the *P*-value to the right of a positive Z value (Z = 2.35) (right-tailed test). Use a significance level of 0.01 ($\alpha = 0.01$).

Computing the *P*-value for a right-tailed Z-test can be achieved employing the general procedure for testing a hypothesis.

Step 1: Specify the null and alternative hypotheses

The two hypotheses (the null and alternative) for a right-tailed test can be written as presented in Eq. (6.2).

$$H_0{:}\mu \leq c \text{ vs } H_1{:}\mu > c \tag{6.2}$$

The hypothesis in Eq. (6.2) represents a one-tailed test because the alternative hypothesis is $H_1{:}\mu > c$ (right-tailed), where c is a given value.

Step 2: Select the significance level (α) for the study

The level of significance is chosen to be 0.01.

Step 3: Use the sample information to calculate the test statistic value

The test statistic value of Z is given to be $Z = 2.35$, otherwise we have to calculate it using a formula.

Step 4: Calculate the *P*-value and identify the critical and noncritical regions for the study

The *P*-value for a Z-test can be computed easily by employing the standard normal table (Table A in the Appendix) to calculate the probability of Z (2.35) or more which represents the required *P*-value. The area to the right of 2.35 is 0.0094 ($1 - 0.9906 = 0.0094$) which represents the *P*-value. The shaded area in Fig. 6.2 represents the required *P*-value.

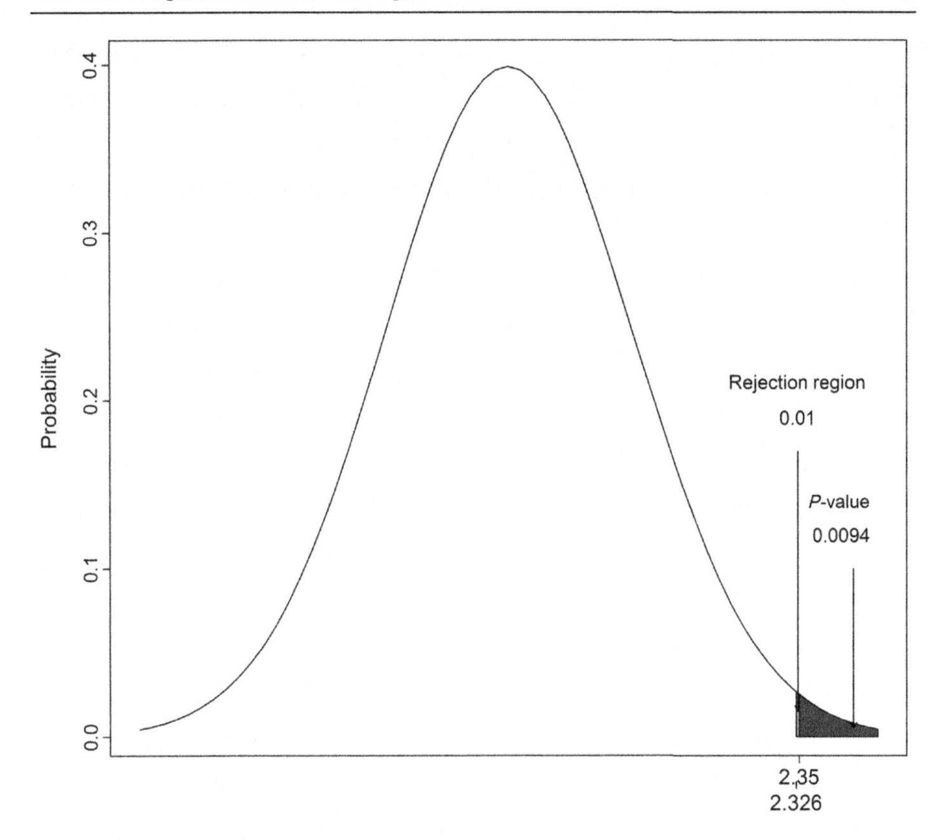

Figure 6.2 The *P*-value to the right of a Z value (2.35).

Step 5: Make a decision using the *P*-value and interpret the results

The null hypothesis should be rejected because the *P*-value is small and less than the level of significance (0.01), ($0.0094 < 0.01$). We can say that the sample data supplied sufficient proof to make a decision to reject the null hypothesis in favor of the alternative hypothesis.

Example 6.3: Compute the *P*-value on the two tails: Compute the *P*-value to the left of a negative Z value ($Z = -1.96$) and to the right of a positive Z value ($Z = 1.96$) (two-tailed test). Use a significance level of 0.01 ($\alpha = 0.01$).

Computing the *P*-value for a two-tailed Z-test can be achieved employing the general procedure for testing a hypothesis.

Step 1: Specify the null and alternative hypotheses

The two hypotheses (the null and alternative) for a two-tailed test can be written as presented in Eq. (6.3).

$$H_0{:}\mu = c \text{ vs } H_1{:}\mu \neq c \tag{6.3}$$

The hypothesis in Eq. (6.3) represents a two-tailed test because the alternative hypothesis is $H_1{:}\mu \neq c$ (two-tailed), where c is a given value.

Step 2: Select the significance level (α) for the study

The level of significance is chosen to be 0.01.

Step 3: Use the sample information to calculate the test statistic value

The test statistic value of Z is given to be $Z = 1.96$, otherwise we have to calculate it using a formula.

Step 4: Calculate the *P*-value and identify the critical and noncritical regions for the study

The *P*-value for a Z-test can be computed easily by employing the standard normal table (Table A in the Appendix) to calculate the probability of Z (-1.96) or less, of Z (1.96) or more which represents the required *P*-value. The area to the right of 1.96 is 0.025 ($1 - 0.9750 = 0.025$), each side (the left and right) receives 0.025. Because the test is two-tailed, the *P*-value should be multiplied by $2 \times (0.025) = 0.05$ to get the required *P*-value. The shaded area in Fig. 6.3 represents the required *P*-value.

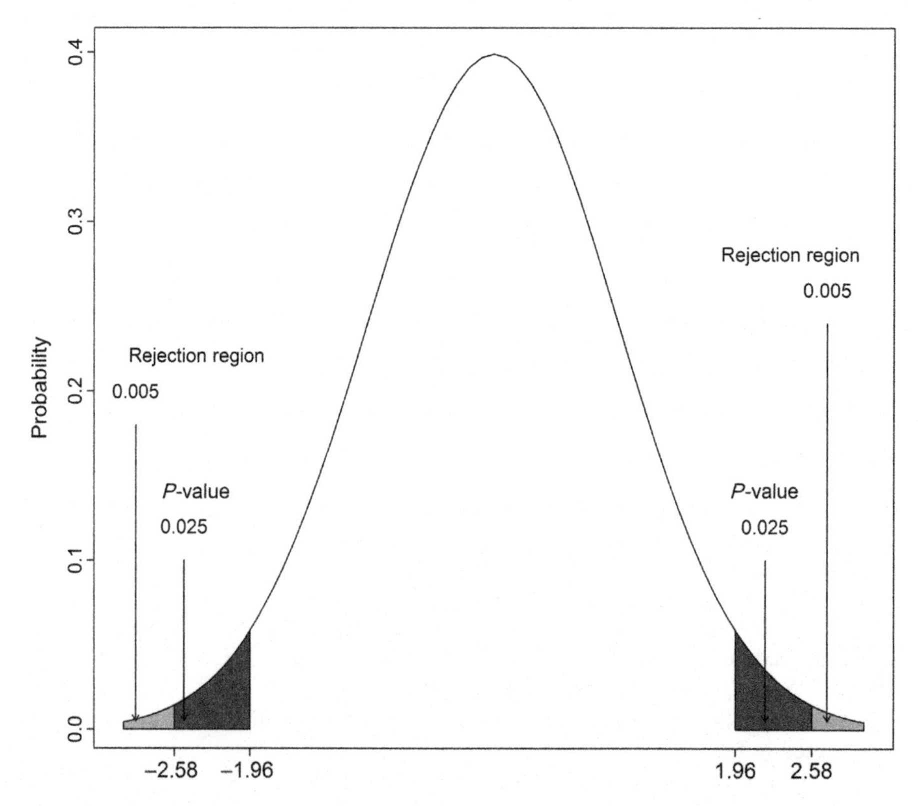

Figure 6.3 The *P*-value to the left of a negative Z value (-1.96) and to the right of a positive Z value (1.96).

Step 5: Make a decision using the *P*-value and interpret the results

The null hypothesis should not be rejected because the *P*-value is large and greater than the level of significance (0.01), $(0.05 > 0.01)$. We can say that the sample data supplied sufficient proof to make a decision not to reject the null hypothesis.

6.4 Testing one sample mean when the variance is known: *P*-value

We have studied hypothesis testing for one sample mean when the sample size is large using the critical value (traditional) procedure. The *P*-value procedure for one sample mean using a Z-test will be illustrated employing the same examples presented in Chapter 2, Z-test for one-sample mean, and solved using the critical value procedure. The three situations of hypothesis testing are covered and the *P*-value are calculated.

Example 6.4: The concentration of cadmium of surface water: Example 2.5 is reproduced "A professor at an environmental section wanted to verify the claim that the mean concentration of cadmium (Cd) of surface water in Juru River is 1.4 (mg/L). He selected 35 samples and tested for cadmium concentration. The collected data showed that the mean concentration of cadmium is 1.6 and the standard deviation of the population is 0.4. A significance level of $\alpha = 0.01$ is chosen to test the claim. Assume that the population is normally distributed."

The five steps for conducting hypothesis testing employing the *P*-value procedure can be used to test the hypothesis regarding the mean concentration of cadmium of surface water in Juru River. The results of the *P*-value procedure will be compared with the critical value (traditional) procedure.

Step 1: Specify the null and alternative hypotheses

The two hypotheses regarding the mean concentration of cadmium of surface water in Juru River are presented in Eq. (6.4).

$$H_0{:}\mu = 1.4 \text{ vs } H_1{:}\mu \neq 1.4 \tag{6.4}$$

Step 2: Select the significance level (α) for the study

The level of significance is chosen to be 0.01. The Z critical values for a two-tailed test with $\alpha = 0.01$ are ± 2.58.

The two procedures will be used to solve this problem; namely the critical value and *P*-value procedures.

Step 3: Use the sample information to calculate the test statistic value

The test statistic value for the Z-test is used to make a decision regarding the mean concentration of cadmium of surface water in Juru River. The test statistic value using the Z-test formula was calculated to be 2.96.

Step 4: Calculate the *P*-value and identify the critical and noncritical regions for the study

The rejection and nonrejection regions using the critical values for testing the mean concentration of cadmium of surface water in Juru River are presented in Fig. 6.4 for the standard normal curve (shaded area: blue represents the *P*-value and orange represents the significance level).

The hypothesis in Eq. (6.4) represents a two-tailed Z-test. The *P*-value represents the probability of observing the test statistic value (Z) or greater or less if the null hypothesis is correct. The area under the standard normal curve to the right of $Z = 2.96$ is 0.0015 $(1 - 0.9984 = 0.0015)$ (blue), and the area to the left of $Z = -2.96$ is 0.0015 (blue). Because the test is two-tailed, the *P*-value is $(2 \times 0.0015 = 0.003)$. The *P*-value is distributed on the right and left tails as shown in Fig. 6.4.

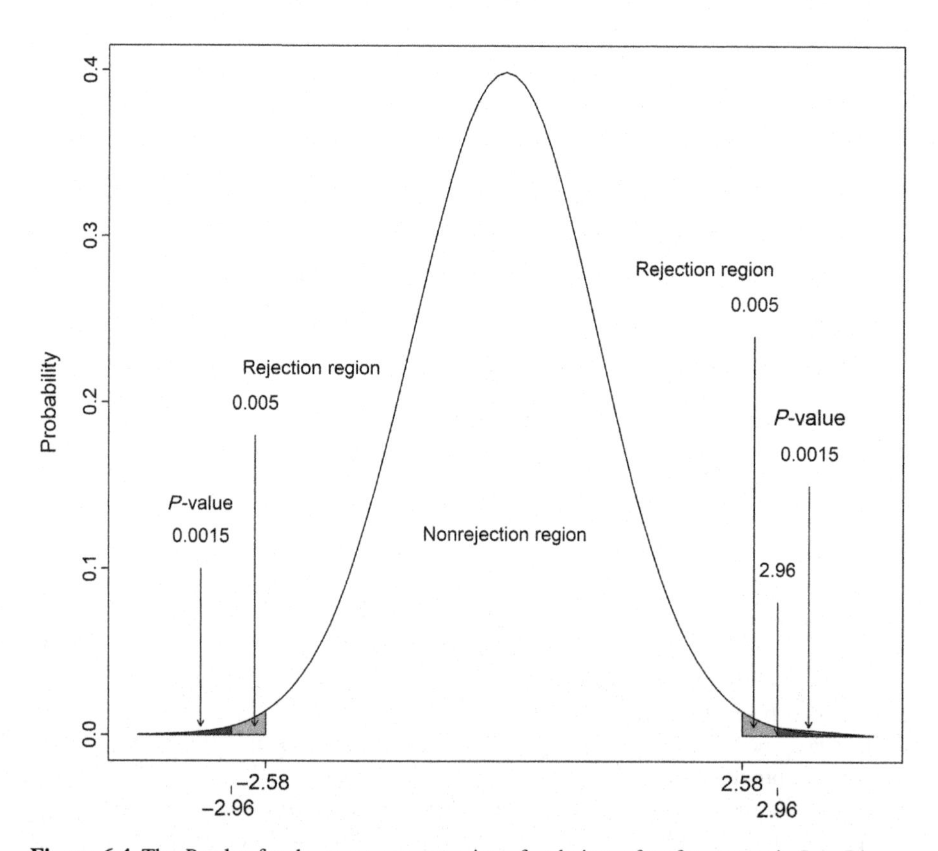

Figure 6.4 The *P*-value for the mean concentration of cadmium of surface water in Juru River.

Step 5: Make a decision using the P-value and interpret the results

1. The decision reached by the critical value procedure: reject the null hypothesis, see Example 2.5.
2. The decision reached by the P-value procedure: reject the null hypothesis because the P-value (0.0015) is smaller than the level of significance (0.01), $(0.0015 < 0.01)$.

The decision reached using the two procedures: rejected the null hypothesis in favor of the alternative hypothesis and we believe that the mean concentration of cadmium of surface water in Juru River is not equal to 1.4.

Example 6.5: The concentration of lead in sediment: Example 2.6 is reproduced "A researcher at a research institution wanted to verify the claim that the mean concentration of lead (Pb) in sediment of Jejawi River is less than 0.23 (mg/L). He selected 33 samples and tested for lead concentration. The collected data showed that the mean concentration of lead is 0.235 and the standard deviation of the population is 0.106. A significance level of $\alpha = 0.01$ is chosen to test the claim. Assume that the population is normally distributed."

The five steps for conducting hypothesis testing employing the P-value procedure can be used to test the hypothesis regarding the mean concentration of lead in sediment of Jejawi River. The results of the P-value procedure will be compared with the critical value (traditional) procedure.

Step 1: Specify the null and alternative hypotheses

The two hypotheses regarding the mean concentration of lead in sediment of Jejawi River are presented in Eq. (6.5).

$$H_0{:}\mu \geq 0.23 \text{ vs } H_1{:}\mu < 0.23 \tag{6.5}$$

Step 2: Select the significance level (α) for the study

The level of significance is chosen to be 0.01. The Z critical value for a left-tailed test with $\alpha = 0.01$ is -2.326.

The two procedures will be used to solve this problem; namely the critical value and P-value procedures.

Step 3: Use the sample information to calculate the test statistic value

The test statistic value for the Z-test is used to make a decision regarding the mean concentration of lead in sediment of Jejawi River. The test statistic value using the Z-test formula was calculated to be 0.27.

Step 4: Calculate the P-value and identify the critical and noncritical regions for the study

The rejection and nonrejection regions using the critical value for testing the mean concentration of lead in sediment of Jejawi River are presented in Fig. 6.5 for the standard normal curve (shaded area: blue represents the P-value and orange represents the significance level).

The hypothesis in Eq. (6.5) represents a left-tailed Z-test. The P-value represents the probability of observing the test statistic value (Z) or less if the null hypothesis is correct. The area under the standard normal curve to the left of $Z = 0.27$ is 0.6064 (blue). The P-value falls on the left tail as shown in Fig. 6.5.

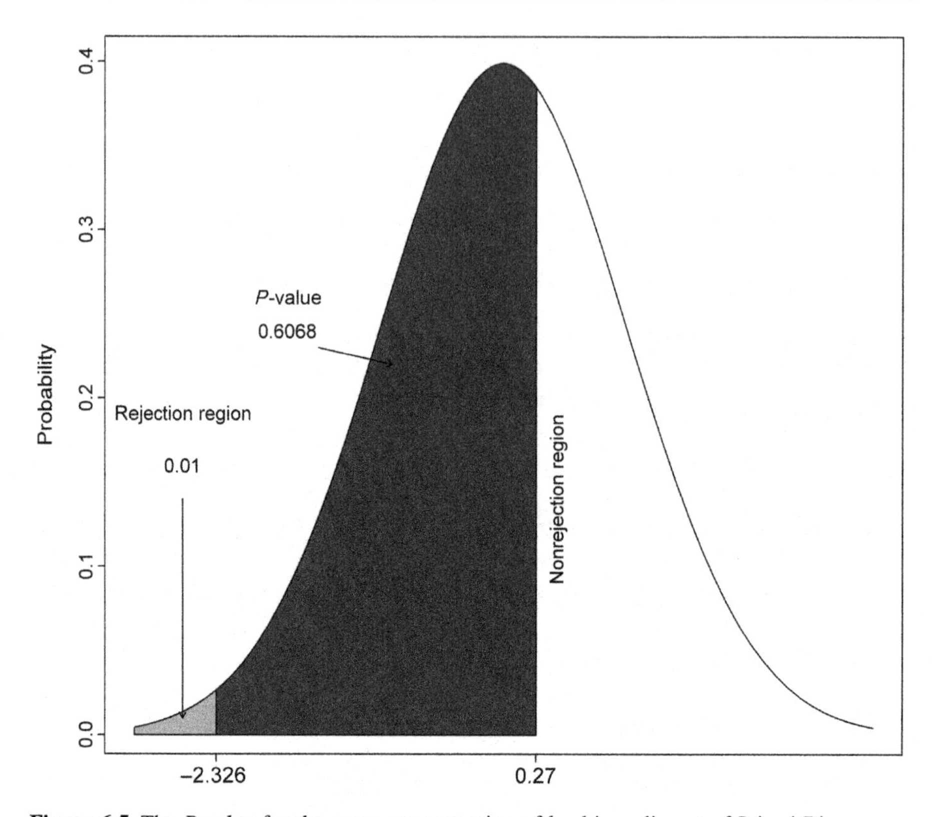

Figure 6.5 The *P*-value for the mean concentration of lead in sediment of Jejawi River.

Step 5: Make a decision using the *P*-value and interpret the results

1. The decision reached using the critical value procedure: fail to reject the null hypothesis, see Example 2.6.
2. The decision reached by the *P*-value procedure: fail to reject the null hypothesis because the *P*-value (0.6064) is greater than the level of significance (0.01), $(0.6064 > 0.01)$.

The decision reached using the two procedures: fail to reject the null hypothesis and believe that the mean concentration of lead in sediment of Jejawi River is more than or equal to 0.23.

Example 6.6: The concentration of sodium in particulate matter in the air: Example 2.7 is reproduced "The concentration of sodium (Na) as an inorganic element in particulate matter (PM_{10}) in the air environment of an equatorial urban coastal location during the summer season was investigated. A researcher wants to verify that the concentration of sodium in the air during the summer season is less than or equal to 6.20. He selected 34 samples and the concentration of sodium was measured. The average concentration of sodium provided by the sample data is 7.46 $(\mu g\, m^{-3})$ and the population standard deviation is 2.18. A significance level of $\alpha = 0.01$ is chosen to test the claim. Assume that the population is normally distributed."

The five steps for conducting hypothesis testing employing the *P*-value procedure can be used to test the hypothesis regarding the mean concentration of sodium in the air environment of an equatorial urban coastal location during the summer season. The results of the *P*-value procedure will be compared with the critical value (traditional) procedure.

Step 1: Specify the null and alternative hypotheses

The two hypotheses regarding the mean concentration of sodium in the air environment of an equatorial urban coastal location during the summer season are presented in Eq. (6.6).

$$H_0:\mu \leq 6.20 \text{ vs } H_1:\mu > 6.20 \tag{6.6}$$

Step 2: Select the significance level (α) for the study

The level of significance is chosen to be 0.01. The Z critical value for a right-tailed test with $\alpha = 0.01$ is 2.326.

The two procedures will be used to solve this problem; namely the critical value and *P*-value procedures.

Step 3: Use the sample information to calculate the test statistic value

The test statistic value for the Z-test is used to make a decision regarding the mean concentration of sodium in the air environment of an equatorial urban coastal location during the summer season. The test statistic value using the Z-test formula was calculated to be 3.37.

Step 4: Calculate the *P*-value and identify the critical and noncritical regions for the study

The rejection and nonrejection regions using the critical value for testing the mean concentration of sodium in the air environment of an equatorial urban coastal location during the summer season are presented in Fig. 6.6 for the standard normal curve (shaded area: blue represents the *P*-value and orange represents the significance level).

The hypothesis in Eq. (6.6) represents a right-tailed Z-test. The *P*-value represents the probability of observing the test statistic value (Z) or greater if the null hypothesis is correct. The area under the standard normal curve to the right of $Z = 3.37$ is $0.0003755911 = 0.00038$ (blue). The *P*-value falls on the right tail as shown in Fig. 6.6.

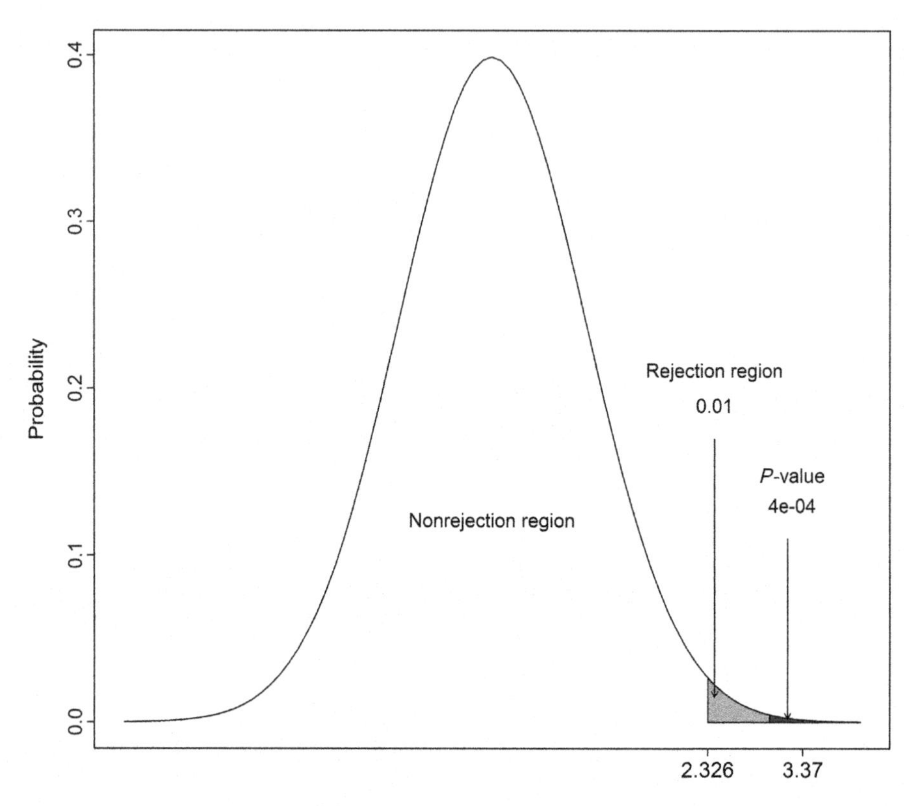

Figure 6.6 The *P*-value for the mean concentration of sodium (Na) in the air environment during the summer season.

Step 5: Make a decision using the *P*-value and interpret the results

1. The decision reached by the critical value procedure: reject the null hypothesis, see Example 2.7.
2. The decision reached by the *P*-value procedure: reject the null hypothesis because the *P*-value (0.00038) is smaller than the level of significance (0.01), $(0.00038 < 0.01)$.

The decision reached using the two procedures: reject the null hypothesis in favor of the alternative hypothesis and we believe that the mean concentration of sodium in the air environment is more than 6.20.

6.5 Computing the *P*-value for a t-test

This section presents the steps for computing the *P*-value for a t-test including one-sided and two-sided tests. The procedure will be illustrated step by step with examples using the general procedure for hypothesis testing.

Example 6.7: Compute the *P*-value to the left of a t value: Compute the *P*-value to the left of a negative t value ($t = -2.02$) and sample size 16, (left-tailed test). Use a significance level of 0.01 ($\alpha = 0.01$).

Computing the *P*-value for a left-tailed t-test can be achieved employing the general procedure for testing a hypothesis.

Step 1: Specify the null and alternative hypotheses

The two hypotheses (the null and alternative) for a left-tailed test can be written as presented in Eq. (6.7).

$$H_0{:}\mu \geq c \text{ vs } H_1{:}\mu < c \tag{6.7}$$

The hypothesis in Eq. (6.7) represents a one-tailed test because the alternative hypothesis is $H_1{:}\mu < c$ (left-tailed), where c is a given value.

Step 2: Select the significance level (α) for the study

The level of significance is chosen to be 0.01.

Step 3: Use the sample information to calculate the test statistic value

The test statistic value for the t-test is given to be $t = -2.02$, otherwise we have to calculate it using a formula.

Step 4: Calculate the *P*-value and identify the critical and noncritical regions for the study

The *P*-value for a t-test can be computed easily by employing the t table (Table B in the Appendix) which depends on two values: the degrees of freedom ($d.f$) and the level of significance (α). The *P*-value for -2.02 with $d.f = 15$ falls somewhere in the interval $0.025 < P\text{-value} < 0.05$ as illustrated below.

We can use the symmetry property of t distribution to compute the *P*-value to the left of a t value because all values in the table are for the positive value. The position of the required value (2.02) with degrees of freedom ($d.f = 15$) falls between 1.753 and 2.131, these two values correspond with the significance level of 0.025 and 0.05 for 1.753 and 2.131, respectively. Thus, one can say that the *P*-value for $t = -2.02$ and $d.f = 15$ is somewhere in the interval $0.025 < P\text{-value} < 0.05$. The exact *P*-value ($0.03147346 = 0.03$) using a statistical software is presented in Fig. 6.7.

The blue shaded area in Fig. 6.7 represents the required *P*-value for the value -2.02 and the level of significance (0.01) represents the orange shaded area.

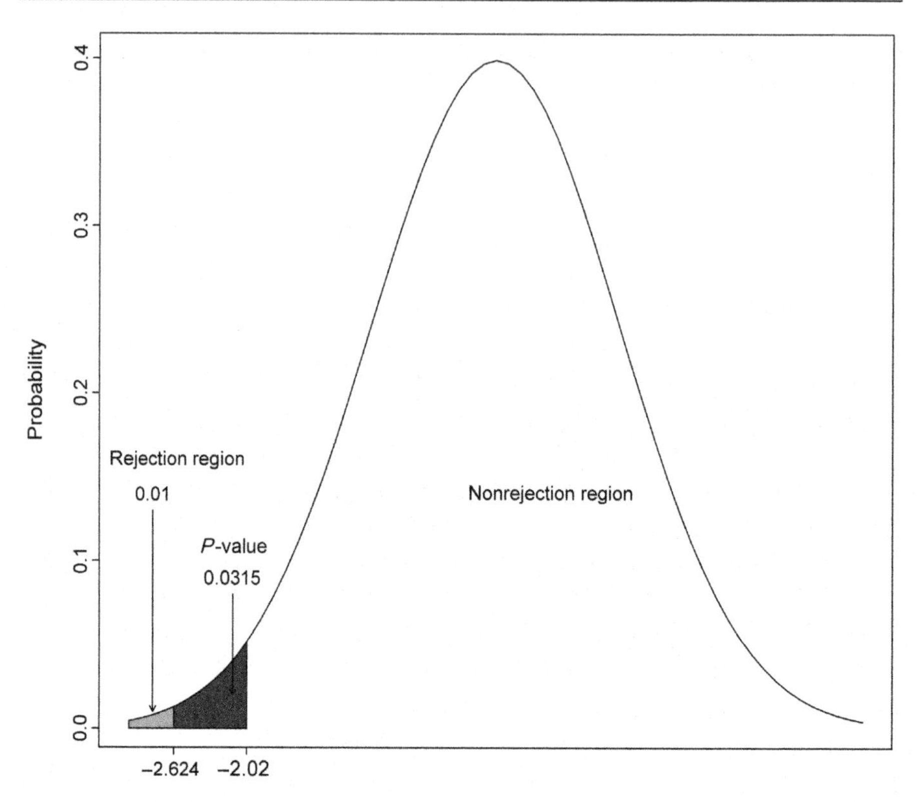

Figure 6.7 The *P*-value to the left of a negative t value ($t = -2.02$).

Step 5: Make a decision using the *P*-value and interpret the results

The null hypothesis should not be rejected because the *P*-value is greater than the level of significance (0.01), ($0.03 > 0.01$). We can say that the sample data supplied sufficient proof to make a decision not to reject the null hypothesis.

Example 6.8: Compute the *P*-value to the right of a t value: Compute the *P*-value to the right of a positive t value ($t = 2.91$) and sample size 20 (right-tailed test). Use a significance level of 0.01 ($\alpha = 0.01$).

Computing the *P*-value for a right-tailed t-test can be achieved employing the general procedure for testing a hypothesis.

Step 1: Specify the null and alternative hypotheses

The two hypotheses (the null and alternative) for a right-tailed test can be written as presented in Eq. (6.8).

$$H_0{:}\mu \leq c \text{ vs } H_1{:}\mu > c \tag{6.8}$$

The hypothesis in Eq. (6.8) represents a one-tailed test because the alternative hypothesis is $H_1{:}\mu > c$ (right-tailed), where c is a given value.

Step 2: Select the significance level (α) for the study

The level of significance is chosen to be 0.01.

Step 3: Use the sample information to calculate the test statistic value

The test statistic value for a t-test is given to be $t = 2.91$, otherwise we have to calculate it using a formula.

Step 4: Calculate the *P*-value and specify the critical and noncritical regions for the study

The *P*-value for a t-test can be computed easily by employing the t table (Table B in the Appendix) which depends on two values: the degrees of freedom ($d.f$) and the level of significance (α). The *P*-value for 2.91 with $d.f = 19$ is less than 0.005 because the value 2.91 falls beyond the *P*-value < 0.005 as illustrated below.

The position of required value (2.91) with degrees of freedom ($d.f = 19$) falls beyond the value of 2.861 (Table B in the Appendix), this value corresponds with the significance level of 0.005. Thus, one can say that the *P*-value for $t = 2.91$ and $d.f = 19$ is somewhere less than *P*-value < 0.005. The exact *P*-value ($0.004489562 = 0.0045$) using statistical software is presented in Fig. 6.8.

The shaded area (blue) in Fig. 6.8 represents the required *P*-value for the value 2.91 and the orange shaded area represents the level of significance (0.01).

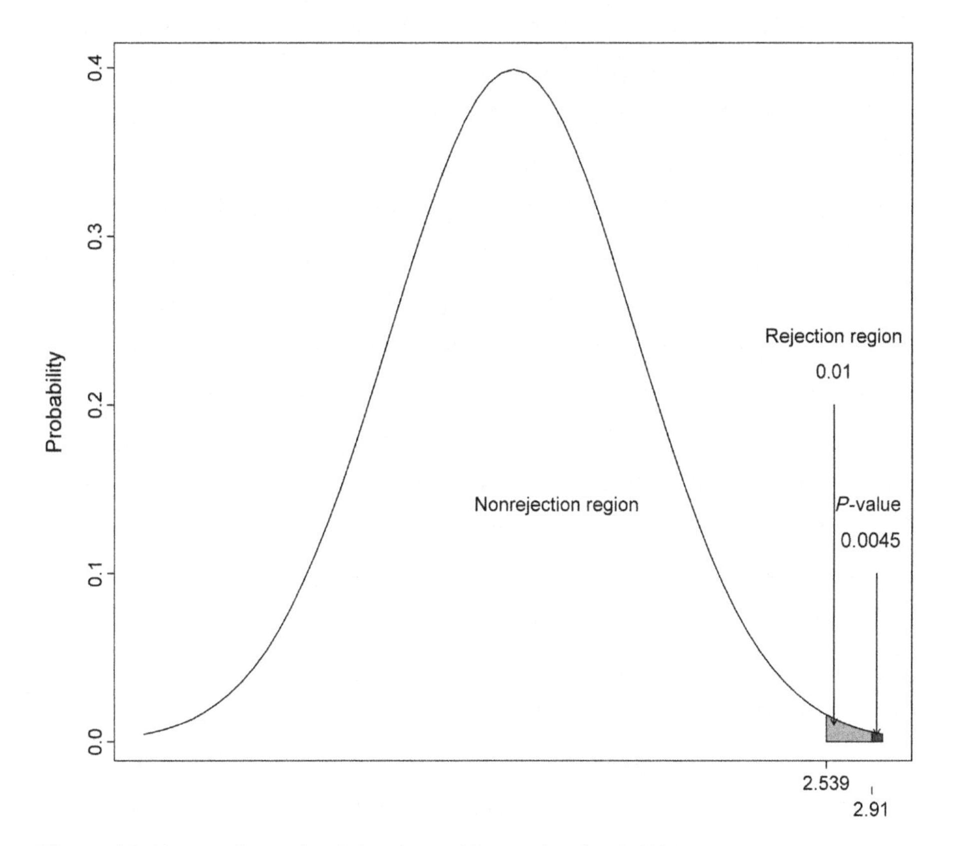

Figure 6.8 The *P*-value to the right of a positive t value ($t = 2.91$).

Step 5: Make a decision using the P-value and interpret the results

The null hypothesis should be rejected because the P-value is smaller than the level of significance (0.01), ($0.0045 < 0.05$). We can say that the sample data supplied sufficient proof to make a decision to reject the null hypothesis.

Example 6.9: Compute the *P*-value on the two tails: Compute the P-value to the left of a negative t value ($t = -2.01$) and to the right of a positive t value $t = 2.01$ (two-tailed test, the t-test statistic value is 2.01) and sample size equal to 28. Use a significance level of 0.01 ($\alpha = 0.01$).

Computing the P-value for a two-tailed t-test can be achieved employing the general procedure for testing a hypothesis.

Step 1: Specify the null and alternative hypotheses

The two hypotheses (the null and alternative) for a two-tailed test can be written as presented in Eq. (6.9).

$$H_0{:}\mu = c \text{ vs } H_1{:}\mu \neq c \tag{6.9}$$

The hypothesis in Eq. (6.9) represents a two-tailed test because the alternative hypothesis is $H_1{:}\mu \neq c$ (two-tailed), where c is a given value.

Step 2: Select the significance level (α) for the study

The level of significance is chosen to be 0.01.

Step 3: Use the sample information to calculate the test statistic value

The test statistic value for a t-test is given to be $t = 2.01$, otherwise we have to calculate it using a formula.

Step 4: Calculate the P-value and identify the critical and noncritical regions for the study

The P-value for a t-test can be computed easily by employing the t table (Table B in the Appendix) which depends on two values: the degrees of freedom ($d.f$) and the level of significance (α). The P-value for 2.01 with $d.f = 27$ falls somewhere in the interval $0.025 < P\text{-value} < 0.05$ as illustrated below.

The position of the required value (2.01) with degrees of freedom ($d.f = 27$) falls between 1.703 and 2.052, and these two values correspond with the significance level of 0.025 and 0.05 for 1.703 and 2.052, respectively. Thus, one can say that the P-value for $t = 2.01$ and $d.f = 27$ is somewhere in the interval $0.025 < P\text{-value} < 0.05$. The exact P-value is presented in Fig. 6.9.

The shaded area (blue) in Fig. 6.9 represents the required P-value for the value 2.01 and the level of significance (0.01) is represented by the shaded area (orange). The exact P-value using statistical software is 0.0545, each tail receives 0.0272.

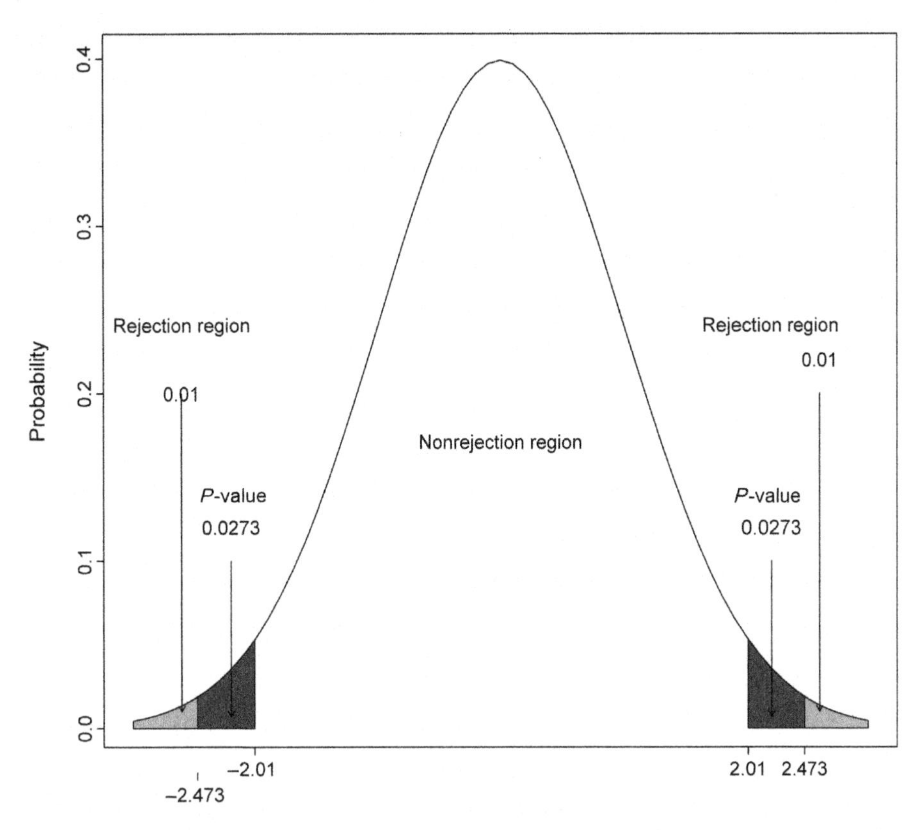

Figure 6.9 The *P*-value to the left of a negative t value ($t = -2.01$) and to the right of a positive t value $t = 2.01$.

Step 5: Make a decision using the *P*-value and interpret the results

The null hypothesis should not be rejected because the *P*-value is greater than the level of significance (0.01), ($0.0545 > 0.01$). We can say that the sample data supplied sufficient proof to make a decision not to reject the null hypothesis.

6.6 Testing one sample mean when the variance is unknown: *P*-value

We have studied hypothesis testing for one sample mean when the population variance is unknown and the sample size is small using the critical value (traditional) procedure for a t-test. The *P*-value procedure for one sample mean using a t-test will be illustrated employing the same examples presented in Chapter 3, t-Test for one-sample mean, and solved using the critical value procedure.

The three situations of hypothesis testing are covered and the *P*-values are calculated.

Example 6.10: The concentration of trace metal iron in the air: Example 3.4 is reproduced "A researcher at an environmental section wishes to verify the claim that the mean concentration of trace metal iron (Fe) on coarse particulate matter (PM_{10}) in the air is 0.36 $(\mu g/m^3)$ during the summer season of an equatorial urban coastal location. Twelve samples were selected, and the concentration of iron was measured. The collected data showed that the mean concentration of iron is 0.44 and the standard deviation is 0.34. A significance level of $\alpha = 0.01$ is chosen to test the claim. Assume that the population is normally distributed."

The five steps for conducting hypothesis testing employing the *P*-value procedure can be used to test the hypothesis regarding the mean concentration of trace metal iron on coarse particulate matter in the air during the summer season of an equatorial urban coastal location. The results of the *P*-value procedure will be compared with the critical value (traditional) procedure.

Step 1: Specify the null and alternative hypotheses

The two hypotheses regarding the mean concentration of iron on coarse particulate matter in the air are presented in Eq. (6.10).

$$H_0{:}\mu = 0.36 \text{ vs } H_1{:}\mu \neq 0.36 \tag{6.10}$$

Step 2: Select the significance level (α) for the study

The level of significance is chosen to be 0.01. The t critical values for a two-tailed test with $\alpha = 0.01$ and $d.f = 12 - 1 = 11$, are $t_{(\frac{\alpha}{2}, d.f)} = t_{(\frac{0.01}{2}, 12-1)} = \mp 3.106$.

The two procedures will be used to solve this problem: namely the critical value and *P*-value procedures.

Step 3: Use the sample information to calculate the test statistic value

The test statistic value for the t-test is used to make a decision regarding the mean concentration of iron in the air. The test statistic value using the t-test formula was calculated to be 0.82.

Step 4: Calculate the *P*-value and identify the critical and noncritical regions for the study

The rejection and nonrejection regions using the critical values for testing the mean concentration of iron in the air are presented in Fig. 6.10 for a t distribution curve (shaded area: blue represents the *P*-value and orange represents the significance level).

The hypothesis in Eq. (6.10) represents a two-tailed t-test. The *P*-value represents the probability of observing the test statistic value (t) or greater or less if the null hypothesis is correct. The area under the t distribution curve to the right of $t = 0.82$ is $0.2161593 = 0.216$, and the area to the left of $t = -0.82$ is $0.2161593 = 0.216$. Because the test is two-tailed, the *P*-value is $(2 \times 0.2161593 = 0.432)$. The exact *P*-values on the right and left tails are calculated using statistical software as shown in Fig. 6.10.

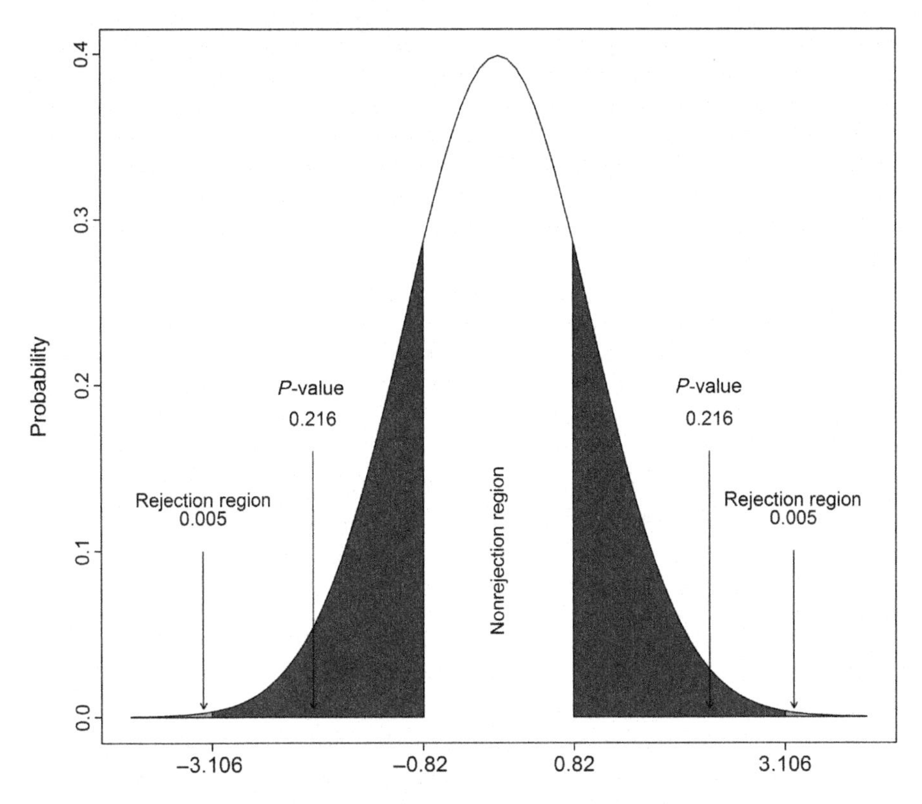

Figure 6.10 The *P*-value for the mean concentration of iron in the air.

Step 5: Make a decision using the *P*-value and interpret the results

1. The decision reached by the critical value procedure: fail to reject the null hypothesis, see Example 3.4.
2. The decision reached by the *P*-value procedure: fail to reject the null hypothesis because the *P*-value (0.432) is greater than the level of significance (0.01), $(0.432 > 0.01)$.

The decision reached using the two procedures: fail to reject the null hypothesis and we believe that the mean concentration of iron on coarse particulate matter in the air is equal to 0.36.

Example 6.11: The concentration of chemical oxygen demand of surface water: Example 3.5 is reproduced "An environmentalist at a research center wishes to verify the claim that the mean concentration of the chemical oxygen demand (COD) in surface water of Juru River is less than 650 (mg/L). Fourteen samples were selected and tested for chemical oxygen demand concentration. The collected data showed that the mean concentration of chemical oxygen demand is 470 and the standard deviation is 245. A significance level of $\alpha = 0.01$ is chosen to test the claim. Assume that the population is normally distributed."

The five steps for conducting hypothesis testing employing the *P*-value procedure can be used to test the hypothesis regarding the mean concentration of chemical oxygen demand in surface water of Juru River. The results of the *P*-value procedure will be compared with the critical value (traditional) procedure.

Step 1: Specify the null and alternative hypotheses

The two hypotheses regarding the mean concentration of chemical oxygen demand in surface water of Juru River are presented in Eq. (6.11).

$$H_0{:}\mu \geq 650 \text{ vs } H_1{:}\mu < 650 \tag{6.11}$$

Step 2: Select the significance level (α) for the study

The level of significance is chosen to be 0.01. The t critical value for a left-tailed test with $\alpha = 0.01$ and $d.f = 14 - 1 = 13$ is $t_{(\alpha,d.f)} = t_{(0.01,13)} = -2.65$.

The two procedures will be used to solve this problem; namely the critical value and *P*-value procedures.

Step 3: Use the sample information to calculate the test statistic value

The test statistic value for the t-test is used to make a decision regarding the mean concentration of chemical oxygen demand in surface water of Juru River. The test statistic value using the t-test formula was calculated to be −2.75.

Step 4: Calculate the *P*-value and identify the critical and noncritical regions for the study

The rejection and nonrejection regions using the critical values for testing the mean concentration of chemical oxygen demand in surface water of Juru River are presented in Fig. 6.11 for the t distribution curve (shaded area: blue represents the *P*-value and orange represents the significance level).

The hypothesis in Eq. (6.11) represents left-tailed t-test. The *P*-value represents the probability of observing the test statistic value (t) or less if the null hypothesis is correct. The area under the t distribution curve to the left of $t = -2.75$ is $0.008285168 = 0.008$. The exact *P*-value (0.008) on the left tail is calculated using a statistical software as shown in Fig. 6.11.

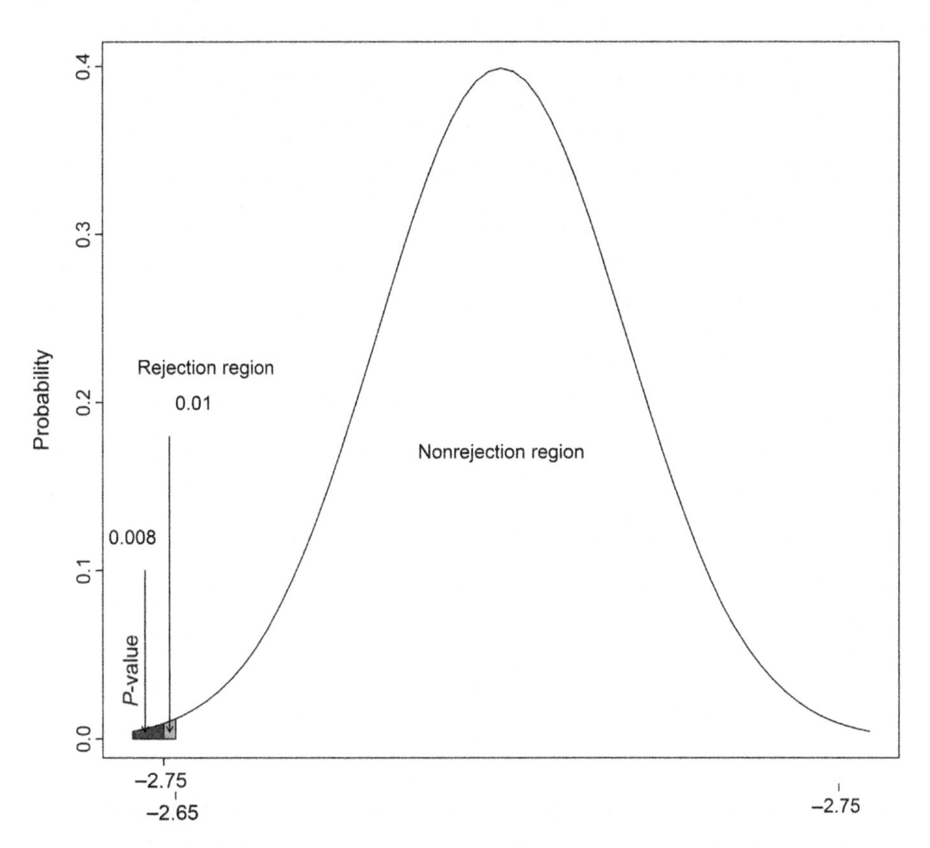

Figure 6.11 The *P*-value for the mean concentration of chemical oxygen demand in surface water of Juru River.

Step 5: Make a decision using the *P*-value and interpret the results

1. The decision reached by the critical value procedure: reject the null hypothesis, see Example 3.5.
2. The decision reached by the *P*-value procedure: reject the null hypothesis because the *P*-value (0.008) is smaller than the level of significance (0.01), $(0.008 < 0.01)$.

The decision reached using the two procedures: reject the null hypothesis and believe that the mean concentration of chemical oxygen demand in surface water of Juru River is less than 650.

Example 6.12: The concentration of total phosphate in the surface water: Example 3.6 is reproduced "The concentration of total phosphate of surface water of Juru River was studied. An environmentalist wishes to verify that the concentration of total phosphate of surface water of Juru River is less than or equal to 0.85 (mg/L). Thirteen samples were selected and the concentration of total phosphate of surface water of Juru River was measured. The average concentration of total phosphate provided by the sample data is 0.97 and the standard deviation is 0.38. A significance level of $\alpha = 0.01$ is chosen to test the claim. Assume that the population is normally distributed."

The five steps for conducting hypothesis testing employing the *P*-value procedure can be used to test the hypothesis regarding the mean concentration of total phosphate in surface water of Juru River. The results of the *P*-value procedure will be compared with the critical value (traditional) procedure.

Step 1: Specify the null and alternative hypotheses

The two hypotheses regarding the mean concentration of total phosphate in surface water of Juru River are presented in Eq. (6.12).

$$H_0{:}\mu \leq 0.85 \text{ vs } H_1{:}\mu > 0.85 \tag{6.12}$$

Step 2: Select the significance level (α) for the study

The level of significance is chosen to be 0.01. The t critical value for a right-tailed test with $\alpha = 0.01$ and $d.f = 13 - 1 = 12$ is $t_{(\alpha,d.f)} = t_{(0.01,12)} = 2.681$.

The two procedures will be used to solve this problem; namely the critical value and *P*-value procedures.

Step 3: Use the sample information to calculate the test statistic value

The test statistic value for the t-test is used to make a decision regarding the mean concentration of total phosphate in surface water of Juru River. The test statistic value using the t-test formula was calculated to be 1.14.

Step 4: Calculate the *P*-value and identify the critical and noncritical regions for the study

The rejection and nonrejection regions using the critical value for testing the mean concentration of total phosphate in surface water of Juru River are presented in Fig. 6.12 for the t distribution curve (shaded area: blue represents the *P*-value and orange represents the level of significance).

The hypothesis in Eq. (6.12) represents a right-tailed t-test. The *P*-value represents the probability of observing the test statistic value (t) or more if the null hypothesis is correct. The area under the t distribution curve to the right of $t = 1.14$ is $0.1385465 = 0.139$. The *P*-value on the right tail is calculated using statistical software as shown in Fig. 6.12.

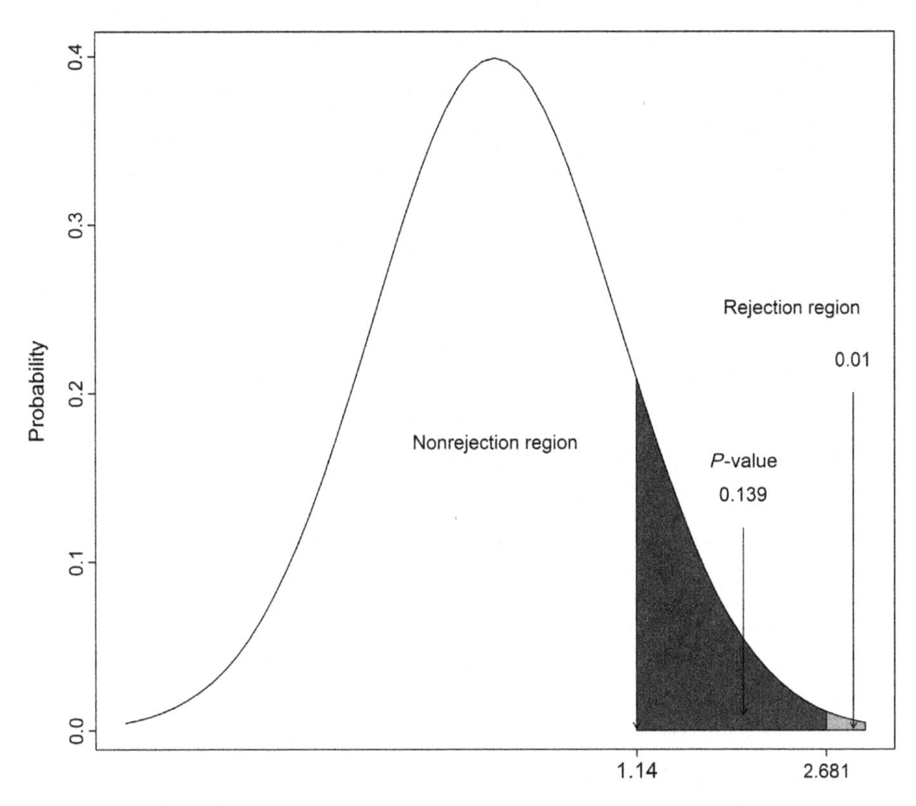

Figure 6.12 The P-value for the mean concentration of total phosphate in surface water of Juru River.

Step 5: Make a decision using the P-value and interpret the results

1. The decision reached by the critical procedure: fail to reject the null hypothesis, see Example 3.6.
2. The decision reached by the P-value procedure: fail to reject the null hypothesis because the P-value (0.139) is greater than the level of significance (0.01), $(0.139 > 0.01)$.

The decision reached using the two procedures: fail to reject the null hypothesis and we believe that the mean concentration of total phosphate in surface water of Juru River is less than or equal to 0.85.

6.7 Testing one sample proportion: *P*-value

Hypothesis testing for one sample proportion using the critical value (traditional) procedure was studied in Chapter 4, Z-test for one sample proportion. The P-value procedure for one sample proportion using a Z-test will be illustrated employing examples to cover various situations of hypothesis testing.

Example 6.13: Mineral water: Example 4.4 is reproduced "A research group at a university wishes to verify the claim announced by a specific factory for mineral water which says that 98% of people are satisfied with the quality of mineral water. The research group selected 700 people randomly and record their responses, they found that 26 were not satisfied. Use the sample information to make a decision whether to support the claim or not. A significance level of $\alpha = 0.01$ is chosen to test the claim. Assume that the population is normally distributed."

The five steps for conducting hypothesis testing employing the *P*-value procedure can be used to test the hypothesis regarding the proportion of people who feel satisfied with the mineral water produced by the factory. The results of the *P*-value procedure will be compared with the critical value (traditional) procedure.

Step 1: Specify the null and alternative hypotheses

The two hypotheses regarding the proportion of people who feel satisfied with the mineral water produced by the factory are presented in Eq. (6.13)

$$H_0{:}p = 0.98 \text{ vs } H_1{:}p \neq 0.98 \tag{6.13}$$

Step 2: Select the significance level (α) for the study

The level of significance is chosen to be 0.01. The Z critical values for a two-tailed test with $\alpha = 0.01$ are ± 2.58.

The two procedures will be used to solve this problem; namely the critical value and *P*-value procedures.

Step 3: Use the sample information to calculate the test statistic value

The test statistic value for the Z-test is used to make a decision regarding the proportion of people who feel satisfied with the mineral water produced by the factory. The test statistic value using the Z-test formula was calculated to be -3.78.

Step 4: Calculate the *P*-value and specify the critical and noncritical regions for the study

The rejection and nonrejection regions using the critical values for testing the proportion of people who feel satisfied with the mineral water produced by the factory are presented in Fig. 6.13 for the standard normal curve (shaded area: blue represents the *P*-value and orange represents the significance level).

The hypothesis in Eq. (6.13) represents a two-tailed Z-test. The *P*-value represents the probability of observing the test statistic value (Z) or greater or less if the null hypothesis is correct. The area under the standard normal curve to the right of $Z = 3.78$ is $7.852614e - 05 = 0.0$, and the area to the left of $Z = -3.78$ is $7.852614e - 05 = 0.0$. Because the test is two-tailed, the *P*-value is $(2 \times 7.852614e - 05 = 0.0)$. The *P*-value on the right and left tails is calculated using a statistical software as shown in Fig. 6.13.

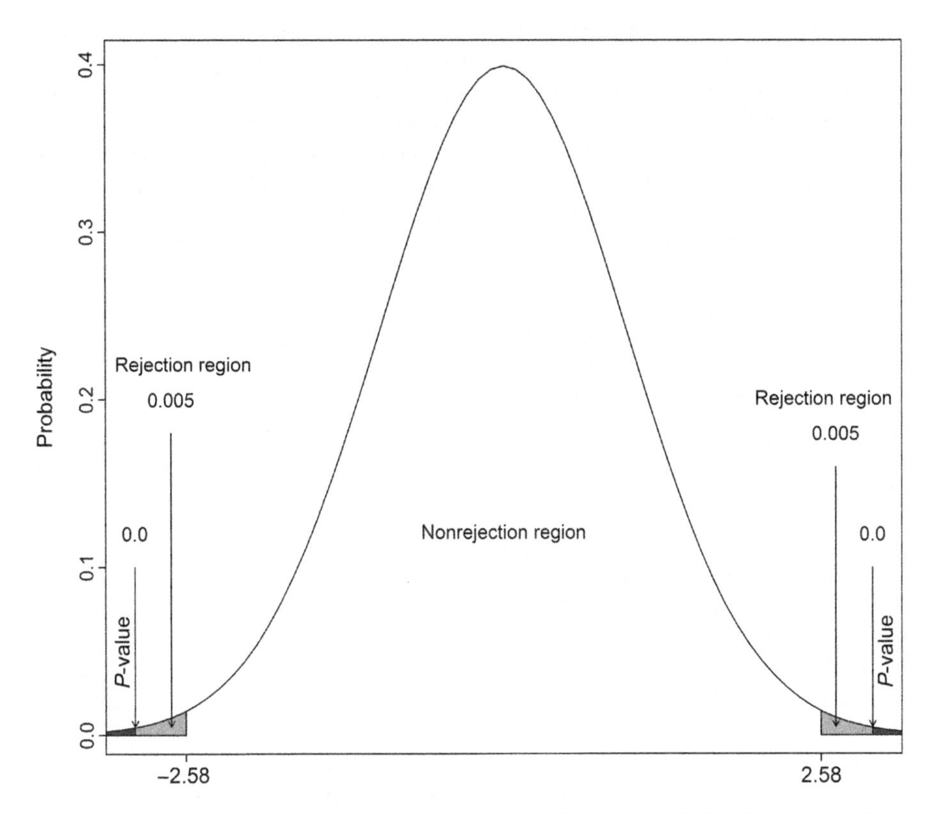

Figure 6.13 The P-value for the proportion of people who feel satisfied with the mineral water produced by the factory.

Step 5: Make a decision using the P-value and interpret the results

1. The decision reached by the critical value procedure: reject the null hypothesis, see Example 4.4.
2. The decision reached by the P-value procedure: reject the null hypothesis because the P-value (0.0) is smaller than the level of significance (0.01), $(0.0 < 0.01)$,

The decision reached using the two procedures: reject the null hypothesis in favor of the alternative hypothesis and believe that the proportion of people who are satisfied with the factory differs from the announced proportion of 0.98.

Example 6.14: Lab skills: Example 4.5 is reproduced "A department of the environment in a university announced that less than or equal to 0.03 of the students have low lab skills. A professor wishes to evaluate her student's skills in the lab work. She selected a group of 200 students and the results were checked. It was found that seven students had wrong result. Use the sample information to make a decision as to whether to support the claim or not. A significance level of $\alpha = 0.01$ is chosen to test the claim. Assume that the population is normally distributed."

The five steps for conducting hypothesis testing employing the *P*-value procedure can be used to test the hypothesis regarding the proportion of students who have low lab skills. The results of the *P*-value procedure will be compared with the critical value (traditional) procedure.

Step 1: Specify the null and alternative hypotheses

The two hypotheses regarding the proportion of students who have low lab skills are presented in Eq. (6.14).

$$H_0{:}p \leq 0.03 \text{ vs } H_1{:}p > 0.03 \tag{6.14}$$

Step 2: Select the significance level (α) for the study

The level of significance is chosen to be 0.01. The Z critical value for a right-tailed test with $\alpha = 0.01$ is 2.326.

The two procedures will be used to solve this problem; namely the critical value and *P*-value procedures.

Step 3: Use the sample information to calculate the test statistic value

The test statistic value for the Z-test is used to make a decision regarding the proportion of students who have low lab skills. The test statistic value using the Z-test formula was calculated to be 0.83.

Step 4: Calculate the *P*-value and identify the critical and noncritical regions for the study

The rejection and nonrejection regions using the critical value for testing the proportion of students who have low lab skills are presented in Fig. 6.14 for the standard normal curve (shaded area: blue represents the *P*-value and orange represents the significance level).

The hypothesis in Eq. (6.14) represents a right-tailed Z-test. The *P*-value represents the probability of observing the test statistic value ($Z = 0.83$) or greater if the null hypothesis is correct. The area under the standard normal curve to the right of $Z = 0.83$ is $0.2035447 = 0.20$. The *P*-value on the right tail was calculated using technology as shown in Fig. 6.14.

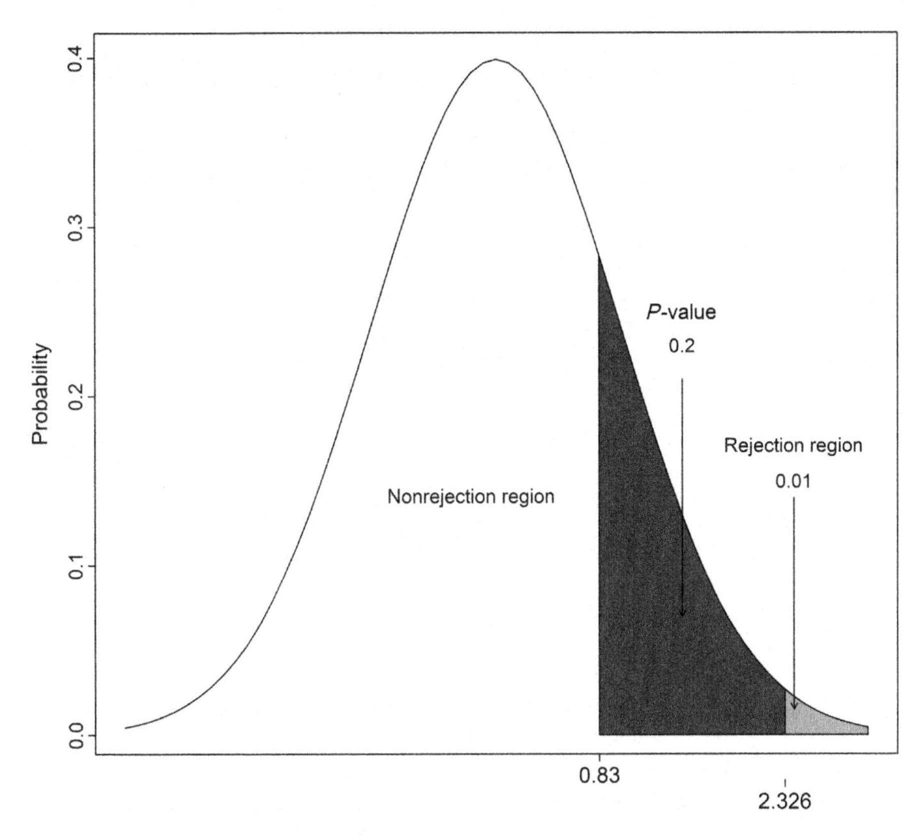

Figure 6.14 The P-value for the proportion of students who have low lab skills.

Step 5: Make a decision using the P-value and interpret the results

1. The decision reached by the critical procedure: fail to reject the null hypothesis, see Example 4.5.
2. The decision reached by the P-value procedure: fail to reject the null hypothesis because the P-value (0.20) is greater than the level of significance (0.01), $(0.20 > 0.01)$.

The decision reached using the two procedures: fail to reject the null hypothesis and we believe that the proportion of students with low lab skills is less than or equal to 0.03.

Example 6.15: Beyond permissible limits: Example 4.6 is reproduced "A study showed that at least 10% of water samples selected from a specific company exceed the permissible limit of heavy metals in water. A research team wishes to investigate the truth regarding the announced proportion. The team selected 130 samples and tested for the heavy metals. The sample data showed three samples with heavy metals beyond the permissible limits. Use the sample information to make a decision as to whether to support the claim or not. A significance level of $\alpha = 0.01$ is chosen to test the claim. Assume that the population is normally distributed."

The five steps for conducting hypothesis testing employing the *P*-value procedure can be used to test the hypothesis regarding the proportion of water samples that exceed the permissible limit of heavy metal produced by the company. The results of the *P*-value procedure will be compared with the critical value (traditional) procedure.

Step 1: Specify the null and alternative hypotheses

The two hypotheses regarding the proportion of water samples that exceed the permissible limit of heavy metal produced by the company are presented in Eq. (6.15).

$$H_0{:}p \geq 0.10 \text{ vs } H_1{:}p < 0.10 \tag{6.15}$$

Step 2: Select the significance level (α) for the study

The level of significance is chosen to be 0.01. The Z critical value for a left-tailed test with $\alpha = 0.01$ is -2.326.

The two procedures will be used to solve this problem; namely the critical value and *P*-value procedures.

Step 3: Use the sample information to calculate the test statistic value

The test statistic value for the Z-test is used to make a decision regarding the proportion of water samples that exceed the permissible limit of heavy metals. The test statistic value using the Z-test formula was calculated to be -3.04.

Step 4: Calculate the *P*-value and identify the critical and noncritical regions for the study

The rejection and nonrejection regions using the critical value for testing the proportion of water samples that exceed the permissible limit of heavy metals are presented in Fig. 6.15 for the standard normal curve (shaded area: blue represents the *P*-value and orange represents the significance level).

The hypothesis in Eq. (6.15) represents a left-tailed Z-test. The *P*-value represents the probability of observing the test statistic value of $Z = -3.04$ or less if the null hypothesis is correct. The area under the standard normal curve to the left of $Z = -3.04$ is $0.001181055 = 0.001$. The *P*-value falls on the left tail as shown in Fig. 6.15.

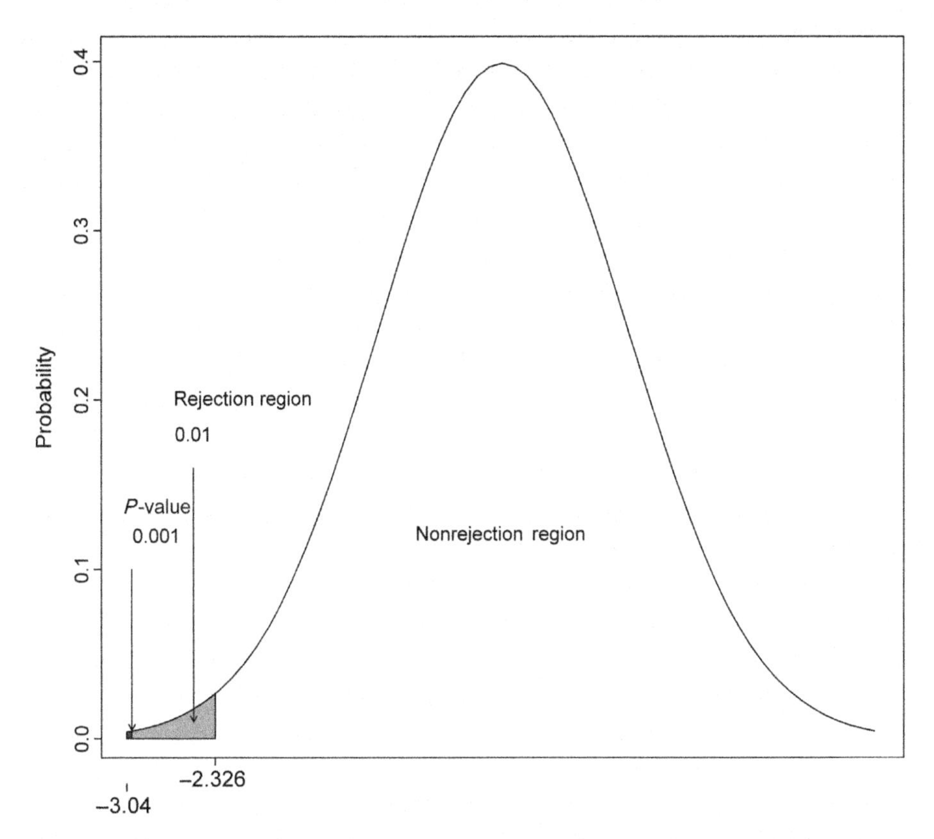

Figure 6.15 The P-value for the proportion of water samples that exceed the permissible limit of heavy metal.

Step 5: Make a decision using the P-value then interpret the results

1. The decision reached by the critical value procedure: reject the null hypothesis, see Example 4.6.
2. The decision reached by the P-value procedure: reject the null hypothesis because the P-value (0.001) is smaller than the level of significance (0.01), $(0.001 < 0.01)$,

The decision reached using the two procedures: reject the null hypothesis in favor of the alternative hypothesis and we believe that the proportion of water samples that exceed the permissible limit of heavy metals is less than 0.10.

6.8 Compute the *P*-value for a chi-square test

We have studied hypothesis testing for one sample variance (standard deviation) using the critical value (traditional) procedure. The P-value procedure for one sample variance or standard deviation using a chi-square test will be illustrated employing examples to cover various situations of hypothesis testing.

Example 6.16: Compute the *P*-value to the left of a χ^2 value: Compute the *P*-value to the left of a chi-square value $\left(\chi^2 = 3.92\right)$ and sample size 12 (left-tailed test). Use a significance level of 0.01 ($\alpha = 0.01$).

Computing the *P*-value for a left-tailed chi-square test can be achieved employing the general procedure for testing a hypothesis.

Step 1: Specify the null and alternative hypotheses

The two hypotheses (the null and alternative) for left-tailed test can be written as presented in (6.16).

$$H_0{:}\sigma^2 \geq c \text{ vs } H_1{:}\sigma^2 < c \tag{6.16}$$

The hypothesis in Eq. (6.16) represents a one-tailed test because the alternative hypothesis is $H_1{:}\mu < c$ (left-tailed), where c is a given value.

Step 2: Select the significance level (α) for the study

The level of significance is chosen to be 0.01.

Step 3: Use the sample information to calculate the test statistic value

The test statistic value of χ^2 is given to be $\chi^2 = 3.92$, otherwise we have to calculate it using a formula.

Step 4: Calculate the *P*-value and identify the critical and noncritical regions for the study

The *P*-value for a chi-square test can be computed easily by employing the chi-square table (Table C in the Appendix) which depends on two values: the degrees of freedom ($d.f$) and the level of significance (α). The χ^2 critical value for the left-tailed is $\chi^2_{(1-\alpha,d.f)} = \chi^2_{(1-0.01,11)} = 3.053$. The *P*-value for 3.92 with $d.f = 11$ falls somewhere in the interval $0.025 < P\text{-}value < 0.05$ as illustrated below.

The position of the required value (3.92) with degrees of freedom ($d.f = 11$) falls between 3.81575 and 4.57481, these two values correspond with the significance level of 0.025 and 0.05 for 3.81575 and 4.57481, respectively. Thus, one can say that the *P*-value for $\chi^2 = 3.92$ and $d.f = 11$ is somewhere in the interval $0.025 < P\text{-value} < 0.05$. The exact *P*-value ($0.02780102 = 0.028 = 0.028$) is calculated using technology as presented in Fig. 6.16.

The blue shaded area in Fig. 6.16 represents the required *P*-value for the value 3.92 and the level of significance (0.01) is represented by the orange shaded area.

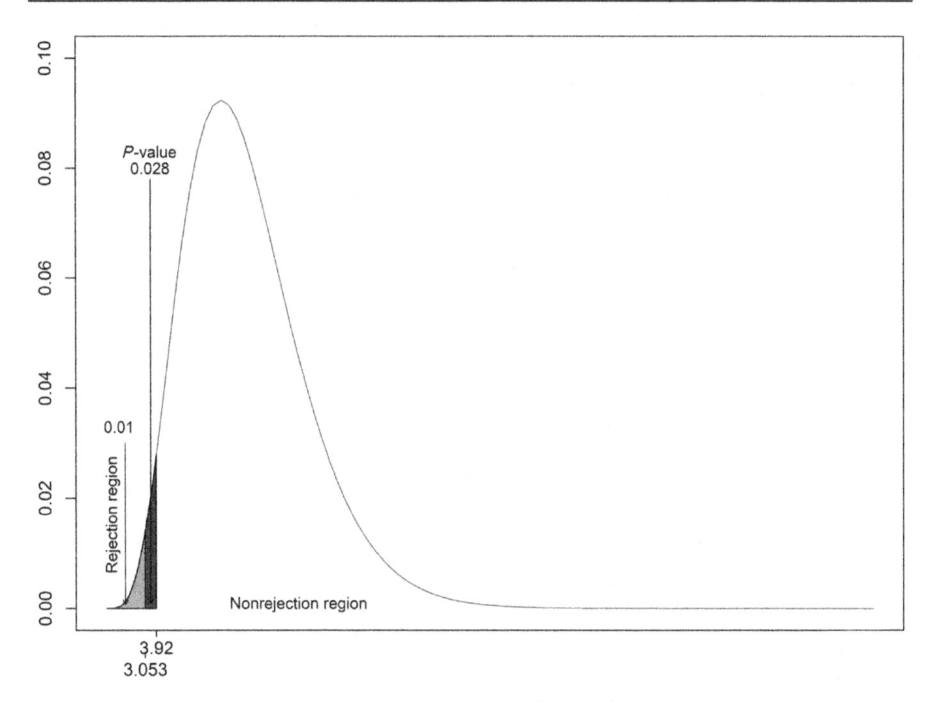

Figure 6.16 The *P*-value to the left of a χ^2 value $(\chi^2 = 3.92)$.

Step 5: Make a decision using the *P*-value then interpret the results

The null hypothesis should not be rejected because the *P*-value is greater than the level of significance (0.01), $(0.028 > 0.01)$. We can say that the sample data supplied sufficient proof to make a decision not to reject the null hypothesis.

Example 6.17: Compute the *P*-value to the right of a χ^2 value: Compute the *P*-value to the right of a chi-square value $(\chi^2 = 18.51)$ and sample size 12 (right-tailed test). Use a significance level of 0.01 $(\alpha = 0.01)$.

Computing the *P*-value for a right-tailed chi-square test can be achieved employing the general procedure for testing a hypothesis.

Step 1: Specify the null and alternative hypotheses

The two hypotheses (the null and alternative) for a right-tailed test can be written as presented in Eq. (6.17).

$$H_0{:}\sigma^2 \le c \text{ vs } H_1{:}\sigma^2 > c \tag{6.17}$$

The hypotheses in Eq. (6.17) represent a one-tailed test because the alternative hypothesis is $H_1{:}\mu > c$ (right-tailed), where c is a given value.

Step 2: Select the significance level (α) for the study

The level of significance is chosen to be 0.01.

Step 3: Use the sample information to calculate the test statistic value

The test statistic value of χ^2 is given to be $\chi^2 = 18.51$, otherwise we have to calculate it using a formula.

Step 4: Calculate the *P*-value and identify the critical and noncritical regions for the study

The *P*-value for chi-square test can be computed easily by employing the chi-square table (Table C in the Appendix) which depends on two values: the degrees of freedom ($d.f$) and the level of significance (α). The χ^2 critical value for a right-tailed test is $\chi^2_{(\alpha,d.f)} = \chi^2_{(0.01,11)} = 24.725$. The *P*-value for 18.51 with $d.f = 11$ falls somewhere in the interval $0.05 < P - value < 0.10$ as illustrated below.

The position of the required value (18.51) with degrees of freedom ($d.f = 11$) falls between 17.275 and 19.6751, these two values correspond with the significance levels of 0.10 and 0.05 for 17.275 and 19.6751, respectively. Thus, one can say that the *P*-value for $\chi^2 = 18.51$ and $d.f = 11$ is somewhere in the interval $0.05 < P\text{-}value < 0.10$. The exact *P*-value ($0.07047653 = 0.07$) is calculated using statistical software presented in Fig. 6.17.

The shaded area (blue) in Fig. 6.17 represents the required *P*-value for the value 18.51 and the level of significance (0.01) is represented by the orange shaded area.

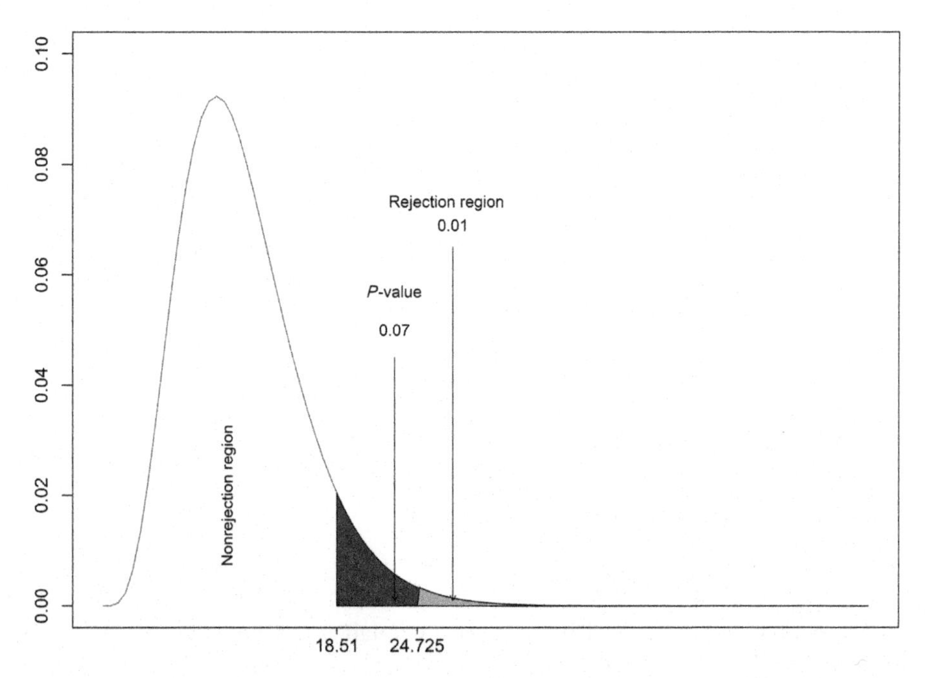

Figure 6.17 The *P*-value to the right of a χ^2 value $(\chi^2 = 18.51)$.

Step 5: Make a decision using the *P*-value then interpret the results

The null hypothesis should not be rejected because the *P*-value is greater than the level of significance (0.01), $(0.07 > 0.01)$. We can say that the sample data supplied sufficient proof to make a decision not to reject the null hypothesis.

Example 6.18: Compute the *P*-value on the two tails: *Compute the* P-*value on the two tails*: Compute the *P*-value for a two-tailed test, the calculated chi-square value is $\left(\chi^2 = 29.15\right)$ and sample size is 12. Use a significance level of 0.01 $(\alpha = 0.01)$.

Computing the *P*-value for a two-tailed chi-square test can be achieved employing the general procedure for testing a hypothesis.

Step 1: Specify the null and alternative hypotheses

The two hypotheses (the null and alternative) for a two-tailed test can be written as presented in Eq. (6.18).

$$H_0{:}\sigma^2 = c \text{ vs } H_1{:}\sigma^2 \neq c \tag{6.18}$$

The hypothesis in Eq. (6.18) represents a two-tailed test because the alternative hypothesis is $H_1{:}\mu \neq c$, where c is a given value.

Step 2: Select the significance level (α) for the study

The level of significance is chosen to be 0.01.

Step 3: Use the sample information to calculate the test statistic value

The test statistic value of χ^2 is given to be $\chi^2 = 29.15$, otherwise we have to calculate it using a formula.

Step 4: Calculate the *P*-value and specify the critical and noncritical regions for the study

The χ^2 critical value for the right-tailed test is $\chi^2_{(\frac{\alpha}{2},d.f)} = \chi^2_{(\frac{0.01}{2},11)} = 26.757$ and the χ^2 critical value for the left-tailed test is $\chi^2_{(1-\frac{\alpha}{2},d.f)} = \chi^2_{(1-\frac{0.01}{2},11)} = 2.603$. The *P*-value for a chi-square test can be computed easily by employing the chi-square table (Table C in the Appendix) which depends on two values: the degrees of freedom ($d.f$) and the level of significance (α). The *P*-value for 29.15 with $d.f = 11$ is less than the $P\text{-value} < 0.005$ as illustrated below.

The position of the required value (29.15) with degrees of freedom ($d.f = 11$) falls beyond the value of 26.7585, this value corresponds with the significance level of 0.005. Thus, one can say that the *P*-value for $\chi^2 = 29.15$ and $d.f = 11$ is less than $P\text{-value} < 0.005$. The exact *P*-value $(0.004302823 = 0.004)$ each side receives (0.002) is calculated using a statistical software as presented in Fig. 6.18.

The blue shaded area in Fig. 6.18 represents the required *P*-value for the value 29.51 and the level of significance (0.01) is represented by the orange shaded area.

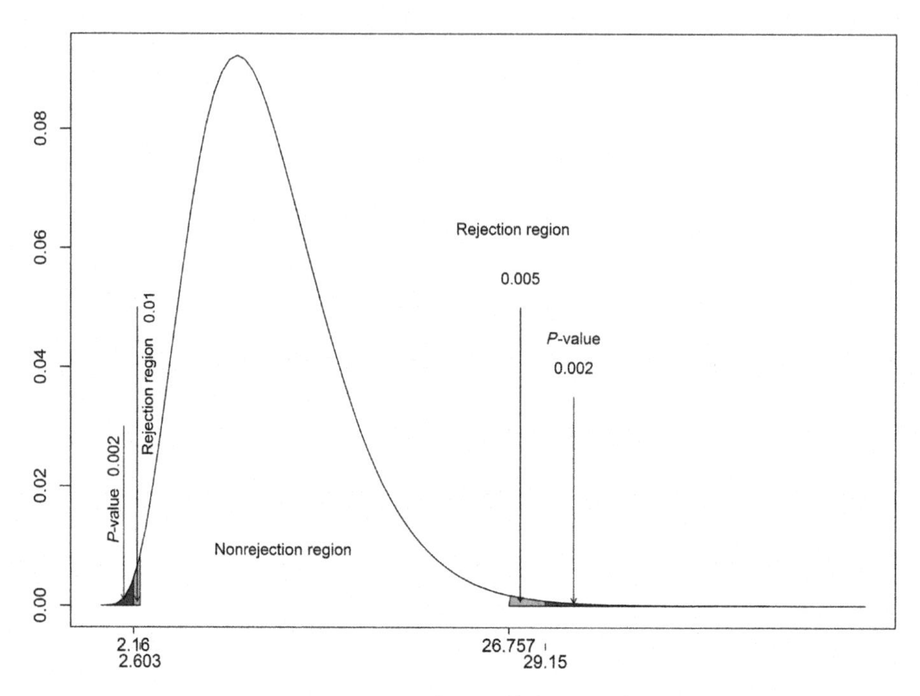

Figure 6.18 The *P*-value for a two-tailed χ^2 value $\left(\chi^2 = 29.51\right)$.

Step 5: Make a decision using the *P*-value then interpret the results

The null hypothesis should be rejected because the *P*-value is smaller than the level of significance (0.01), $(0.004 > 0.01)$. We can say that the sample data supplied sufficient proof to make a decision to reject the null hypothesis.

6.9 Testing one-sample population variance or standard deviation: *P*-value

We have studied hypothesis testing for one sample variance (standard deviation) using the critical value (traditional) procedure. The *P*-value procedure for one sample variance (standard deviation) using a chi-square test will be illustrated employing the same examples presented in Chapter 5, Chi -square test for one sample variance, and solved using the critical value procedure. The three situations of hypothesis testing are covered, and the *P*-values are calculated.

Example 6.19: The concentration of total suspended solids of surface water: Example 5.4 is reproduced "A researcher at an environmental section wishes to verify the claim that the variance of total suspended solids concentration (TSS) of Beris dam surface water is 1.25 (mg/L). Twelve samples were selected and the total suspended solids concentration was measured. The collected data showed that the standard deviation of total suspended solids concentration is 1.80. A significance level of $\alpha = 0.01$ is chosen to test the claim. Assume that the population is normally distributed."

The five steps for conducting hypothesis testing employing the *P*-value procedure can be used to test the hypothesis regarding the variance of total suspended solids concentration in the surface water of Beris dam. The results of the *P*-value procedure will be compared with the critical value (traditional) procedure.

Step 1: Specify the null and alternative hypotheses

The two hypotheses regarding the variance of total suspended solid concentration in the surface water of Beris dam are presented in Eq. (6.19).

$$H_0{:}\sigma^2 = 1.25 \text{ vs } H_1{:}\sigma^2 = 1.25 \tag{6.19}$$

Step 2: Select the significance level (α) for the study

The level of significance is chosen to be 0.01. The χ^2 critical values for a two-tailed test with $\alpha = 0.01$ are:

The left-tailed value: The chi-square critical value for a left-tailed test for $d.f = 11$ and $1 - \frac{\alpha}{2} = 0.995$ is 2.603, $\chi^2_{\left(1-\frac{\alpha}{2},d.f\right)} = \chi^2_{(0.995,11)} = 2.603$

The right-tailed value: The chi-square critical value for a right-tailed test for $d.f = 11$ and $\frac{\alpha}{2} = 0.005$ is 26.757, $\chi^2_{\left(\frac{\alpha}{2},d.f\right)} = \chi^2_{(0.005,11)} = 26.757$

The two procedures will be used to solve this problem; namely the critical value and *P*-value procedures.

Step 3: Use the sample information to calculate the test statistic value

The test statistic value for the χ^2-test is used to make a decision regarding the variance of total suspended solids concentration in the surface water of Beris dam. The test statistic value using the chi-square formula was calculated to be $= 28.512$.

Step 4: Calculate the *P*-value and identify the critical and noncritical regions for the study

The rejection and nonrejection regions using the critical values for testing the variance of total suspended solids concentration in the surface water of Beris are presented in Fig. 6.19 for χ^2 distribution curve (shaded area: blue represents the *P*-value and orange represents the significance level).

The *P*-value for a chi-square test can be computed easily by employing the chi-square table (Table C in the Appendix) which depends on two values: the degrees of freedom ($d.f$) and the level of significance (α). The *P*-value for 28.512 with $d.f = 11$ is less than $P\text{-value} < 0.005$ as illustrated below.

The position of the required value (28.512) with degrees of freedom ($d.f = 11$) falls beyond the value of 26.7585, this value corresponds with the significance level of 0.005. Thus, one can say that the *P*-value for $\chi^2 = 28.512$ and $d.f = 11$ is less than $P\text{-value} < 0.005$. The exact *P*-value [(0.005401711 = 0.0054), each side receives 0.0027] is calculated using statistical software as presented in Fig. 6.19.

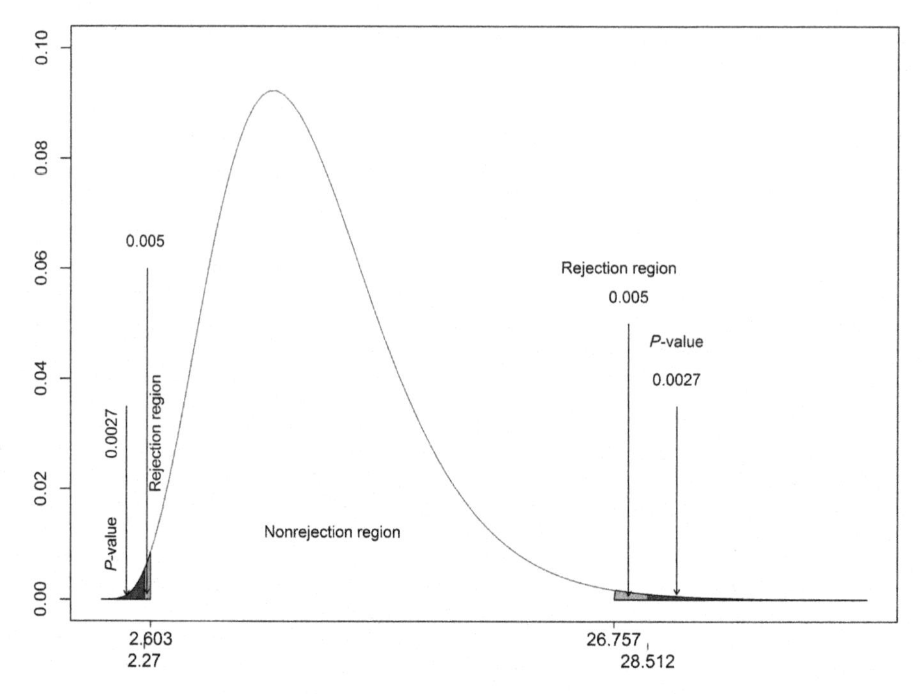

Figure 6.19 The *P*-value for total suspended solids of surface water.

Step 5: Make a decision using the *P*-value and interpret the results

1. The decision reached by the critical value procedure: reject the null hypothesis, see Example 5.4.
2. The decision reached by the *P*-value procedure: reject the null hypothesis because the *P*-value is smaller than the level of significance (0.01) $(0.0054 < 0.01)$.

The decision reached by the two procedures: reject the null hypothesis in favor of the alternative hypothesis and believe that the variance of total suspended solids of surface water differs from 1.25.

Example 6.20: The level of particulate matter in a palm oil mill: Example 5.6 is reproduced "A research group wishes to verify that the standard deviation of the particulate matter (PM_1) level in the air of a palm oil mill is less 23.50 $(\mu g/m^3)$. Twelve samples were selected and the PM_1 level was measured. The sample data showed that the standard deviation of PM_1 level is 26.89. A significance level of $\alpha = 0.01$ is chosen to test the claim. Assume that the population is normally distributed."

The general procedure for conducting hypothesis testing can be used to make the decision regarding the variance of PM_1 level in the air of the palm oil mill.

The five steps for conducting hypothesis testing employing the *P*-value procedure can be used to test the hypothesis regarding the variance of PM_1 level in the air of the palm oil mill. The results of the *P*-value procedure will be compared with the critical value (traditional) procedure.

Step 1: Specify the null and alternative hypotheses

The two hypotheses regarding the variance of PM_1 level in the air of palm oil mill are presented in Eq. (6.20).

$$H_0{:}\sigma \geq 23.50 \text{ vs } H_1{:}\sigma < 23.50 \tag{6.20}$$

Step 2: Select the significance level (α) for the study

The level of significance is chosen to be 0.01. The χ^2 critical value for a left-tailed test with $\alpha = 0.01$ and $d.f = 11$ is 3.053 ($\chi^2_{(1-0.01,d.f)} = \chi^2_{(0.99,11)} = 3.053$).

The two procedures will be used to solve this problem; namely the critical value and *P*-value procedures.

Step 3: Use the sample information to calculate the test statistic value

The test statistic value for the χ^2-test is used to make a decision regarding the variance of PM_1 level in the air of the palm oil mill. The test statistic value using a chi-square formula was calculated to be $= 14.40$.

Step 4: Calculate the *P*-value and identify the critical and noncritical regions for the study

The rejection and nonrejection regions using the critical value for testing the variance of PM_1 level in the air of palm oil mill are presented in Fig. 6.20 for the χ^2 distribution curve (shaded area: blue represents the *P*-value and orange represents the significance level).

The *P*-value for a chi-square test can be computed easily by employing the chi-square table (Table C in the Appendix) which depends on two values: the degrees of freedom ($d.f$) and the level of significance (α). The *P*-value for 14.40 with $d.f = 11$ is somewhere in the interval $010 < P\text{-value} < 0.25$ as illustrated below.

The position of required value (14.40) with degrees of freedom ($d.f = 11$) falls between 13.70069 and 17.275, these two values correspond with the significance levels of 0.250 and 0.10. Thus, one can say that the *P*-value for $\chi^2 = 14.40$ and $d.f = 11$ falls somewhere in the interval $010 < P\text{-value} < 0.25$. The exact *P*-value using statistical software is $0.2115148 = 0.212$ as presented in Fig. 6.20.

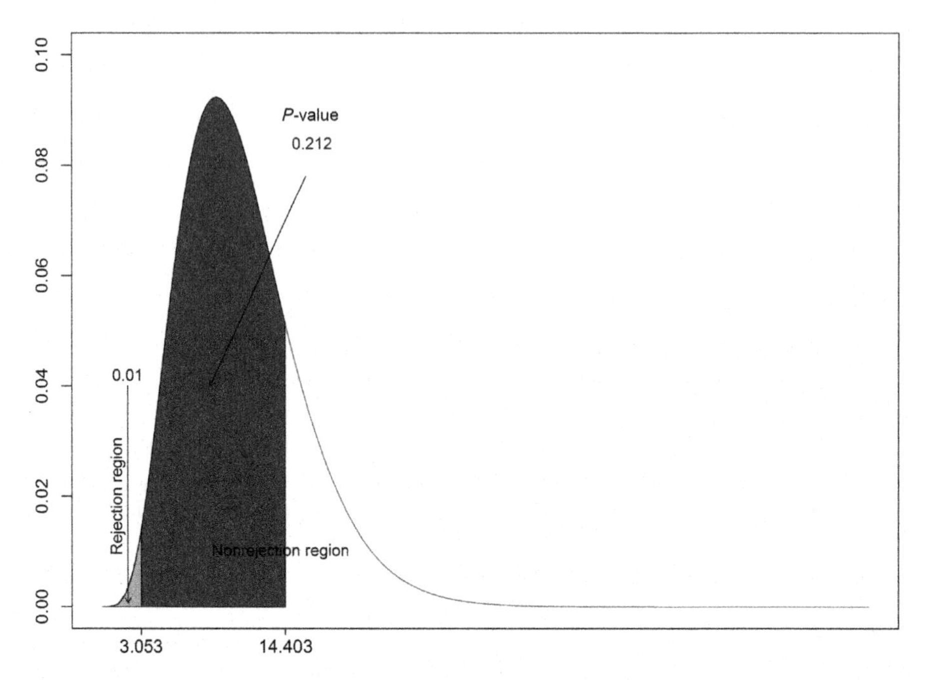

Figure 6.20 The P-value for the level of particulate matter in the air of a palm oil mill.

Step 5: Make a decision using the P-value and interpret the results

1. The decision reached by the critical procedure: fail to reject the null hypothesis, see Example 5.6.
2. The decision reached by the P-value procedure: fail to reject the null hypothesis because the P-value is greater than the level of significance (0.01), $(0.212 > 0.01)$.

The decision reached by the two procedures: fail to reject the null hypothesis and we believe that the variance of the level of PM_1 in the air of a palm oil mill is more than or equal to 23.50.

Example 6.21: The concentration of lead in sediment: Example 5.5 is reproduced "A professor at a research center wishes to verify the claim that the variance of lead (Pb) concentration in sediment of Juru River is less than or equal to 0.20 (mg/L). Ten samples were selected and tested for lead concentration. The collected data showed that the standard deviation of lead concentration is 0.23. A significance level of $\alpha = 0.01$ is chosen to test the claim. Assume that the population is normally distributed."

The general procedure for conducting hypothesis testing can be used to make the decision regarding the variance of lead concentration in sediment of Juru River.

The five steps for conducting hypothesis testing employing the *P*-value procedure can be used to test the hypothesis regarding the variance of lead concentration in sediment of Juru River. The results of the *P*-value procedure will be compared with the critical value (traditional) procedure.

Step 1: Specify the null and alternative hypotheses

The two hypotheses regarding the variance of lead concentration in sediment of Juru River are presented in Eq. (6.21).

$$H_0{:}\sigma^2 \le 0.20 \text{ vs } H_1{:}\sigma^2 > 0.20 \tag{6.21}$$

Step 2: Select the significance level (α) for the study

The level of significance is chosen to be 0.01. The χ^2 critical value for a right-tailed test with $\alpha = 0.01$ and $d.f = 9$ is 21.666 ($\chi^2_{(\alpha,d.f)} = \chi^2_{(0.01,9)} = 21.666$).

The two procedures will be used to solve this problem; namely the critical value and *P*-value procedures.

Step 3: Use the sample information to calculate the test statistic value

The test statistic value for the χ^2-test is used to make a decision regarding the variance of lead concentration in sediment of Juru River. The test statistic value using a chi-square formula was calculated to be $= 2.381$.

Step 4: Calculate the *P*-value and identify the critical and noncritical regions for the study

The rejection and nonrejection regions using the critical values for testing the variance of lead concentration in sediment of Juru River are presented in Fig. 6.21 for a χ^2 distribution curve (shaded area: blue represents the *P*-value and orange represents the significance level).

The *P*-value for a chi-square test can be computed easily by employing the chi-square table (Table C in the Appendix) which depends on two values: the degrees of freedom ($d.f$) and the level of significance (α). The *P*-value for 2.381 with $d.f = 9$ falls somewhere in the interval $0.975 < P\text{-value} < 0.990$ as illustrated below.

The position of the required value (2.381) with degrees of freedom ($d.f = 9$) falls between 2.0879 and 2.70039, these two values correspond with the significance levels of 0.990 and 0.975, respectively. Thus, one can say that the *P*-value for $\chi^2 = 2.38$ and $d.f = 11$ falls somewhere in the interval $0975 < P\text{-value} < 0.990$. The exact *P*-value ($0.9839264 = 0.9839$) is calculated using statistical software as presented in Fig. 6.21.

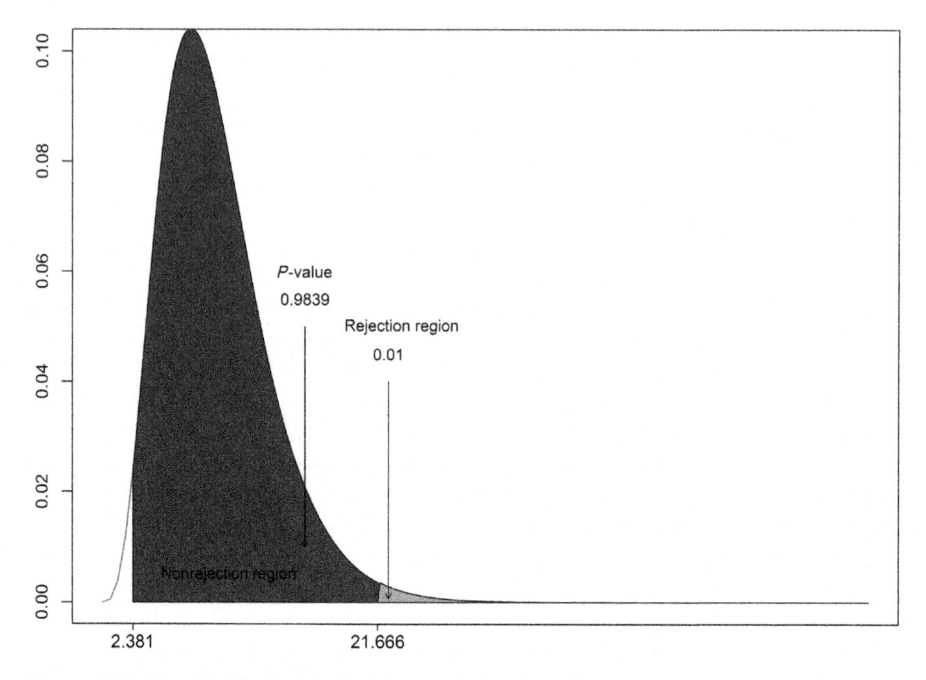

Figure 6.21 The *P*-value for the lead concentration in sediment.

Step 5: Make a decision using the *P*-value and interpret the results

1. The decision reached by the critical value procedure: fail to reject the null hypothesis, see Example 5.5.
2. The decision reached by the *P*-value procedure: fail to reject the null hypothesis because the *P*-value is greater than the level of significance (0.01), $(0.9839 > 0.01)$.

The decision reached by the two procedures: fail to reject the null hypothesis and believe that the variance of lead concentration in sediment is less than or equal to 0.20.

Note:

- Statistical software usually provides the *P*-value for each problem.

Further reading

Alkarkhi, A. F. M., Ahmad, A., & Easa, A. M. (2009a). Assessment of surface water quality of selected estuaries of Malaysia: Multivariate statistical techniques. *Environmentalist*, *29*, 255–262.

Alkarkhi, A. F. M., Ismail, N., Ahmed, A., & Easa, A. m (2009b). Analysis of heavy metal concentrations in sediments of selected estuaries of Malaysia—A statistical assessment. *Environmental Monitoring and Assessment*, *153*, 179–185.

Bluman, A. G. (1998). *Elementary statistics: A step by step approach* (3rd ed.). Boston: WCB, McGraw-Hill.

Eberly College of Science-Department of Statistics (2020). Hypothesis testing (P-value approach). In, Vol. 2020. Pennstate: The Pennsylvania State University.

Editor, M.B. (2014). How to correctly interpret P values. In, Vol. 2020. Minitab.

Rahman, N. N. N. A., Chard, N. C., Al-Karkhi, A. F. M., Rafatullah, M., & Kadir, M. O. A. (2017). *Analysis of particulate matters in air of palm oil mills – A statistical assessment*, *16*(11), 2537–2543.

Weiss, N. A. (2012). *Introductory statistics* (9th ed.). Pearson.

Yusup, Y., & Alkarkhi, A. F. M. (2011). Cluster analysis of inorganic elements in particulate matter in the air environment of an equatorial urban coastal location. *Chemistry and Ecology*, *27*(3), 273–286.

Interval estimation for one population

7

Abstract

Estimation of the population parameter using sample data can be expressed statistically either by a single value (point estimate) or as an interval (confidence interval). The concepts of confidence interval, confidence level, and standard error are delivered in a simple and enjoyable way. The confidence interval for one population parameter is given for normal distribution including the mean value and proportion, t distribution for the mean value, and chi-square distribution for the variance value. Furthermore, examples from the fields of environmental science and engineering are used to explain the procedure for building the confidence interval with sufficient demonstrations of each step matched to the field of study.

Keywords: Confidence interval; confidence level; mean value; proportion value; variance value

Learning outcomes

After completing this chapter, readers will be able to:

- Understand the meaning and importance of interval estimation (confidence interval);
- Know the general procedure for building confidence interval;
- Describe the steps for computing confidence interval for one population mean;
- Describe the steps for computing confidence interval for one population proportion;
- Describe the steps for computing confidence interval for one population variance;
- Choose a suitable formula to compute the confidence interval for the selected population parameter;
- Know two-tailed and one-tailed confidence intervals for one population parameter;
- Understand the steps used to interpret the confidence interval;
- Write a report for the results.

7.1 Introduction

Two types of estimation can be recognized in statistics, the first type is called point estimation, such as the mean, median, variance, and others, and the second type of estimation is called interval estimation, which represents the estimation in a range (interval). Summarizing the collected data can be represented by a single value (point estimate) or by an interval which includes lower and upper limits. As an example, an

Applications of Hypothesis Testing for Environmental Science. DOI: https://doi.org/10.1016/B978-0-12-824301-5.00008-3

environmentalist wants to investigate the mean concentration of chemical oxygen demand (COD) of surface water of a river, 20 samples were selected and tested for COD concentration, the results of 20 samples can be represented by a single value such as the mean concentration of COD or represent the results between two values (upper and lower limits) to show the range of the mean concentration of COD. If a new sample is selected then we expect that the concentration of COD will be between these two values. The concept of interval estimation, how to build a confidence interval, and other terms related to confidence interval will be delivered in this chapter.

7.2 What is interval estimation?

The sample data are usually used to estimate a population parameter either by a single point or an interval. The concept of point estimation and the concept of confidence interval are given first.

Suppose we would like to estimate the mean concentration of the pH value of surface water (population mean μ), in this case we should select a number of sampling points and measure the pH value of the surface water at each point and then represent the data by a single value, $\overline{X}$ (sample mean), because the sample mean ($\overline{X}$) is the point estimate of the population mean (μ). This estimation is called point estimation.

How confident are we that the sample mean is close to the population mean? To answer this question lets us think of a range of values that we expect to include the true population mean. Thus, the mean concentration of pH of surface water could be any value within two values [the lower (L) and upper (U) limits], in this case, it can be said that we are confident that the mean value falls in the interval with a confidence level $(1-\alpha)100\%$:

$$L \leq \mu \leq U$$

This estimation is called the interval estimate or confidence interval. Another important concept that is associated with confidence interval is called the confidence level; the confidence level, $(1-\alpha)100\%$, can be defined as the probability of all possible outcomes that the interval estimate will contain the true population parameter. The confidence level is usually selected to be 0.95 or 0.99, but may be another value.

We can interpret the confidence level as "we are confident (at confident level) that the population parameter falls in the interval $L \leq \mu \leq U$."

Other names are given to the confidence level such as the confidence coefficient or degree of confidence.

We can build confidence intervals for population parameters using the general procedure with four steps.

Step 1: Use the sample data to compute the sample statistic

Step 2: Select the significance level (α) for the study and find the critical values

Step 3: Compute the confidence interval

Step 4: Interpret the results

7.3 Confidence interval for one population mean

We have noted that hypothesis testing for one population mean includes two cases, the first case when the sample size is large, and the second case when the sample size is small. We can use these two cases to calculate the confidence intervals for one population mean employing the general procedure for calculating the confidence interval presented earlier.

7.3.1 When the sample size is large

Consider a large random sample ($n \geq 30$) that is selected from a normally distributed population and Y represents a random variable of interest. A confidence interval regarding the population mean of the variable of interest can be built using the sample data. The mathematical formula for computing the confidence interval for one sample mean is presented in Eq. (7.1).

$$\overline{Y} \pm Z_{\frac{\alpha}{2}}\left[\frac{\sigma}{\sqrt{n}}\right] \tag{7.1}$$

The confidence interval can be written using another form:

$$\overline{Y} - Z_{\frac{\alpha}{2}}\left[\frac{\sigma}{\sqrt{n}}\right] \leq \mu \leq \overline{Y} + Z_{\frac{\alpha}{2}}\left[\frac{\sigma}{\sqrt{n}}\right]$$

where

$\overline{Y}$ represents the sample mean;
μ represents the population mean;
σ represents the population standard deviation;
$Z_{\frac{\alpha}{2}}$ represents the Z critical value;
n represents the sample size;
$\frac{\sigma}{\sqrt{n}}$ represents the standard error; and
$Z_{\frac{\alpha}{2}}\left[\frac{\sigma}{\sqrt{n}}\right]$ represents the margin of error (E).

Note:

- When the sample size is large and the population standard deviation (σ) is not provided, then we can use the sample standard deviation (S);
- The margin of error (E) can be used to write the confidence interval $\overline{Y} - E \leq \mu \leq \overline{Y} + E$;
- The lower one-tailed confidence interval is given in Eq. (7.2).

$$\overline{Y} - Z_{\alpha}\left[\frac{\sigma}{\sqrt{n}}\right] \leq \mu \tag{7.2}$$

or it can be written as the interval $\left[\overline{Y} - Z_{\alpha}\left[\frac{\sigma}{\sqrt{n}}\right], \infty\right)$

- The upper one-tailed confidence interval is given in Eq. (7.3).

$$\mu \leq \overline{Y} + Z_{\alpha}\left[\frac{\sigma}{\sqrt{n}}\right] \tag{7.3}$$

or it can be written as the interval $\left(-\infty, \overline{Y} + Z_{\alpha}\left[\frac{\sigma}{\sqrt{n}}\right]\right]$

Example 7.1: The biochemical oxygen demand concentration of water before the dam: A researcher wants to investigate the range of biochemical oxygen demand (BOD) of surface water before Beris dam in Malaysia. He selected 65 samples and the concentration of BOD was measured. The results showed that the average concentration of BOD is 5.48 (mg/L) and the standard deviation is 1.01. Calculate the confidence interval for the average concentration of BOD using 0.01 as the level of significance. Assume that the population is normally distributed.

The general procedure for calculating the confidence interval, including the four steps, will be used to calculate the confidence interval for the BOD concentration of water before the dam.

Step 1: Use the sample data to compute the sample statistic

The mean concentration of BOD for water before the dam is already calculated to be 5.48 (mg/L).

Step 2: Select the significance level (α) for the study and find the critical values

The level of significance is chosen to be 0.01 ($\alpha = 0.01$) and $\frac{\alpha}{2} = \frac{0.01}{2} = 0.005$. The Z critical value is $Z_{\frac{\alpha}{2}} = 2.58$ (Table A in the Appendix).

Step 3: Compute the confidence interval

The entries for the confidence interval formula as presented in Eq. (7.1) (two-sided confidence interval) are already provided, the mean concentration of BOD for water before the dam is 5.48, the standard deviation is 1.01, and the sample size is 65.

Because it is a two-sided confidence interval, the formula in Eq. (7.1) is used to calculate the confidence interval for the mean concentration of BOD of water before the dam.

$$\overline{Y} - Z_{\frac{\alpha}{2}}\left[\frac{\sigma}{\sqrt{n}}\right] \leq \mu \leq \overline{Y} + Z_{\frac{\alpha}{2}}\left[\frac{\sigma}{\sqrt{n}}\right]$$

$$5.48 - 2.58\left[\frac{1.01}{\sqrt{65}}\right] \leq \mu \leq 5.48 + 2.58\left[\frac{1.01}{\sqrt{65}}\right]$$

$$5.158 \leq \mu \leq 5.802$$

The confidence interval for the mean concentration of BOD of water falls in the interval (5.518, 5.802).

Step 4: Interpret the results

It can be said that we are 99% confident that the mean concentration of BOD of surface water falls somewhere between 5.158 and 5.802.

Example 7.2: The concentration of zinc in particulate matter in the air: An environmentalist wants to investigate the range of zinc (Zn) concentration as an inorganic element in particulate matter (PM_{10}) in the air environment of an equatorial urban coastal location during the winter season. He selected 75 samples and the concentration of zinc was measured. The results showed that the average concentration of zinc is 2.84 (μgm^{-3}) and the standard deviation is 0.81. Calculate the confidence interval for the average concentration of zinc using 0.01 as the level of significance. Assume that the population is normally distributed.

The general procedure for calculating the confidence interval is used to calculate the confidence interval for the concentration of zinc in particulate matter in the air environment of an equatorial urban coastal location during the winter season.

Step 1: Use the sample data to compute the sample statistic

The mean concentration of zinc in particulate matter in the air environment of an equatorial urban coastal location during the winter season is already calculated to be 2.48.

Step 2: Select the significance level (α) for the study and find the critical values

The level of significance is chosen to be 0.01 ($\alpha = 0.01$) and $\frac{\alpha}{2} = \frac{0.01}{2} = 0.005$. The Z critical value is $Z_{\frac{\alpha}{2}} = 2.58$ (Table A in the Appendix).

Step 3: Compute the confidence interval

The entries for the confidence interval formula as presented in Eq. (7.1) (two-sided confidence interval) are already provided, the mean concentration of zinc in particulate matter in the air environment of an equatorial urban coastal location during the winter season is 2.84, the standard deviation is 0.81, and the sample size is 75.

Because it is a two-sided confidence interval, the formula in Eq. (7.1) is used to calculate the confidence interval for the mean concentration of zinc in particulate matter in the air environment of an equatorial urban coastal location during the winter season.

$$\overline{Y} - Z_{\frac{\alpha}{2}}\left[\frac{\sigma}{\sqrt{n}}\right] \leq \mu \leq \overline{Y} + Z_{\frac{\alpha}{2}}\left[\frac{\sigma}{\sqrt{n}}\right]$$

$$2.48 - 2.58\left[\frac{0.81}{\sqrt{75}}\right] \leq \mu \leq 2.48 + 2.58\left[\frac{0.81}{\sqrt{75}}\right]$$

$$2.599 \leq \mu \leq 3.081$$

The confidence interval for the mean concentration of zinc in particulate matter in the air environment of an equatorial urban coastal location during the winter season falls in the interval (2.599, 3.081).

Step 4: Interpret the results

It can be said that we are 99% confident that the mean concentration of zinc in particulate matter in the air environment of an equatorial urban coastal location during the winter season falls somewhere between 2.599 and 3.081.

7.3.2 When the sample size is small

Consider a small random sample ($n < 30$) that is selected from a normally distributed population and Y represents a random variable of interest. A confidence interval regarding the population mean value of the variable of interest can be calculated using the sample data. The mathematical formula for computing the confidence interval for one population mean is presented in Eq. (7.4).

$$\overline{Y} \pm t_{\frac{\alpha}{2}}\left[\frac{S}{\sqrt{n}}\right] \tag{7.4}$$

The confidence interval can be written using another form:

$$\overline{Y} - t_{\frac{\alpha}{2}}\left[\frac{S}{\sqrt{n}}\right] \leq \mu \leq \overline{Y} + t_{\frac{\alpha}{2}}\left[\frac{S}{\sqrt{n}}\right]$$

where

$\overline{Y}$ represents the sample mean;
μ represents the population mean;
S represents the sample standard deviation;
$t_{\frac{\alpha}{2}}$ represents the t critical value;
n represents the sample size; and
$t_{\frac{\alpha}{2}}\left[\frac{S}{\sqrt{n}}\right]$ represents the margin of error (E).

Note:

- The margin of error (E) can be used to write the confidence interval $\overline{Y} - E \leq \mu \leq \overline{Y} + E$;
- The lower one-tailed confidence interval is

$$\overline{Y} - t_{\alpha}\left[\frac{S}{\sqrt{n}}\right] \leq \mu$$

- The upper one-tailed confidence interval is

$$\mu \leq \overline{Y} + t_{\alpha}\left[\frac{S}{\sqrt{n}}\right]$$

Example 7.3: The concentration of arsenic in the water: A researcher wants to investigate the range of arsenic (As) in the surface water of a river. He selected 10 samples and the concentration of arsenic was measured. The results showed that the average concentration of arsenic is 3.38 (mg/L) and the standard deviation is 1. Calculate the confidence interval for the average concentration of arsenic using 0.01 as the level of significance. Assume that the population is normally distributed.

The general procedure for calculating the confidence interval is used to calculate the confidence interval for the concentration of arsenic in the surface water.

Step 1: Use the sample data to compute the sample statistic

The mean concentration of arsenic in the surface water is already calculated to be 3.38.

Step 2: Select the significance level (α) for the study and find the critical values

The level of significance is chosen to be 0.01 ($\alpha = 0.01$) and $\frac{\alpha}{2} = \frac{0.01}{2} = 0.005$ and the sample size is 10. The t critical value is $t_{\frac{\alpha}{2},n-1} = 3.25$ (Table B in the Appendix).

Step 3: Compute the confidence interval

The entries for the confidence interval formula as presented in Eq. (7.4) (two-sided confidence interval) are already provided, the mean concentration of arsenic in the surface water is 3.38, the standard deviation is 1, and the sample size is 10.

Because it is a two-sided confidence interval, the formula in Eq. (7.4) is used to calculate the confidence interval for the mean concentration of arsenic in the surface water.

$$\overline{Y} - t_{\frac{\alpha}{2}}\left[\frac{\sigma}{\sqrt{n}}\right] \leq \mu \leq \overline{Y} + t_{\frac{\alpha}{2}}\left[\frac{\sigma}{\sqrt{n}}\right]$$

$$3.38 - 3.25\left[\frac{1}{\sqrt{10}}\right] \leq \mu \leq 3.38 + 3.25\left[\frac{1}{\sqrt{10}}\right]$$

$$2.352 \leq \mu \leq 4.408$$

The confidence interval for the mean concentration of arsenic in the surface water falls in the interval (2.352, 4.408).

Step 4: Interpret the results

It can be said that we are 99% confident that the mean concentration of arsenic (As) in the surface water falls somewhere between 2.352 and 4.408.

Example 7.4: The concentration of cadmium in sediment: A professor at a research center wants to know the range of cadmium (Cd) in sediment. She selected 12 samples and the concentration of cadmium was measured. The results showed that the average concentration of cadmium is 1.95 (mg/L) and the standard deviation is 0.04. Calculate the confidence interval for the average concentration of cadmium in sediment using 0.01 as the level of significance. Assume that the population is normally distributed.

The general procedure for calculating the confidence interval is used to calculate the confidence interval for the concentration of cadmium in the sediment.

Step 1: Use the sample data to compute the sample statistic

The mean concentration of cadmium in sediment is already calculated to be 1.95.

Step 2: Select the significance level (α) for the study and find the critical values

The level of significance is chosen to be 0.01 ($\alpha = 0.01$) and $\frac{\alpha}{2} = \frac{0.01}{2} = 0.005$ and the sample size is 12. The t critical value is $t_{\frac{\alpha}{2},(n-1)} = 3.11$ (Table B in the Appendix).

Step 3: Compute the confidence interval

The entries for the confidence interval formula as presented in Eq. (7.4) (two-sided confidence interval) are already provided, the mean concentration of cadmium in sediment is 1.95, the standard deviation is 0.04, and the sample size is 12.

Because it is a two-sided confidence interval, the formula in Eq. (7.4) is used to calculate the confidence interval for the mean concentration of cadmium in sediment.

$$\overline{Y} - t_{\frac{\alpha}{2}}\left[\frac{\sigma}{\sqrt{n}}\right] \leq \mu \leq \overline{Y} + t_{\frac{\alpha}{2}}\left[\frac{\sigma}{\sqrt{n}}\right]$$

$$1.95 - 3.11\left[\frac{0.04}{\sqrt{12}}\right] \leq \mu \leq 1.95 + 3.11\left[\frac{0.04}{\sqrt{12}}\right]$$

$$1.914 \leq \mu \leq 1.986$$

The confidence interval for the mean concentration of cadmium in sediment falls in the interval (1.914, 1.986).

Step 4: Interpret the results

It can be said that we are 99% confident that the mean concentration of cadmium in sediment falls somewhere between 1.914 and 1.986.

7.4 Confidence interval for one population proportion

Hypothesis testing for one population proportion was studied in Chapter 4, Z-test for one sample proportion. Calculating the confidence interval for one population proportion can be achieved by employing the general procedure for calculating the confidence interval presented earlier.

Consider a random sample that is selected from a normally distributed population and p represents the proportion of a characteristic in the population. A confidence interval regarding the population proportion can be calculated using binomial distribution. When a large sample size is used, then binomial distribution can be approximated to normal distribution; the probability of success is small and $np > 5$ and $nq \geq 5$. The mathematical formula for computing the confidence interval for one population proportion is presented in Eq. (7.5).

$$\hat{p} \pm Z_{\frac{\alpha}{2}}\sqrt{\frac{\hat{p}(1-\hat{p})}{n}} \tag{7.5}$$

The confidence interval can be written using another form:

$$\hat{p} - Z_{\frac{\alpha}{2}}\sqrt{\frac{\hat{p}(1-\hat{p})}{n}} \leq p \leq \hat{p} + Z_{\frac{\alpha}{2}}\sqrt{\frac{\hat{p}(1-\hat{p})}{n}}$$

where

$\hat{p}$ represents the sample proportion;
p represents the population proportion;
$Z_{\frac{\alpha}{2}}$ represents the Z critical value;
n represents the sample size; and
$Z_{\frac{\alpha}{2}}\sqrt{\frac{\hat{p}(1-\hat{p})}{n}}$ represents the margin of error (E).

- The margin of error (E) can be used to write the confidence interval $\hat{p} - E \leq p \leq p + E$.
- The lower one-tailed confidence interval is

$$\hat{p} - Z_{\alpha}\sqrt{\frac{\hat{p}(1-\hat{p})}{n}} \leq \mu$$

- The upper one-tailed confidence interval is

$$\mu \leq \hat{p} + Z_{\alpha}\sqrt{\frac{\hat{p}(1-\hat{p})}{n}}$$

Example 7.5: The concentration of chromium in sediment: A research center wants to know the range of high chromium (Cr) concentration in sediment. Thirty-five samples were selected and the concentration of chromium was measured. The results showed that 40% of the samples have a high chromium concentration. Calculate the confidence interval for the proportion of samples that have a high chromium concentration using 0.01 as the level of significance. Assume that the population is normally distributed.

The general procedure for calculating the confidence interval is used to calculate the confidence interval for the proportion of samples that have a high chromium concentration in sediment.

Step 1: Use the sample data to compute the sample statistic

The proportion of high chromium concentration in sediment is already calculated to be 0.40.

Step 2: Select the significance level (α) for the study and find the critical values

The level of significance is chosen to be 0.01 ($\alpha = 0.01$) and $\frac{\alpha}{2} = \frac{0.01}{2} = 0.005$. The Z critical value is $Z_{\frac{\alpha}{2}} = 2.58$ (Table A in the Appendix).

Step 3: Compute the confidence interval

The entries for the confidence interval formula as presented in Eq. (7.5) (two-sided confidence interval) are already provided, the proportion of samples that have a high chromium concentration in sediment is 0.40, and the sample size is 35.

Because it is a two-sided confidence interval, the formula in Eq. (7.5) is used to calculate the confidence interval for the proportion of samples that have a high chromium concentration in sediment.

$$\hat{p} - Z_{\frac{\alpha}{2}}\sqrt{\frac{\hat{p}(1-\hat{p})}{n}} \le p \le \hat{p} + Z_{\frac{\alpha}{2}}\sqrt{\frac{\hat{p}(1-\hat{p})}{n}}$$

$$0.40 - 2.58\sqrt{\frac{(0.40)(0.60)}{35}} \le p \le 0.40 + 2.58\sqrt{\frac{(0.40)(0.60)}{35}}$$

$$0.187 \le p \le 0.613$$

Thus, the 99% confidence interval for the proportion of samples that have a high chromium concentration in sediment is (0.187, 0.613).

Step 4: Interpret the results

It can be said that we are 99% confident that the proportion of samples that have a high chromium concentration in sediment falls somewhere between 0.187 and 0.613.

Example 7.6: The concentration of lead in river water: An environmentalist wishes to know the range of lead (Pb) in the surface water of a river. Fifty samples were selected, and the concentration of lead was measured. The results showed that in 15% of the samples, the lead concentration exceeds the permissible limits. Calculate the confidence interval for the proportion of samples that the lead concentration exceeds the permissible limit using 0.01 as the level of significance. Assume that the population is normally distributed.

The general procedure for calculating the confidence interval is used to calculate the confidence interval for the proportion of samples that the lead concentration exceeds the permissible limit.

Step 1: Use the sample data to compute the sample statistic

The proportion of lead concentration that exceeds the permissible limit in river water is already calculated to be 0.15.

Step 2: Select the significance level (α) for the study and find the critical values

The level of significance is chosen to be 0.01 ($\alpha = 0.01$) and $\frac{\alpha}{2} = \frac{0.01}{2} = 0.005$. The Z critical value is $Z_{\frac{\alpha}{2}} = 2.58$ (Table A in the Appendix).

Step 3: Compute the confidence interval

The entries for the confidence interval formula as presented in Eq. (7.5) (two-sided confidence interval) are already provided, the proportion of samples that the lead concentration exceeds the permissible limit in river water is 0.15, and the sample size is 50.

Because it is a two-sided confidence interval, the formula in Eq. (7.5) is used to calculate the confidence interval for the proportion of samples that the lead concentration exceeds the permissible limit in river water.

$$\hat{p} - Z_{\frac{\alpha}{2}}\sqrt{\frac{\hat{p}(1-\hat{p})}{n}} \le p \le \hat{p} + Z_{\frac{\alpha}{2}}\sqrt{\frac{\hat{p}(1-\hat{p})}{n}}$$

$$0.15-2.58\sqrt{\frac{(0.15)(0.85)}{50}}\le p\le 0.15+2.58\sqrt{\frac{(0.15)(0.85)}{50}}$$

$$0.02\le p\le 0.28$$

Thus, the 99% confidence interval for the proportion of lead concentration that exceeds the permissible limit in river water is (0.02, 0.28).

Step 4: Interpret the results

It can be said that we are 99% confident that the proportion of lead concentration that exceeds the permissible limit in river water falls somewhere between 0.02 and 0.28.

7.5 Confidence interval for one population variance

Hypothesis testing for one population variance was studied in Chapter 5, Chi-square test for one sample variance. Calculating the confidence interval for one population variance can be achieved by employing the general procedure for calculating the confidence interval presented earlier.

Consider a random sample that is selected from a normally distributed population and σ^2 represents the variance of the population. A confidence interval regarding the population variance can be calculated using chi-square distribution. The mathematical formula for computing the confidence interval for one population variance is presented in Eq. (7.6).

$$\frac{(n-1)S^2}{\chi_R^2}\le\sigma^2\le\frac{(n-1)S^2}{\chi_L^2} \tag{7.6}$$

Follow a chi-square distribution with $d.f=(n-1)$.

where, $\chi_R^2=\chi^2_{(\frac{\alpha}{2},n-1)}$ which represents the right tail and $\chi_L^2=\chi^2_{(1-\frac{\alpha}{2},n-1)}$ represents the left tail.

The confidence interval for the standard deviation is:

$$\sqrt{\frac{(n-1)S^2}{\chi_R^2}}\le\sigma\le\sqrt{\frac{(n-1)S^2}{\chi_L^2}}$$

- The lower one-tailed confidence interval for the population variance is $\frac{(n-1)S^2}{\chi_R^2}\le\sigma^2$
- The upper one-tailed confidence interval for the population variance is $\sigma^2\le\frac{(n-1)S^2}{\chi_L^2}$

Example 7.7: The wind speed: A professor at a research center wants to know the range of variance of wind speed at the place where she works. She selected 20 days and the wind speed was measured. The results showed that the variance of the wind speed is 0.27 (m s^{-1}). Calculate the confidence interval for the variance of wind speed using 0.01 as the level of significance. Assume that the population is normally distributed.

The general procedure for calculating the confidence interval with the five steps will be used to calculate the confidence interval for the variance of wind speed.

Step 1: Use the sample data to compute the sample statistic

The variance of the wind speed is already calculated to be 0.27.

Step 2: Select the significance level (α) for the study and find the critical values

The level of significance is chosen to be 0.01 ($\alpha = 0.01$) and $\frac{\alpha}{2} = \frac{0.01}{2} = 0.005$. The chi-square critical values for $\alpha = 0.01$ and $d.f = 19$ are $\chi_L^2 = \chi^2_{(1-\frac{\alpha}{2}, n-1)=(1-\frac{0.05}{2}, 19)} = 6.843971$ and $\chi_R^2 = \chi^2_{(\frac{\alpha}{2}, n-1)=(\frac{0.01}{2}, 19)} = 38.58226$ (Table C in the Appendix).

Step 3: Compute the confidence interval

The entries for the confidence interval formula as presented in Eq. (7.6) (two-sided confidence interval) are already provided, the variance of wind speed is 0.27, and the sample size is 20.

Because it is a two-sided confidence interval, the formula in Eq. (7.6) is used to calculate the confidence interval for the variance of wind speed.

$$\frac{(n-1)S^2}{\chi_R^2} \leq \sigma^2 \leq \frac{(n-1)S^2}{\chi_L^2}$$

$$\frac{(20-1)0.27}{38.58226} \leq \sigma^2 \leq \frac{(20-1)0.27}{6.843971}$$

$$0.133 \leq \sigma^2 \leq 0.750$$

Thus, the 99% confidence interval for the variance of wind speed is (0.133, 0.750).

Step 4: Interpret the results

It can be said that we are 99% confident that the variance of wind speed falls somewhere between 0.133 and 0.750.

Example 7.8: The concentration of mercury in cockle: An environmentalist wishes to know the range of variance of mercury concentration (Hg) in cockles obtained from a river. He selected 40 samples and the concentration of mercury was measured. The results showed that the variance of the mercury is 0.006 (mg kg^{-1}). Calculate the confidence interval for the variance of mercury in cockles using 0.01 as the level of significance. Assume that the population is normally distributed.

The general procedure for calculating the confidence interval is used to calculate the confidence interval for the variance of mercury in cockles.

Step 1: Use the sample data to compute the sample statistic

The variance of mercury in cockles is already calculated to be 0.006.

Step 2: Select the significance level (α) for the study and find the critical values

The level of significance is chosen to be 0.01 ($\alpha = 0.01$) and $\frac{\alpha}{2} = \frac{0.01}{2} = 0.005$. The chi-square critical values for $\alpha = 0.01$ and $d.f = 39$ are $\chi_L^2 = \chi^2_{(1-\frac{\alpha}{2}, n-1)=(1-\frac{0.01}{2}, 39)} = 19.99587$ and $\chi_R^2 = \chi^2_{(\frac{\alpha}{2}, n-1)=(\frac{0.01}{2}, 39)} = 65.47557$ (Table C in the Appendix).

Step 3: Compute the confidence interval

The entries for the confidence interval formula as presented in Eq. (7.6) (two-sided confidence interval) are already provided, the variance of mercury in cockles is 0.006, and the sample size is 40.

Because two-sided confidence interval, the formula in Eq. (7.6) is used to calculate the confidence interval for the variance of mercury in cockles.

$$\frac{(n-1)S^2}{\chi_R^2} \le \sigma^2 \le \frac{(n-1)S^2}{\chi_L^2}$$

$$\frac{(40-1)0.006}{65.47557} \le \sigma^2 \le \frac{(40-1)0.006}{19.99587}$$

$$0.0036 \le \sigma^2 \le 0.0117.$$

Thus, the 99% confidence interval for the variance of mercury concentration in cockles is (0.0036, 0.0117).

Step 4: Interpret the results

It can be said that we are 99% confident that the variance of the mercury concentration in cockles falls somewhere between 0.0036 and 0.0117.

Further reading

Alkarkhi, A. F. M., Ahmad, A., & Easa, A. M. (2009). *Assessment of surface water quality of selected estuaries of Malaysia: Multivariate statistical techniques. The Environmentalist* (29, pp. 255–262).

Alkarkhi, A. F. M., & Chin, L. H. (2012). *Elementary statistics for technologist* (1st ed.). Malaysia: Universiti Sains Malaysia.

Alkarkhi, A. F. M., Ismail, N., Ahmed, A., & Easa, Am (2009). Analysis of heavy metal concentrations in sediments of selected estuaries of Malaysia—A statistical assessment. *Environmental Monitoring and Assessment, 153*, 179–185.

Alkarkhi, F. M. A., Ismail, N., & Easa, A. M. (2008). Assessment of arsenic and heavy metal contents in cockles (*Anadara granosa*) using multivariate statistical techniques. *Journal of Hazardous Material, 150*, 783–789.

Bluman, A. G. (1998). *Elementary statistics: A step by step approach* (3rd ed.). Boston: WCB/McGraw-Hill.

Weiss, N. A. (2012). *Introductory statistics* (9th ed.). Pearson.

NIST/SEMATECH (2013). Engineering statistics handbook. In, Vol. 2020. NIST/SEMATECH. https://www.itl.nist.gov/div898/handbook/eda/section3/eda3674.htm.

The interval estimation procedure: hypothesis testing for one population

8

Abstract

This chapter addresses the confidence interval procedure for testing various hypotheses including one-tailed and two-tailed tests for one population parameter. Moreover, the results of confidence interval for hypothesis testing are compared with the results of two other procedures: critical value (traditional) and *P*-value procedures. Three examples for each test are given including a Z-test for the mean when the sample size is large, a Z-test for the proportion, a t-test for the mean when the sample size is small, and a chi-square test for the variance. The three procedures provide the same decision regarding one-tailed and two-tailed tests. The procedure is explained in an easy and enjoyable way to show the comparison between different procedures.

Keywords: Confidence interval; *P*-value; traditional procedure; mean; proportion; variance

Learning outcomes

After completing this chapter, readers should be able to:

- Know the steps on how to use confidence interval for testing various hypothesis;
- Understand the procedure for testing a hypothesis concerning one sample mean when a large sample size is used;
- Understand the procedure for testing a hypothesis concerning one sample mean when a small sample size is used;
- Understand the procedure for testing a hypothesis concerning one sample proportion;
- Understand the procedure for testing a hypothesis concerning one sample variance;
- Understand the procedure for testing a hypothesis concerning one-sided and two-sided tests;
- Describe the procedure for using a confidence interval to make a decision regarding the population parameter;
- Draw smart conclusions regarding the problem under investigation.

8.1 Introduction

The third procedure for testing a hypothesis in statistics is called the interval estimation (confidence interval) procedure. Hypothesis testing for the mean, proportion, and variance values will be tested using the confidence interval procedure. Examples regarding various environmental issues are presented and

Applications of Hypothesis Testing for Environmental Science. DOI: https://doi.org/10.1016/B978-0-12-824301-5.00004-6

tested for various parameters using the confidence interval procedure. The result of the confidence interval will be compared with the critical value and P-value procedures.

8.2 The steps for the confidence interval procedure

We can perform hypothesis testing using the confidence interval procedure employing similar steps used for critical value (traditional) and P-value procedures. The four steps for conducting hypothesis testing employing the confidence interval procedure are presented below.

Step 1: Specify the null and alternative hypotheses

Step 2: Select the significance level (α) for the study and find the critical values

Step 3: Use the sample information to calculate the confidence interval

Step 4: Make a decision using the confidence interval and interpret the results

- Do not reject the null hypothesis if the hypothesized (claimed) value falls in the confidence interval which indicates that the result is statistically insignificant.
- Reject the null hypothesis if the hypothesized (claimed) value does not fall in the confidence interval which indicates that the result is statistically significant.

8.3 Confidence interval for testing one mean value: Z-test

We have studied hypothesis testing for one sample mean when the sample size is large using the critical value (traditional) and P-value procedures. The confidence interval procedure for one sample mean will be illustrated employing examples to cover various situations of hypothesis testing.

Let us present two-tailed and one-tailed confidence intervals for large samples before giving examples to illustrate the procedure.

- The two-tailed confidence interval for the mean value is given in Eq. (8.1).

$$\overline{Y} - Z_{\frac{\alpha}{2}}\left[\frac{\sigma}{\sqrt{n}}\right] \leq \mu \leq \overline{Y} + Z_{\frac{\alpha}{2}}\left[\frac{\sigma}{\sqrt{n}}\right] \tag{8.1}$$

Do not reject the null hypothesis if the hypothesized (claimed) value falls in the confidence interval and reject the null hypothesis if the hypothesized (claimed) value does not fall in the confidence interval.

- The one-tailed confidence interval:

The lower one-tailed confidence interval for the mean value is given in Eq. (8.2).

$$\overline{Y} - Z_{\alpha}\left[\frac{\sigma}{\sqrt{n}}\right] \leq \mu \tag{8.2}$$

- Reject the null hypothesis if the hypothesized value does not fall in the confidence interval $\overline{Y} - Z_{\alpha}[\frac{\sigma}{\sqrt{n}}] \leq \mu$, which means we reject the null hypothesis if $\overline{Y} - Z_{\alpha}[\frac{\sigma}{\sqrt{n}}] > \mu$

and the upper one-tailed confidence interval for the mean value is given in Eq. (8.3).

$$\mu \le \overline{Y} + Z_{\alpha}\left[\frac{\sigma}{\sqrt{n}}\right] \tag{8.3}$$

- Reject the null hypothesis if the hypothesized value does not fall in the confidence interval $\mu \le \overline{Y} + Z_{\alpha}[\frac{\sigma}{\sqrt{n}}]$, which means we reject the null hypothesis if $\overline{Y} - Z_{\alpha}[\frac{\sigma}{\sqrt{n}}] < \mu$.

Example 8.1: The concentration of cadmium of surface water: Example 2.5 is reproduced "A professor at an environmental section wanted to verify the claim that the mean concentration of cadmium (Cd) of surface water in Juru River is 1.4 (mg/L). He selected 35 samples and tested for the cadmium concentration. The collected data showed that the mean concentration of cadmium is 1.6 and the standard deviation of the population is 0.4. A significance level of $\alpha = 0.01$ is chosen to test the claim. Assume that the population is normally distributed."

The four steps for conducting hypothesis testing employing the confidence interval procedure can be used to test the hypothesis regarding the mean concentration of cadmium of surface water in Juru River. The result of the confidence interval procedure will be compared with the critical value (traditional) and *P*-value procedures.

Step 1: Specify the null and alternative hypotheses

The two hypotheses regarding the mean concentration of cadmium of surface water in Juru River are presented in Eq. (8.4).

$$H_0\colon \mu = 1.4 \quad \text{vs} \quad H_1\colon \mu \ne 1.4 \tag{8.4}$$

Step 2: Select the significance level (α) for the study and find the critical values

The level of significance is chosen to be 0.01. The Z critical value for a two-tailed test with $\alpha = 0.01$ is 2.58.

Step 3: Use the sample information to calculate the confidence interval

The hypothesis in Eq. (8.4) represents a two-tailed test. Thus, the two-tailed confidence interval formula presented in Eq. (8.1) should be used to calculate the confidence interval for the mean concentration of cadmium of surface water in Juru River. The formula in Eq. (8.1) represents the alternative hypothesis for a two-tailed test.

$$\overline{Y} - Z_{\frac{\alpha}{2}}\left[\frac{\sigma}{\sqrt{n}}\right] \le \mu \le \overline{Y} + Z_{\frac{\alpha}{2}}\left[\frac{\sigma}{\sqrt{n}}\right]$$

$$1.6 - 2.58\left[\frac{0.4}{\sqrt{35}}\right] \le \mu \le 1.6 + 2.58\left[\frac{0.4}{\sqrt{35}}\right]$$

$$1.443 \le \mu \le 1.757$$

Step 4: Make a decision using the confidence interval and interpret the results

Making a decision using the confidence interval procedure requires checking the confidence interval whether it contains the hypothesized (claimed) value or not. It can be seen that the hypothesized value (1.4) is not in the calculated confidence interval $1.443 \leq \mu \leq 1.757$.

1. The decision reached by the confidence interval procedure: reject the null hypothesis and we believe that the mean concentration of cadmium of surface water in Juru River is not equal to 1.4.
   The result of the confidence interval is compared with:
2. The decision achieved by the critical value (traditional) procedure: reject the null hypothesis as presented in Example 2.5.
3. The decision achieved by the P-value procedure: reject the null hypothesis as presented in Example 6.4.

We can say that the three procedures have reached the same decision, which is to reject the null hypothesis and we believe that the mean concentration of cadmium of surface water in Juru River is not equal to 1.4.

Example 8.2: The concentration of lead in sediment: Example 2.6 is reproduced "A researcher at a research institution wanted to verify the claim that the mean concentration of lead (Pb) in sediment of Jejawi River is less than 0.23 (mg/L). He selected 33 samples and tested for the lead concentration. The collected data showed that the mean concentration of lead is 0.235 and the standard deviation of the population is 0.106. A significance level of $\alpha = 0.01$ is chosen to test the claim. Assume that the population is normally distributed."

The four steps for conducting hypothesis testing employing the confidence interval procedure can be used to test the hypothesis regarding the mean concentration of lead in sediment of Jejawi River. The result of the confidence interval procedure will be compared with the critical value (traditional) and P-value procedures.

Step 1: Specify the null and alternative hypotheses

The two hypotheses regarding the mean concentration of lead in sediment of Jejawi River are presented in Eq. (8.5).

$$H_0: \mu \geq 0.23 \quad \text{vs} \quad H_1: \mu < 0.23 \tag{8.5}$$

Step 2: Select the significance level (α) for the study and find the critical values

The level of significance is chosen to be 0.01. The Z critical value for a one-tailed test with $\alpha = 0.01$ is 2.326.

Step 3: Use the sample information to calculate the confidence interval

The hypothesis in Eq. (8.5) represents a one-tailed test. Thus, the upper confidence interval formula presented in Eq. (8.3) should be used to calculate the confidence interval for the mean concentration of lead in sediment of Jejawi River. The formula in Eq. (8.3) represents the alternative hypothesis for a left-tailed test.

$$\mu \leq \overline{Y} + Z_\alpha \left[\frac{\sigma}{\sqrt{n}}\right]$$

$$\mu \leq 0.235 + (2.326)\left[\frac{0.106}{\sqrt{33}}\right]$$

$$\mu \leq 0.278$$

Step 4: Make a decision using the confidence interval and interpret the results

Making a decision using the confidence interval procedure requires checking the confidence interval whether it contains the hypothesized value or not. It can be seen that the hypothesized value (0.23) is in the calculated confidence interval $\mu \leq 0.278$.

1. The decision reached by the confidence interval procedure: fail to reject the null hypothesis and we believe that the mean concentration of lead in sediment of Jejawi River is more than or equal to 0.23.

   The result of the confidence interval is compared with:
2. The decision achieved by the critical value (traditional) procedure: fail to reject the null hypothesis as presented in Example 2.6.
3. The decision achieved by the *P*-value procedure: fail to reject the null hypothesis as presented in Example 6.5.

We can say that the three procedures have reached the same decision which is to fail to reject the null hypothesis and we believe that the mean concentration of lead in sediment of Jejawi River is more than or equal to 0.23.

Example 8.3: The concentration of sodium in particulate matter in the air: Example 2.7 is reproduced "The concentration of sodium (Na) as an inorganic element in particulate matter (PM_{10}) in the air environment of an equatorial urban coastal location during the summer season was investigated. A researcher wants to verify that the concentration of sodium in the air during the summer season is less than or equal to 6.20. He selected 34 samples and the concentration of sodium was measured. The average concentration of sodium provided by the sample data is 7.46 (μgm^{-3}) and the population standard deviation is 2.18. A significance level of $\alpha = 0.01$ is chosen to test the claim. Assume that the population is normally distributed."

The four steps for conducting hypothesis testing employing the confidence interval procedure can be used to test the hypothesis regarding the mean concentration of sodium in the air environment of an equatorial urban coastal location during the summer season. The results of the confidence procedure will be compared with the critical value (traditional) and *P*-value procedures.

Step 1: Specify the null and alternative hypotheses

The two hypotheses regarding the mean concentration of sodium in the air environment of an equatorial urban coastal location during the summer season are presented in Eq. (8.6).

$$H_0: \mu \leq 6.20 \quad \text{vs} \quad H_1: \mu > 6.20 \tag{8.6}$$

Step 2: Select the significance level (α) for the study and find the critical values

The level of significance is chosen to be 0.01. The Z critical value for a one-tailed test with $\alpha = 0.01$ is 2.326.

Step 3: Use the sample information to calculate the confidence interval

The hypothesis in Eq. (8.6) represents a one-tailed test. Thus, the lower confidence interval formula presented in Eq. (8.2) should be used to calculate the confidence interval for the mean concentration of sodium in the air environment of an equatorial urban coastal location during the summer season. The formula in Eq. (8.2) represents the alternative hypothesis for a right-tailed test.

$$\overline{Y} - Z_{\alpha}\left[\frac{\sigma}{\sqrt{n}}\right] \leq \mu$$

$$7.46 - 2.326\left[\frac{2.18}{\sqrt{34}}\right] \leq \mu$$

$$6.497 \leq \mu$$

Step 4: Make a decision using the confidence interval and interpret the results

Making a decision using the confidence interval procedure requires checking the confidence interval whether it contains the hypothesized value or not. It can be seen that the hypothesized value (6.20) is not in the calculated confidence interval $6.497 \leq \mu$.

1. The decision reached by the confidence interval procedure: reject the null hypothesis and we believe that the mean concentration of sodium in the air environment is more than 6.20. The result of the confidence interval is compared with:
2. The decision achieved by the critical value (traditional) procedure: reject the null hypothesis as presented in Example 2.7.
3. The decision achieved by the *P*-value procedure: reject the null hypothesis as presented in Example 6.6.

We can say that the three procedures have reached the same decision which is to reject the null hypothesis and we believe that the mean concentration of sodium (Na) in the air environment is more than 6.20.

8.4 Confidence interval for testing one mean value: t-test

We have studied hypothesis testing for one sample mean when the sample size is small using two procedures: critical value (traditional) and *P*-value procedures. The confidence interval procedure for one sample mean using a t-test will be illustrated employing examples to cover various situations of hypothesis testing.

Let us present two-tailed and one-tailed confidence intervals for small samples before giving examples to illustrate the procedure.

- The two-tailed confidence interval for the mean value is given in Eq. (8.7).

$$\overline{Y} - t_{\frac{\alpha}{2}}\left[\frac{\sigma}{\sqrt{n}}\right] \leq \mu \leq \overline{Y} + t_{\frac{\alpha}{2}}\left[\frac{\sigma}{\sqrt{n}}\right] \tag{8.7}$$

Do not reject the null hypothesis if the hypothesized (claimed) value falls in the confidence interval and reject the null hypothesis if the hypothesized value does not fall in the confidence interval.

- The one-tailed confidence interval:
  The lower one-tailed confidence interval for the mean value is given in Eq. (8.8).

$$\overline{Y} - t_\alpha\left[\frac{\sigma}{\sqrt{n}}\right] \le \mu \tag{8.8}$$

Reject the null hypothesis if the hypothesized value does not fall in the confidence interval $\overline{Y} - t_\alpha[\frac{\sigma}{\sqrt{n}}] \le \mu$, which means we reject the null hypothesis if $\overline{Y} - t_\alpha[\frac{\sigma}{\sqrt{n}}] > \mu$ and the upper one-tailed confidence interval for the mean value is given in Eq. (8.9).

$$\mu \le \overline{Y} + t_\alpha\left[\frac{\sigma}{\sqrt{n}}\right] \tag{8.9}$$

Reject the null hypothesis if the hypothesized value does not fall in the confidence interval $\mu \le \overline{Y} + t_\alpha[\frac{\sigma}{\sqrt{n}}]$, which means we reject the null hypothesis if $\mu > \overline{Y} + t_\alpha[\frac{\sigma}{\sqrt{n}}]$

Example 8.4: The concentration of trace metal iron in the air: Example 3.4 is reproduced "A researcher at an environmental section wishes to verify the claim that the mean concentration of trace metal iron (Fe) on coarse particulate matter (PM_{10}) in the air is 0.36 ($\mu g/m^3$) during the summer season of an equatorial urban coastal location. Twelve samples were selected, and the concentration of iron was measured. The collected data showed that the mean concentration of iron is 0.44 and the standard deviation is 0.34. A significance level of $\alpha = 0.01$ is chosen to test the claim. Assume that the population is normally distributed."

The four steps for conducting hypothesis testing employing the confidence interval procedure can be used to test the hypothesis regarding the mean concentration of iron on coarse particulate matter (PM_{10}) in the air during the summer season of an equatorial urban coastal location. The result of the confidence interval procedure will be compared with the critical value (traditional) and P-value procedures.

Step 1: Specify the null and alternative hypotheses

The two hypotheses regarding the mean concentration of iron in the air during the summer season are presented in Eq. (8.10).

$$H_0: \mu = 0.36 \quad \text{vs} \quad H_1: \mu \ne 0.36 \tag{8.10}$$

Step 2: Select the significance level (α) for the study and find the critical values

The level of significance is chosen to be 0.01. The t critical value for a two-tailed test with $\alpha = 0.01$ and degrees of freedom 11 is 3.106.

Step 3: Use the sample information to calculate the confidence interval

The hypothesis in Eq. (8.10) represents a two-tailed test. Thus, the two-tailed confidence interval formula presented in Eq. (8.7) should be used to calculate the confidence interval for the mean concentration of iron in the air using the confidence interval procedure. The formula in Eq. (8.7) represents the alternative hypothesis for a two-tailed test.

$$\overline{Y} - t_{\frac{\alpha}{2}}\left[\frac{\sigma}{\sqrt{n}}\right] \leq \mu \leq \overline{Y} + t_{\frac{\alpha}{2}}\left[\frac{\sigma}{\sqrt{n}}\right]$$

$$0.44 - 3.106\left[\frac{0.34}{\sqrt{12}}\right] \leq \mu \leq 0.44 + 3.106\left[\frac{0.34}{\sqrt{12}}\right]$$

$$0.135 \leq \mu \leq 0.745$$

Step 4: Make a decision using the confidence interval and interpret the results

Making a decision using the confidence interval procedure requires checking the confidence interval whether it contains the hypothesized value or not. It can be seen that the hypothesized value (0.36) is somewhere in the calculated confidence interval $0.135 \leq \mu \leq 0.745$.

1. The decision reached by the confidence interval procedure: fail to reject the null hypothesis and we believe that the mean concentration of iron in the air is 0.36.
   The result of the confidence interval is compared with:
2. The decision reached by the critical value (traditional) procedure: fail to reject the null hypothesis as presented in Example 3.4.
3. The decision reached by the *P*-value procedure: fail to reject the null hypothesis as presented in Example 6.10.

We can say that the three procedures have reached the same decision which is fail to reject the null hypothesis and we believe that the mean concentration of iron in the air is 0.36.

Example 8.5: The concentration of chemical oxygen demand of surface water: Example 3.5 is reproduced "An environmentalist at a research center wishes to verify the claim that the mean concentration of the chemical oxygen demand (COD) in surface water of Juru River is less than 650 (mg/L). Fourteen samples were selected and tested for chemical oxygen demand concentration. The collected data showed that the mean concentration of chemical oxygen demand is 470 and the standard deviation is 245. A significance level of $\alpha = 0.01$ is chosen to test the claim. Assume that the population is normally distributed."

The four steps for conducting hypothesis testing employing the confidence interval procedure can be used to test the hypothesis regarding the mean concentration of chemical oxygen demand in surface water of Juru River. The result of the confidence interval procedure will be compared with the critical value (traditional) and *P*-value procedures.

Step 1: Specify the null and alternative hypotheses

The two hypotheses regarding the mean concentration of chemical oxygen demand in surface water of Juru River are presented in Eq. (8.11).

$$H_0: \mu \geq 650 \quad \text{vs} \quad H_1: \mu < 650 \tag{8.11}$$

Step 2: Select the significance level (α) for the study and find the critical values

The level of significance is chosen to be 0.01. The t critical value for a one-tailed test with $\alpha = 0.01$ and degrees of freedom 13 is 2.65.

Step 3: Use the sample information to calculate the confidence interval

The hypothesis in Eq. (8.11) represents a one-tailed test. Thus, the upper confidence interval formula presented in Eq. (8.9) should be used to calculate the confidence interval for the mean concentration of chemical oxygen demand in surface water of Juru River. The formula in Eq. (8.9) represents the alternative hypothesis for a left-tailed test.

$$\mu \leq \overline{Y} + t_{\alpha}\left[\frac{\sigma}{\sqrt{n}}\right]$$

$$\mu \leq 470 + 2.65\left[\frac{245}{\sqrt{14}}\right]$$

$$\mu \leq 643.54$$

Step 4: Make a decision using the confidence interval and interpret the results

Making a decision using the confidence interval procedure requires checking the confidence interval whether it contains the hypothesized value or not. It can be seen that the hypothesized value (650) is not in the calculated confidence interval $\mu \leq 643.54$.

1. The decision reached by the confidence interval procedure: reject the null hypothesis and believe that the mean concentration of chemical oxygen demand of surface water in Juru River is less than 650.
   The result of the confidence interval is compared with:
2. The decision reached by the critical value (traditional) procedure: reject the null hypothesis as presented in Example 3.5.
3. The decision reached by the *P*-value procedure: reject the null hypothesis as presented in Example 6.11.

We can say that the three procedures have reached the same decision which is to reject the null hypothesis and we believe that the mean concentration of chemical oxygen demand of surface water in Juru River is less than 650.

Example 8.6: The concentration of total phosphate in the surface water: Example 3.6 is reproduced "The concentration of total phosphate of surface water of Juru River was studied. An environmentalist wishes to verify that the concentration of total phosphate of surface water of Juru River is less than or equal to 0.85 (mg/L). Thirteen samples were selected and the concentration of total phosphate of surface water of Juru River was measured. The average concentration of total phosphate provided by the sample data is 0.97 and the standard deviation is

0.38. A significance level of $\alpha = 0.01$ is chosen to test the claim. Assume that the population is normally distributed."

The four steps for conducting hypothesis testing employing the confidence interval procedure can be used to test the hypothesis regarding the mean concentration of total phosphate in surface water of Juru River. The result of the confidence interval procedure will be compared with the critical value (traditional) and *P*-value procedures.

Step 1: Specify the null and alternative hypotheses

The two hypotheses regarding the mean concentration of total phosphate in surface water of Juru River are presented in Eq. (8.12).

$$H_0: \mu \leq 0.85 \quad \text{vs} \quad H_1: \mu > 0.85 \tag{8.12}$$

Step 2: Select the significance level (α) for the study and find the critical values

The level of significance is chosen to be 0.01. The t critical value for a one-tailed test with $\alpha = 0.01$ and degrees of freedom 12 is 2.681.

Step 3: Use the sample information to calculate the confidence interval

The hypothesis in Eq. (8.12) represents a one-tailed test. Thus, the lower confidence interval formula presented in Eq. (8.8) should be used to calculate the confidence interval for the mean concentration of total phosphate in surface water of Juru River. The formula in Eq. (8.8) represents the alternative hypothesis for a right-tailed test.

$$\mu \geq \overline{Y} - t_\alpha \left[\frac{\sigma}{\sqrt{n}} \right]$$

$$\mu \geq 0.97 - 2.681 \left[\frac{0.38}{\sqrt{13}} \right]$$

$$\mu \geq 0.687$$

Step 4: Make a decision using the confidence interval and interpret the results

Making a decision using the confidence interval procedure requires checking the confidence interval whether it contains the hypothesized value or not. It can be seen that the hypothesized value (0.85) is somewhere in the calculated confidence interval $\mu \geq 0.687$.

1. The decision reached by the confidence interval procedure: fail to reject the null hypothesis and we believe that the mean concentration of total phosphate in surface water of Juru River is less than or equal to 0.85.

   The result of the confidence interval is compared with:
2. The decision reached by the critical value (traditional) procedure: fail to reject the null hypothesis as presented in Example 3.6.
3. The decision reached by the *P*-value procedure: fail to reject the null hypothesis as presented in Example 6.12.

We can say that the three procedures have reached the same decision which is fail to reject the null hypothesis and we believe that the mean concentration of total phosphate in surface water of Juru River is less than or equal 0.85.

8.5 Confidence interval for testing one proportion value

Hypothesis testing for one sample proportion was studied using two procedures: critical value (traditional) and P-value procedures. The confidence interval procedure for one sample proportion will be illustrated employing examples to cover various situations of hypothesis testing.

Let us present two-tailed and one-tailed confidence intervals for one proportion before giving examples to illustrate the procedure.

- The two-tailed confidence interval for the proportion value is given in Eq. (8.13).

$$\hat{p} - Z_{\frac{\alpha}{2}}\sqrt{\frac{\hat{p}(1-\hat{p})}{n}} \leq p \leq \hat{p} + Z_{\frac{\alpha}{2}}\sqrt{\frac{\hat{p}(1-\hat{p})}{n}} \quad (8.13)$$

Do not reject the null hypothesis if the hypothesized (claimed) value falls in the confidence interval and reject the null hypothesis if the hypothesized value does not fall in the confidence interval.

- The one-tailed confidence interval:

The lower one-tailed confidence interval for the one proportion value is given in Eq. (8.14).

$$\hat{p} - Z_{\alpha}\sqrt{\frac{\hat{p}(1-\hat{p})}{n}} \leq p \quad (8.14)$$

or it can be written as

$$\left[\hat{p} - Z_{\alpha}\sqrt{\frac{\hat{p}(1-\hat{p})}{n}},\ \infty\right)$$

Reject the null hypothesis if the hypothesized value does not fall in the confidence interval $\hat{p} - Z_{\alpha}\sqrt{\frac{\hat{p}(1-\hat{p})}{n}} \leq p$, which means we reject the null hypothesis if $\hat{p} - Z_{\alpha}\sqrt{\frac{\hat{p}(1-\hat{p})}{n}} > p$ and the upper one-tailed confidence interval for the one proportion value is given in Eq. (8.15).

$$p \leq \hat{p} - Z_{\alpha}\sqrt{\frac{\hat{p}(1-\hat{p})}{n}} \quad (8.15)$$

or it can be written as

$$\left(-\infty,\ \hat{p} - Z_{\alpha}\sqrt{\frac{\hat{p}(1-\hat{p})}{n}}\right]$$

Reject the null hypothesis if the hypothesized value does not fall in the confidence interval $p \leq \hat{p} - Z_{\alpha}\sqrt{\frac{\hat{p}(1-\hat{p})}{n}}$, which means we reject the null hypothesis if $p > \hat{p} - Z_{\alpha}\sqrt{\frac{\hat{p}(1-\hat{p})}{n}}$.

Example 8.7: Mineral water: Example 4.4 is reproduced "A research group at a university wishes to verify the claim announced by a specific factory for mineral water which says that 98% of people are satisfied with the quality of mineral water. The research group selected 700 people randomly and recorded their response, they found that 26 were not satisfied. Use the sample information to make a decision as to whether to support the claim or not. A significance level of $\alpha = 0.01$ is chosen to test the claim. Assume that the population is normally distributed."

The four steps for conducting hypothesis testing employing the confidence interval procedure can be used to test the hypothesis regarding the proportion of people who feel satisfied with the mineral water produced by the factory. The result of the confidence interval procedure will be compared with the critical value (traditional) and P-value procedures.

Step 1: Specify the null and alternative hypotheses

The two hypotheses regarding the proportion of people who feel satisfied with the mineral water produced by the factory are presented in Eq. (8.16)

$$H_0: p = 0.98 \quad \text{vs} \quad H_1: p \neq 0.98 \tag{8.16}$$

Step 2: Select the significance level (α) for the study and find the critical values

The level of significance is chosen to be 0.01. The Z critical value for a two-tailed test with $\alpha = 0.01$ is 2.58.

Step 3: Use the sample information to calculate the confidence interval

The hypothesis in Eq. (8.16) represents a two-tailed test. Thus, the two-tailed confidence interval formula presented in Eq. (8.13) should be used to calculate the confidence interval for the proportion of people who feel satisfied with the mineral water produced by the factory. The formula in Eq. (8.13) represents the alternative hypothesis for a two-tailed test. The $\hat{p}$ was calculated in Example 4.4 to be 0.96.

$$\hat{p} - Z_{\frac{\alpha}{2}}\sqrt{\frac{\hat{p}(1-\hat{p})}{n}} \leq p \leq \hat{p} + Z_{\alpha/2}\sqrt{\frac{\hat{p}(1-\hat{p})}{n}}$$

$$0.96 - 2.58\sqrt{\frac{0.96(1-0.96)}{700}} \leq p \leq 0.96 + 2.58\sqrt{\frac{0.96(1-0.96)}{700}}$$

$$0.941 \leq p \leq 0.979$$

Step 4: Make a decision using the confidence interval and interpret the results

Making a decision using the confidence interval procedure requires checking the confidence interval whether it contains the hypothesized value or not. It can be seen that the hypothesized value (0.98) is not in the calculated confidence interval $0.941 \leq p \leq 0.979$.

1. The decision reached by the confidence interval procedure: reject the null hypothesis and we believe that the proportion of people who are satisfied with the mineral water produced by the factory differs from the announced proportion of 0.98.

   The result of the confidence interval is compared with:

2. The decision reached by the critical value (traditional) procedure: reject the null hypothesis as presented in Example 4.4.
3. The decision reached by the P-value procedure: reject the null hypothesis as presented in Example 6.13.

We can say that the three procedures have reached the same decision which is to reject the null hypothesis and we believe that the proportion of people who are satisfied with the factory differs from the announced proportion of 0.98.

Example 8.8: Lab skills: Example 4.5 is reproduced "A department of the environment in a university announced that less than or equal to 0.03 of the students have low lab skills. A professor wishes to evaluate her students' skills in the lab work. She selected a group of 200 students and the results were checked. It was found that seven students had the wrong result. Use the sample information to make a decision as to whether to support the claim or not. A significance level of $\alpha = 0.01$ is chosen to test the claim. Assume that the population is normally distributed."

The four steps for conducting hypothesis testing employing the confidence interval procedure can be used to test the hypothesis regarding the proportion of students who have low lab skills. The results of the confidence interval procedure will be compared with the critical value (traditional) and P-value procedures.

Step 1: Specify the null and alternative hypotheses

The two hypotheses regarding the proportion of students who have low lab skills are presented in Eq. (8.17).

$$H_0: p \leq 0.03 \quad \text{vs} \quad H_1: p > 0.03 \tag{8.17}$$

Step 2: Select the significance level (α) for the study and find the critical values

The level of significance is chosen to be 0.01. The Z critical value for a one-tailed test with $\alpha = 0.01$ is 2.326.

Step 3: Use the sample information to calculate the confidence interval

The hypothesis in Eq. (8.17) represents a one-tailed test. Thus, the lower confidence interval formula presented in Eq. (8.14) should be used to calculate the confidence interval for the proportion of students who have low lab skills. The formula represents the alternative hypothesis for a right-tailed test. The $\hat{p}$ was calculated in Example 4.5 to be 0.04.

$$\hat{p} - Z_{\frac{\alpha}{2}}\sqrt{\frac{\hat{p}(1-\hat{p})}{n}} \leq p$$

$$0.04 - 2.326\sqrt{\frac{0.04(1-0.04)}{200}} \leq p$$

$$0.008 \leq p$$

Step 4: Make a decision using the confidence interval and interpret the results

Making a decision using the confidence interval procedure requires checking the confidence interval whether it contains the hypothesized value or not. It can be seen that the hypothesized value (0.03) is somewhere in the calculated confidence interval $0.008 \leq p$.

1. The decision reached by the confidence interval procedure: fail to reject the null hypothesis and we believe that the proportion of students with low lab skills is less than or equal to 0.03.
   The result of the confidence interval is compared with:
2. The decision reached by the critical value (traditional) procedure: fail to reject the null hypothesis as presented in Example 4.5.
3. The decision reached by the P-value procedure: fail to reject the null hypothesis as presented in Example 6.14.

We can say that the three procedures have reached the same decision which is fail to reject the null hypothesis and we believe that the proportion of students with low lab skills is less than or equal to 0.03.

Example 8.9: Beyond permissible limits: Example 4.6 is reproduced "A study showed that at least 10% of water samples selected from a specific company exceed the permissible limit of heavy metals in water. A research team wishes to investigate the truth regarding the announced proportion. The team selected 130 samples and tested for the heavy metals. The sample data showed that three samples with heavy metals beyond the permissible limits. Use the sample information to make a decision as to whether to support the claim or not. A significance level of $\alpha = 0.01$ is chosen to test the claim. Assume that the population is normally distributed."

The four steps for conducting hypothesis testing employing the confidence interval procedure can be used to test the hypothesis regarding the proportion of water samples that exceed the permissible limit of heavy metal produced by the company. The result of the confidence interval procedure will be compared with the critical value (traditional) and P-value procedures.

Step 1: Specify the null and alternative hypotheses

The two hypotheses regarding the proportion of water samples that exceed the permissible limit of heavy metal produced by the company are presented in Eq. (8.18).

$$H_0: p \geq 0.10 \quad \text{vs} \quad H_1: p < 0.10 \tag{8.18}$$

Step 2: Select the significance level (α) for the study and find the critical values

The level of significance is chosen to be 0.01. The Z critical value for a one-tailed test with $\alpha = 0.01$ is 2.326.

Step 3: Use the sample information to calculate the confidence interval

The hypothesis in Eq. (8.18) represents a one-tailed test. Thus, the upper confidence interval formula presented in Eq. (8.15) should be used to calculate the confidence interval for the proportion of water samples that exceed the permissible limit of heavy metal. The formula in Eq. (8.15) represents the alternative hypothesis for a left-tailed test. The $\hat{p}$ was calculated in Example 4.6 to be 0.02.

$$p \leq \hat{p} + Z_{\alpha}\sqrt{\frac{\hat{p}(1-\hat{p})}{n}}$$

$$p \leq 0.02 + 2.326\sqrt{\frac{0.02(1-0.02)}{130}}$$

$$p \leq 0.049$$

Step 4: Make a decision using the confidence interval and interpret the results

Making a decision using the confidence interval procedure requires checking the confidence interval as to whether it contains the hypothesized value or not. It can be seen that the hypothesized value (0.10) is not in the calculated confidence interval $p \leq 0.049$.

1. The decision reached by the confidence interval procedure: reject the null hypothesis and we believe that the proportion of samples that exceed the permissible limit is less than 0.10.
   The result of the confidence interval is compared with:
2. The decision reached by the critical value (traditional) procedure: reject the null hypothesis as presented in Example 4.6.
3. The decision reached by the *P*-value procedure: reject the null hypothesis as presented in Example 6.14.

We can say that the three procedures have reached the same decision which is to reject the null hypothesis and we believe that the proportion of samples that exceed the permissible limit is less than 0.10.

8.6 Confidence interval for testing one standard deviation value

Hypothesis testing for one sample variance or standard deviation was studied using two procedures: critical value (traditional) and *P*-value procedures. The confidence interval procedure for one sample variance or standard deviation test will be illustrated employing examples to cover various situations of hypothesis testing.

Let us present two-tailed and one-tailed confidence intervals for one variance or standard deviation before giving examples to illustrate the procedure.

- The two-tailed confidence interval for one variance is given in Eq. (8.19).

$$\frac{(n-1)S^2}{\chi_R^2} \leq \sigma^2 \leq \frac{(n-1)S^2}{\chi_L^2} \tag{8.19}$$

with $d \cdot f = (n-1)$

where, $\chi_R^2 = \chi^2_{(\frac{\alpha}{2}, n-1)}$ which represents the right tail and $\chi_L^2 = \chi^2_{(1-\frac{\alpha}{2}, n-1)}$ which represents the left tail, and the confidence interval for the standard deviation is

$$\sqrt{\frac{(n-1)S^2}{\chi_R^2}} \leq \sigma \leq \sqrt{\frac{(n-1)S^2}{\chi_L^2}}$$

Do not reject the null hypothesis if the hypothesized value falls in the confidence interval and reject the null hypothesis if the hypothesized value does not fall in the confidence interval.

- The one-tailed confidence interval:

  The lower one-tailed confidence interval for one variance is given in Eq. (8.20).

$$\frac{(n-1)S^2}{\chi_R^2} \le \sigma^2 \tag{8.20}$$

Reject the null hypothesis if the hypothesized value does not fall in the confidence interval $\frac{(n-1)S^2}{\chi_R^2} \le \sigma^2$, which means we reject the null hypothesis if $\frac{(n-1)S^2}{\chi_R^2} > \sigma^2$ and the upper one-tailed confidence interval for the variance value is given in Eq. (8.21).

$$\sigma^2 \le \frac{(n-1)S^2}{\chi_L^2} \tag{8.21}$$

Reject the null hypothesis if the hypothesized value does not fall in the confidence interval $\sigma^2 \le \frac{(n-1)S^2}{\chi_L^2}$, which means we reject the null hypothesis if $\sigma^2 > \frac{(n-1)S^2}{\chi_L^2}$

Example 8.10: The concentration of lead in sediment: Example 5.5 is reproduced "A professor at a research center wishes to verify the claim that the variance of lead (Pb) concentration in sediment of Juru River is less than or equal 0.20 (mg/L). Ten samples were selected and tested for lead concentration. The collected data showed that the standard deviation of lead concentration is 0.23. A significance level of $\alpha = 0.01$ is chosen to test the claim. Assume that the population is normally distributed."

The four steps for conducting hypothesis testing employing the confidence interval procedure can be used to test the hypothesis regarding the variance of lead concentration in sediment of Juru River. The result of the confidence interval procedure will be compared with the critical value (traditional) and P-value procedures.

Step 1: Specify the null and alternative hypotheses

The two hypotheses regarding the variance of lead concentration in sediment of Juru River are presented in Eq. (8.22).

$$H_0: \sigma^2 \le 0.20 \quad \text{vs} \quad H_1: \sigma^2 > 0.20 \tag{8.22}$$

Step 2: Select the significance level (α) for the study and find the critical values

The level of significance is chosen to be 0.01. The χ^2 critical value for a right-tailed test with $\alpha = 0.01$ and $d \cdot f = 9$ is 21.666 ($\chi^2_{(\alpha, d.f)} = \chi^2_{(0.01,9)} = 21.666$).

Step 3: Use the sample information to calculate the confidence interval

The hypothesis in Eq. (8.22) represents a one-tailed chi-square test. Thus, the lower confidence interval formula presented in Eq. (8.20) should be used to calculate the confidence interval for the variance of lead concentration in sediment of

Juru River. The formula in Eq. (8.20) represents the alternative hypothesis for a right-tailed test.

$$\sigma^2 \geq \frac{(n-1)S^2}{\chi_R^2}$$

$$\sigma^2 \geq \frac{(n-1)S^2}{\chi_R^2} = \frac{(10-1)(0.23)^2}{21.666}$$

$$\sigma^2 \geq 0.02197453 = 0.022$$

Step 4: Make a decision using the confidence interval and interpret the results

Making a decision using the confidence interval procedure requires checking the confidence interval whether it contains the hypothesized value or not. It can be seen that the hypothesized value (0.20) is somewhere in the calculated confidence interval $\sigma^2 \geq 0.022$.

1. The decision reached by the confidence interval procedure: fail to reject the null hypothesis and we believe that the variance of lead concentration in sediment is less than or equal 0.20.

   The result of the confidence interval is compared with:
2. The decision reached by the critical value (traditional) procedure: fail to reject the null hypothesis as presented in Example 5.5.
3. The decision reached by the *P*-value procedure: fail to reject the null hypothesis as presented in Example 6.21.

We can say that the three procedures have reached the same decision which is fail to reject the null hypothesis and we believe that the variance of lead concentration in sediment is less than or equal to 0.20.

Example 8.11: The level of particulate matter in a palm oil mill: Example 5.6 is reproduced "A research group wishes to verify that the standard deviation of the particulate matter (PM_1) level in the air of a palm oil mill is less 23.50 ($\mu g/m^3$). Twelve samples were selected and the PM_1 level was measured. The sample data showed that the standard deviation of PM_1 level is 26.89. A significance level of $\alpha = 0.01$ is chosen to test the claim. Assume that the population is normally distributed".

The four steps for conducting hypothesis testing employing the confidence interval procedure can be used to test the hypothesis regarding the standard deviation of PM_1 level in the air of the palm oil mill. The result of the confidence interval procedure will be compared with the critical value (traditional) and *P*-value procedures.

Step 1: Specify the null and alternative hypotheses

The two hypotheses regarding the standard deviation of PM_1 level in the air of the palm oil mill are presented in Eq. (8.23).

$$H_0: \sigma \geq 23.50 \quad \text{vs} \quad H_1: \sigma < 23.50 \tag{8.23}$$

Step 2: Select the significance level (α) for the study and find the critical values

The level of significance is chosen to be 0.01. The χ^2 critical value for a one-tailed test with $\alpha = 0.01$ and $d \cdot f = 11$ is 3.053 ($\chi^2_{(1-0.01,12-1)} = \chi^2_{(0.99,11)} = 3.053$).

Step 3: Use the sample information to calculate the confidence interval

The hypothesis in Eq. (8.22) represents a one-tailed chi-square test. Thus, the upper confidence interval formula presented in Eq. (8.21) should be used to calculate the confidence interval for the standard deviation of PM_1 level in the air of palm oil mill. The formula in Eq. (8.21) represents the alternative hypothesis for a left-tailed test.

$$\sigma^2 \leq \frac{(n-1)S^2}{\chi^2_L}$$

$$\sigma^2 \leq \frac{(12-1)(26.89)^2}{3.053484}$$

$$\sigma^2 \leq 2604.825$$

$$\sqrt{\sigma^2} \leq 51.03749 = 51.037$$

Step 4: Make a decision using the confidence interval and interpret the results

Making a decision using the confidence interval procedure requires checking the confidence interval whether it contains the hypothesized value or not. It can be seen that the hypothesized value (23.50) is somewhere in the calculated confidence interval $\sqrt{\sigma^2} \leq 51.037$.

1. The decision reached by the confidence interval procedure: fail to reject the null hypothesis and we believe that the variance of PM_1 level is more than or equal to 23.50.
   The result of the confidence interval is compared with:
2. The decision reached by the critical value (traditional) procedure: fail to reject the null hypothesis as presented in Example 5.6.
3. The decision reached by the *P*-value procedure: fail to reject the null hypothesis as presented in Example 6.20.

We can say that the three procedures have reached the same decision which is fail to reject the null hypothesis and we believe that the standard deviation of PM_1 level is more than or equal to 23.50.

Example 8.12: The concentration of total suspended solids of surface water: Example 5.4 is reproduced "A researcher at an environmental section wishes to verify the claim that the variance of total suspended solids concentration (TSS) of Beris dam surface water is 1.25 (mg/L). Twelve samples were selected and the total suspended solids concentration was measured. The collected data showed that the standard deviation of total suspended solids concentration is 1.80. A significance level of $\alpha = 0.01$ is chosen to test the claim. Assume that the population is normally distributed."

The four steps for conducting hypothesis testing employing the confidence interval procedure can be used to test the hypothesis regarding the variance of total suspended solids in the surface water of Beris dam. The result of the confidence interval procedure will be compared with the critical value (traditional) and *P*-value procedures.

Step 1: Specify the null and alternative hypotheses

The two hypotheses regarding the variance of total suspended solids concentration in the surface water of Beris dam are presented in Eq. (8.24).

$$H_0\colon \sigma^2 = 1.25 \quad \text{vs} \quad H_1\colon \sigma^2 = 1.25 \tag{8.24}$$

Step 2: Select the significance level (α) for the study and find the critical values

The level of significance is chosen to be 0.01. The χ^2 critical values for a two-tailed test with $\alpha = 0.01$ are:

The left-tailed value: The chi-square critical value for the left-tailed test for $d.f = 11$ and $1 - \frac{\alpha}{2} = 0.995$ is 2.603, $\chi^2_{(1-\frac{\alpha}{2}, d\cdot f)} = \chi^2_{(0.995, 11)} = 2.603$.

The right-tailed value: The chi-square critical value for the right-tailed test for $.fd \cdot f = 11$ and $\frac{\alpha}{2} = 0.005$ is 26.757, $\chi^2_{(\frac{0.01}{2}, 12-1)} = \chi^2_{0.005, 11} = 26.757$.

Step 3: Use the sample information to calculate the confidence interval

The hypothesis in Eq. (8.24) represents a two-tailed chi-square test. Thus, the two-tailed confidence interval formula presented in Eq. (8.19) should be used to calculate the confidence interval for the variance of total suspended solids concentration in the surface water of Beris dam. The formula in Eq. (8.19) represents the alternative hypothesis for a two-tailed test.

$$\frac{(n-1)S^2}{\chi^2_R} \le \sigma^2 \le \frac{(n-1)S^2}{\chi^2_L}$$

$$\frac{(12-1)(1.80)^2}{26.757} \le \sigma^2 \le \frac{(12-1)(1.80)^2}{2.603}$$

$$1.331995 \le \sigma^2 \le 13.69073$$

$$1.332 \le \sigma^2 \le 13.691$$

Step 4: Make a decision using the confidence interval and interpret the results

Making a decision using the confidence interval procedure requires checking the confidence interval whether it contains the hypothesized value or not. It can be seen that the hypothesized value (1.25) is not in the calculated confidence interval $1.332 \le \sigma^2 \le 13.691$.

1. The decision reached by the confidence interval procedure: reject the null hypothesis and we believe that the variance of total suspended solids concentration in surface water differs from 1.25.

   The result of the confidence interval is compared with:

2. The decision reached by the critical value (traditional) procedure: reject the null hypothesis as presented in Example 5.4.
3. The decision reached by the P-value procedure: reject the null hypothesis as presented in Example 6.19.

We can say that the three procedures have reached the same decision which is to reject the null hypothesis and we believe that the variance of total suspended solids concentration in surface water differs from 1.25.

Further reading

Alkarkhi, A. F. M., Ahmad, A., & Easa, A. M. (2009a). Assessment of surface water quality of selected estuaries of Malaysia: Multivariate statistical techniques. *Environmentalist*, *29*, 255–262.

Alkarkhi, A. F. M., Ismail, N., Ahmed, A., & Easa, A. M. (2009b). Analysis of heavy metal concentrations in sediments of selected estuaries of Malaysia—A statistical assessment. *Environmental Monitoring and Assessment*, *153*, 179–185.

Eberly College of Science-Department of Statistics. (2020). *Confidence intervals & hypothesis testing* In: (Vol. 2020). Pennstate: The Pennsylvania State University.

NIST/SEMATECH. (2013). *Engineering statistics handbook* In: (Vol. 2020). NIST/SEMATECH.

Rahman, N. N. N. A., Chard, N. C., Al-Karkhi, A. F. M., Rafatullah, M., & Kadir, M. O. A. (2017). Analysis of particulate matters in air of palm oil mills – A statistical assessment. *Environmental Engineering and Management Journal*, *16*(11), 2537–2543.

Yusup, Y., & Alkarkhi, A. F. M. (2011). Cluster analysis of inorganic elements in particulate matter in the air environment of an equatorial urban coastal location. *Chemistry and Ecology*, *27*(3), 273–286.

Hypothesis testing for the difference between two populations

9

Abstract

This chapter addresses hypothesis testing for the difference between two population parameters including the concept of null and alternative hypotheses for the difference between two population parameters. Moreover, the general procedure for carrying out hypothesis testing for two population parameters is given in a detailed and enjoyable presentation. Three examples for each test are delivered to cover various situations of hypothesis testing, the tests are *Z*-tests for the mean when the sample size is large, *Z*-tests for the proportion, *t*-tests for the mean when the sample size is small covering two cases (dependent and independent samples), and *F*-tests for the ratio of two variances.

Keywords
Z-test; *t*-test; *F*-test; dependent samples; independent samples; mean; proportion; variance

Learning outcomes

After completing this chapter, readers should be able to:

- Describe the common steps for testing a hypothesis concerning two population means: large sample size;
- Describe the common steps for testing a hypothesis concerning two population means: small sample size;
- Describe the common steps for testing a hypothesis concerning two population proportions;
- Describe the common steps for testing a hypothesis concerning two population variances;
- Describe the common steps for testing a hypothesis concerning two dependent samples;
- Compute the *Z*-test statistic value for two population means and two population proportions;
- Compute the *t*-test statistic value for two population independent samples;
- Compute the *t*-test statistic value for two dependent samples (paired samples);
- Differentiate between two important concepts, independent and paired samples;
- Know the procedure for interpreting the results and match it to the field of study;
- Report on smart conclusions.

9.1 Introduction

Hypothesis testing regarding one population parameter was studied and investigated properly using three procedures for testing various hypotheses. Hypothesis testing

Applications of Hypothesis Testing for Environmental Science. DOI: https://doi.org/10.1016/B978-0-12-824301-5.00009-5

can be applied for the difference between two population means, proportions and variances for the comparison purpose. The general procedure for hypothesis testing can be used for testing the difference between two populations parameters with a simple change in the formulas to meet the purpose of testing for two population parameters. As an example, for two populations, consider that the objective of a study is to compare the concentration of total suspended solids in two rivers or two regions which represent two populations.

Hypothesis testing for the difference between two population means, proportions and variances, is covered in this chapter including setting up the two hypotheses (null and alternative) for the difference between two parameters and the general procedure for hypothesis testing in detail. Furthermore, the concept of two important terms related to hypothesis testing for two population parameters is delivered; the two concepts are dependent and independent samples.

9.2 The general procedure for testing two samples

The concept of dependent and independent samples should be given before delivering the general procedure for hypothesis testing for the difference between two population parameters because these two concepts are important and related to the hypothesis testing procedure for two population parameters.

Independent samples: Consider there are two samples, the two samples are called independent samples if the data values of the first sample have no link (not connected) with the data values of the second sample.

Dependent samples: Consider there are two samples, the two samples are called dependent samples if the data values of the first sample have a link (connected, matched, paired) with the data values of the second sample.

The general procedure for conducting hypothesis testing for two population parameters can be summarized by the steps given below.

Step 1: Specify the null and alternative hypotheses for the difference between two populations

Step 2: Select the significance level (α) for the study

Step 3: Use the sample information to calculate the test statistic value

Step 4: Identify the critical and noncritical regions for the study

Step 5: Make a decision and interpret the results

We will discuss step 1 to clarify how to specify the null and alternative hypotheses for the difference between two population parameters, other steps are similar to the general procedure for one population parameter.

Step 1: Specify the null and alternative hypotheses

Three steps can be employed to define the null and alternative hypotheses for the difference between two population parameters. The three steps are:

1. Find out the two hypotheses concerning the issue under investigation;
2. Translate the words of the null hypothesis into a mathematical expression;
3. Translate the words of the alternative hypothesis into a mathematical expression.

Example 9.1: Specify the two hypotheses for a two-samples test (two-sided): An environmentalist wishes to investigate a claim regarding the mean concentration of cadmium (Cd) of surface water in two rivers. The environmentalist wants to examine the claim that the mean concentration of cadmium of surface water of Juru River is the same as the mean concentration of cadmium of Jejawi River.

The three steps for setting up the null and alternative hypotheses can be used to write the two hypotheses.

1. *Find out the two hypotheses concerning the issue under investigation.*
   One can see that the hypothesis is given in words, thus we need to figure out the hypothesis. The hypothesis says, "the mean concentration of cadmium of surface water of Juru River is the same as (equal to) Jejawi River."
2. *Translate the words of the null hypothesis into a mathematical expression.*
   The next step in specifying the null and alternative hypotheses is to translate the words of the null hypothesis into a mathematical form. The null hypothesis states that the mean concentration of cadmium of surface water of Juru River (μ_1) is the same as the mean concentration of cadmium of surface water of Jejawi River (μ_2). This statement can be translated into a mathematical form as given below.

   $$H_0: \mu_1 = \mu_2$$

   One can see that the null hypothesis can be expressed as a difference between the two averages.

   $$H_0: \mu_1 - \mu_2 = 0$$

   where μ_1 and μ_2 refer to the population mean for the first and second populations, respectively.
   One can see that *equality* sign is under the null hypothesis.
3. *Translate the words of the alternative hypothesis into a mathematical expression.*
   The null hypothesis states that the mean concentration of cadmium of surface water of Juru River (μ_1) is equal to the mean concentration of cadmium of surface water of Jejawi River (μ_2). The opposite of the null hypothesis represents the alternative hypothesis, which is not equal. Thus if the mean concentration of cadmium of surface water of Juru River is not equal to the mean concentration of cadmium of surface water of Jejawi River, then the mean concentration of cadmium of surface water of Juru River will be either greater than or less than the mean concentration of cadmium of surface water of Jejawi River. We can recognize two directions for the alternative hypothesis, the first direction, (μ_1) is greater than (μ_2), and the second direction, (μ_1) is less than (μ_2). The two directions can be represented in a mathematical form as $\neq$. The alternative hypothesis in terms of a mathematical form is:

   $$H_1: \mu_1 \neq \mu_2$$

   One can see that the null hypothesis can be expressed as a difference between the two averages.

   $$H_1: \mu_1 - \mu_2 \neq 0$$

   One can see that the null and alternative hypotheses are *mutually exclusive* (cannot occur at the same time).

Example 9.2: Specify the two hypotheses for a two-samples test (right-sided): An environmentalist wishes to investigate a claim regarding the mean concentration of cadmium (Cd) of surface water in two rivers. The environmentalist wants to examine the claim that the mean concentration of cadmium of surface water of Juru River is greater than the mean concentration of cadmium of surface water of Jejawi River.

The three steps for setting up the null and alternative hypotheses can be used to write the two hypotheses.

1. *Find out the two hypotheses concerning the issue under investigation.*
   One can see that the hypothesis is given in words, thus we need to figure out the hypothesis. The claim says "the mean concentration of cadmium of surface water of Juru River is greater than the mean concentration of cadmium of surface water of Jejawi River."
2. *Translate the words of the null hypothesis into a mathematical expression.*
   The next step in specifying the null and alternative hypotheses is to translate the words of the null hypothesis into a mathematical form. The claim states that the mean concentration of cadmium of surface water of Juru River (μ_1) is greater than the mean concentration of cadmium of surface water of Jejawi River (μ_2). Thus if the mean concentration of cadmium of surface water of Juru River is not greater than the mean concentration of cadmium of surface water of Jejawi River, then the mean concentration of cadmium of surface water of Juru River will be either less than or equal to the mean concentration of cadmium of surface water of Jejawi River. Thus we can use the two directions to represent the null hypothesis as $\leq$. This statement can be translated into a mathematical form as given below.

   $$H_0: \mu_1 \leq \mu_2$$

   One can see that the null hypothesis can be expressed as a difference between the two averages.

   $$H_0: \mu_1 - \mu_2 \leq 0$$

   where μ_1 and μ_2 refer to the population mean for the first and second populations, respectively.
3. *Translate the words of the alternative hypothesis into a mathematical expression.*
   The null hypothesis states that the mean concentration of cadmium of surface water of Juru River is less than or equal to the mean concentration of cadmium of surface water of Jejawi River. The opposite of the null hypothesis represents the alternative hypothesis, thus if the mean concentration of cadmium of surface water of Juru River is not less than or equal to the mean concentration of cadmium of surface water of Jejawi River, then the mean concentration of cadmium of surface water of Juru River will be greater than the mean concentration of cadmium of surface water of Jejawi River. Thus we can recognize only one direction for the alternative hypothesis which is (μ_1) greater than (μ_2). This direction can be represented in a mathematical form as $>$. The alternative hypothesis in terms of a mathematical form is:

   $$H_1: \mu_1 > \mu_2$$

One can see that the null hypothesis can be expressed as a difference between the two averages.

$$H_1: \mu_1 - \mu_2 > 0$$

where μ_1 and μ_2 refer to the population mean for the first and second populations, respectively.

One can see that the *equality* sign is under the null hypothesis. One can see that the null and alternative hypotheses are *mutually exclusive* (cannot occur at the same time).

Example 9.3: Specify the two hypotheses for a two-samples test (left-sided): An environmentalist wishes to investigate a claim regarding the mean concentration of cadmium (Cd) of surface water in two rivers. The environmentalist wants to examine the claim that the mean concentration of cadmium of surface water of Juru River is less than the mean concentration of cadmium of surface water of Jejawi River.

The three steps for setting up the null and alternative hypotheses can be used to write the two hypotheses.

1. *Find out the two hypotheses concerning the issue under investigation.*

   One can see that the hypothesis is given in words, thus we need to figure out the hypothesis. The claim says "the mean concentration of cadmium of surface water of Juru River is less than the mean concentration of cadmium of surface water of Jejawi River."
2. *Translate the words of the null hypothesis into a mathematical expression.*

   The next step in specifying the null and alternative hypotheses is to translate the words of the null hypothesis into a mathematical form. The claim states that the mean concentration of cadmium of surface water of Juru River (μ_1) is less than the mean concentration of cadmium of surface water of Jejawi River (μ_2). Thus if the mean concentration of cadmium of surface water of Juru River is not less than the mean concentration of cadmium of surface water of Jejawi River, then the mean concentration of cadmium of surface water of Juru River will be either greater than or equal to the mean concentration of cadmium of surface water of Jejawi River. Thus we can use the two directions to represent the null hypothesis as $\geq$. This statement can be translated into a mathematical form as given below.

   $$H_0: \mu_1 \geq \mu_2$$

   One can see that the null hypothesis can be expressed as a difference between the two averages.

   $$H_0: \mu_1 - \mu_2 \geq 0$$

   where μ_1 and μ_2 refer to the population mean for the first and second populations, respectively.
3. *Translate the words of the alternative hypothesis into a mathematical expression.*

   The null hypothesis states that the mean concentration of cadmium of surface water of Juru River is greater than or equal to the mean concentration of cadmium of surface water of Jejawi River. The opposite of the null hypothesis represents the alternative hypothesis, thus if the mean concentration of cadmium of surface water of Juru River is not greater than or equal to the mean concentration of cadmium of surface water of Jejawi River, then the mean concentration of cadmium of surface water of Juru River will be less than the mean concentration of cadmium of surface water of Jejawi River. Thus we can recognize only one direction for the alternative hypothesis which is (μ_1) is greater than (μ_2). This direction can be represented in a mathematical form as $<$. The alternative hypothesis in terms of a mathematical form is:

$$H_1: \mu_1 < \mu_2$$

One can see that the alternative hypothesis can be expressed as a difference between the two averages.

$$H_1: \mu_1 - \mu_2 < 0$$

where μ_1 and μ_2 refer to the population mean for the first and second populations, respectively.

One can see that the *equality* sign is under the null hypothesis. One can see that the null and alternative hypotheses are *mutually exclusive* (they cannot occur at the same time).

Note:

We can summarize the type of hypothesis into two types, the first type is called a two-sided test and the second type is called a one-sided test, as shown below.

a. Two-sided or two-tailed test.

The two hypotheses for this type of test take the form given in Eq. (9.1)

$$H_0: \mu_1 = \mu_2 \quad H_1 = \mu_1 \neq \mu_2 \tag{9.1}$$

Or the two hypotheses can be expressed as a difference between the two averages.

$$H_0: \mu_1 - \mu_2 = 0 \quad H_1: \mu_1 - \mu_2 \neq 0$$

b. One-sided or one-tailed test.

The two hypotheses for this type of test take the form given in Eqs. (9.2) and (9.3) for the right-tailed and left-tailed tests, respectively.

1. Right-tailed

$$H_0: \mu_1 \leq \mu_2 \quad H_1: \mu_1 > \mu_2 \tag{9.2}$$

Or the two hypotheses can be expressed as a difference between the two averages.

$$H_0: \mu_1 - \mu_2 \leq 0 \quad H_1: \mu_1 - \mu_2 > 0$$

2. Left-tailed

$$H_0: \mu_1 \geq \mu_2 \quad H_1: \mu_1 < \mu_2 \tag{9.3}$$

Or the two hypotheses can be expressed as a difference between the two averages.

$$H_0: \mu_1 - \mu_2 \geq 0 \quad H_1: \mu_1 - \mu_2 < 0$$

where μ_1 and μ_2 refer to the population mean for the first and second populations, respectively.

9.3 Testing the difference between two means when the sample size is large

Consider two large random samples ($n \geq 30$) are selected, each sample is selected from a normally distributed population. Consider Y_1 and Y_2 are two variables of interest, n_1 and n_2 represent the sample sizes selected from population 1 and population 2, respectively, with $\overline{Y}_1$ and $\overline{Y}_2$ to represent the average value and σ_1^2 and σ_2^2 to represent the variance for populations 1 and 2, respectively. A claim regarding the difference between two population means of the variable of interest can be tested employing a Z-test for two sample means to make a decision regarding the hypothesis. The mathematical formula for computing the test statistic value for two samples Z-test is presented in Eq. (9.4).

$$Z = \frac{\overline{Y}_1 - \overline{Y}_2 - (\mu_1 - \mu_2)}{\sqrt{\frac{\sigma_1^2}{n_1} + \frac{\sigma_1^2}{n_2}}} \tag{9.4}$$

where,
$\overline{Y}_1 - \overline{Y}_2$ represents the observed difference between the two-sample means,
$\mu_1 - \mu_2$ represents the expected difference which equals zero ($\mu_1 - \mu_2 = 0$).
σ_1^2 and σ_2^2 can be replaced by the sample variances S_1^2 and S_2^2 when the sample size is large.

The computed test statistic value obtained from the sample data (9.4) is used to make a decision to reject or fail to reject the null hypothesis regarding the difference between the difference between two sample means. The procedure to make a decision is to compare the absolute value of the test statistic with the theoretical value (critical value) of the normal distribution or using a normal distribution curve, large test statistic value leads to reject the null hypothesis ($Z_\alpha < |Computed\ Z|$).

Example 9.4: The concentration of dissolved oxygen (DO) before and after the dam: A professor at an environmental section wanted to verify the claim that the mean concentration of DO (mg/L) of water before Beris dam is the same as the mean concentration of DO of water after the dam. He selected 35 samples from each region and tested for DO concentration. The collected data showed that the mean concentration of DO before the dam is 5.20 and the standard deviation of the population is 0.41, while the collected data after the dam showed that the mean concentration of DO is 2.50 and the standard deviation of the population is 1.10. A significance level of $\alpha = 0.01$ is chosen to test the claim. Assume that the two populations are normally distributed.

The Z-test for two samples can be used to make a decision regarding the equality of the mean concentration of DO of water before the dam μ_1 and the mean concentration of DO of water after the dam μ_2. Two procedures will be used to test the hypothesis regarding DO of water before and after the dam, the two procedures are critical value (traditional) and P-value.

Step 1: Specify the null and alternative hypotheses

The mean concentration of DO of water before the dam μ_1 is equal to the mean concentration of DO of water after the dam μ_2, this claim should be under the null hypothesis because the claim represents equality (=). If the mean concentration of DO of water before the dam is not equal to the mean concentration of DO of water after the dam, then two directions should be considered, the first direction could be $\mu_1 > \mu_2$, and the second direction could be $\mu_1 < \mu_2$. The two directions (greater than and less than) can be represented mathematically as $\neq$ (represents the alternative hypothesis). Thus we can write the two hypotheses (null and alternative) as presented in Eq. (9.5).

$$H_0: \mu_1 = \mu_2 \quad \text{vs} \quad H_1: \mu_1 \neq \mu_2 \tag{9.5}$$

We should make a decision regarding the null hypothesis whether the mean concentration of DO of water before the dam μ_1 is equal to the mean concentration of DO of water after the dam μ_2, or μ_1 differs from μ_2.

Step 2: Select the significance level (α) for the study

The significance level of 0.01 ($\alpha = 0.01$) is selected to test the hypothesis. We divide the value of significance level (0.01) by 2 to represent the two-tailed test (more or less) of the alternative hypothesis, thus $\frac{\alpha}{2} = \frac{0.01}{2} = 0.005$ which represents the rejection region in each tail of the standard normal curve (the left and right tails). The Z critical value for the two-tailed test with $\alpha = 0.01$ is 2.58 as appeared in the standard normal table (Table A in the Appendix). Thus the Z critical values for both sides are ± 2.58, we use the two Z critical values to make a decision whether to reject or fail to reject the null hypothesis.

a. The critical value procedure

Step 3: Use the sample information to calculate the test statistic value

The entries for the Z-test statistic formula as presented in Eq. (9.4) are already provided. The collected data showed that the mean concentration of DO of water before the dam is 5.20, the standard deviation is 0.41 and the sample size is 35, the mean concentration of DO of water after the dam is 2.50, the standard deviation is 1.10 and the sample size is 35.

We apply the formula presented in Eq. (9.4) to test the hypothesis regarding the equality of the mean concentration of DO of water before the dam and the mean concentration of DO of water after the dam.

$$Z = \frac{\overline{Y}_1 - \overline{Y}_2}{\sqrt{\frac{\sigma_1^2}{n_1} + \frac{\sigma_1^2}{n_2}}} = \frac{5.20 - 2.50}{\sqrt{\frac{(0.41)^2}{35} + \frac{(1.10)^2}{35}}}$$

$$Z = 13.60684 = 13.61$$

The test statistic value for the mean concentration of DO of water before and after the dam is found to be 13.61.

Step 4: Identify the critical and noncritical regions for the study

We can easily identify the rejection and nonrejection regions for the two-tailed test employing the Z critical values found in step 2. The rejection and nonrejection regions for the difference between the mean concentration of dissolved oxygen of water before the dam μ_1 and the mean concentration of dissolved oxygen of water after the dam μ_2, are presented in Fig. 9.1 for the standard normal curve (orange shaded area).

Step 5: Make a decision and interpret the results

One can observe that the null hypothesis should be rejected because the test statistic value of the Z-test calculated from the sample data is 13.61, which is greater than the Z critical value for a two-tailed test with $\alpha = 0.01$ (2.58). Moreover, one can use the standard normal curve to decide, the same decision (reject the null hypothesis) can be reached using the normal curve and it can be seen that the test statistic value falls in the rejection region on the right tail as shown in Fig. 9.1. The null hypothesis is rejected in favor of the alternative hypothesis and one should believe that the mean concentration of dissolved oxygen of water before the dam μ_1 differs from the mean concentration of dissolved oxygen of water after the dam μ_2.

We can conclude at a 1% significance level, that there is sufficient evidence provided by the collected data to believe that the mean concentration of dissolved oxygen of water before the dam μ_1 differs from the mean concentration of dissolved oxygen of water after the dam μ_2.

b. The *P*-value procedure

The first three steps of the general procedure are the same for critical value and *P*-value procedures. Thus we should start from step 4 to calculate the *P*-value.

Step 4: Calculate the P-value and specify the critical and noncritical regions for the study

Consider that the null hypothesis is correct, calculating the *P*-value for a two-tailed test requires computing the probability of observing a value of *Z* of 13.61 or greater or a value of *Z* (−13.61) or less. The probability to the right of *Z* value (13.61) is 0.0001 ($1 - 0.9999 = 0.0001$) and the area to the left of *Z* (−13.61) is 0.0001. Because the problem is two-tailed the *P*-value should be multiplied by 2 to calculate for a two-sided test ($2 \times 0.0001 = 0.0002$). The exact *P*-value (blue shaded area) using technology is very small and is equal to 0 ($1.823257e - 42 = 0$) as presented in Fig. 9.1.

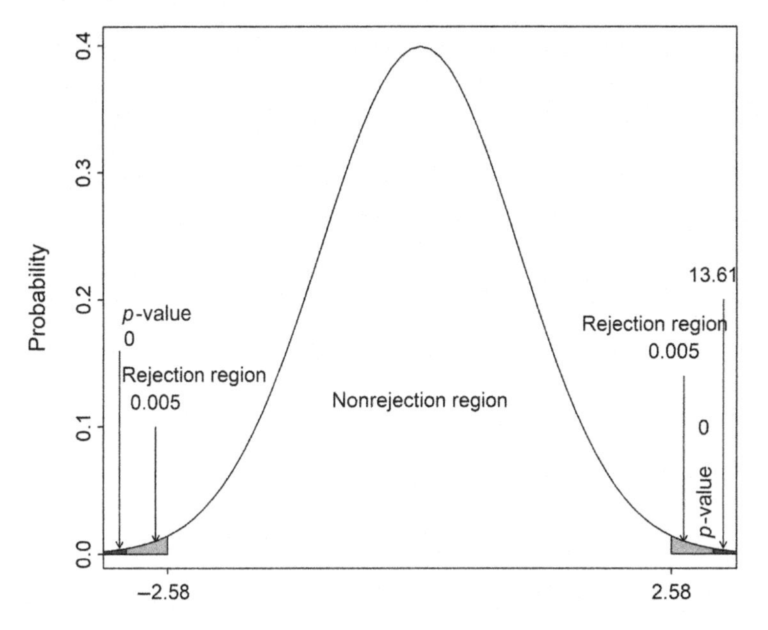

Figure 9.1 The rejection and nonrejection regions for the mean concentration of dissolved oxygen of water before and after the dam.

Step 5: Make a decision using P-value and interpret the results

A small *P*-value leads to rejecting the null hypothesis in favor of the alternative hypothesis. The decision is made by comparing the *P*-value with the significance level (α), one can see that the calculated *P*-value is almost zero (0) which is less than the significance level of 0.01, $0 < 0.01$.

The same decision is reached by using the *P*-value and critical value procedures: reject the null hypothesis and one should believe that the mean concentration of dissolved oxygen of water before the dam differs from the mean concentration of dissolved oxygen of water after the dam.

Example 9.5: The concentration of mercury in river water: A researcher at a research institution wanted to verify the claim that the mean concentration of mercury (Hg) (mg/L) in water of Juru River is less than the mean concentration of mercury (Hg) in water of Jejawi River. He selected 33 samples from each river and tested for mercury concentration. The collected data showed that the mean concentration of mercury of Juru River is 0.13 and the standard deviation is 0.02, while the collected data for Jejawi River showed that the mean concentration of mercury is 0.20 and the standard deviation is 0.07. A significance level of $\alpha = 0.01$ is chosen to test the claim. Assume that the two populations are normally distributed.

The *Z*-test for two samples can be used to make a decision regarding the mean concentration of mercury in water of Juru River μ_1 and the mean concentration of mercury (Hg) in water of Jejawi River μ_2. Two procedures will be used to test the hypothesis regarding the mean concentration of mercury in water of Juru and Jejawi Rivers, the two procedures are critical value (traditional) and *P*-value.

Step 1: Specify the null and alternative hypotheses

The mean concentration of mercury in water of Juru River μ_1 is less than the mean concentration of mercury in water of Jejawi River μ_2, this claim should be under the alternative hypothesis because the claim represents only one direction which is less than. If the mean concentration of mercury in water of Juru River is not less than the mean concentration of mercury in water of Jejawi River, then two directions should be considered, the first direction could be $\mu_1 > \mu_2$ and the second direction is $\mu_1 = \mu_2$. The two directions (greater than and equal to) can be represented mathematically as $\geq$ (represents the null hypothesis). Thus we can write the two hypotheses (null and alternative) as presented in Eq. (9.6).

$$H_0: \mu_1 \geq \mu_2 \quad \text{vs} \quad H_1: \mu_1 < \mu_2 \tag{9.6}$$

We should make a decision regarding the null hypothesis as to whether the mean concentration of mercury in water of Juru River μ_1 is greater than or equal to the mean concentration of mercury in water of Jejawi River or the mean concentration of mercury in water of Juru River is less than that of Jejawi River.

Step 2: Select the significance level (α) for the study

The significance level of 0.01 ($\alpha = 0.01$) is selected to test the hypothesis. Because the test is a one-tailed test as presented by the alternative hypothesis (less than), we should represent the rejection region on the left tail of the standard normal curve. The *Z* critical value for a one-tailed test with $\alpha = 0.01$ is 2.326 as appeared in the standard normal table (Table A in the Appendix). Thus the Z critical value for the left-tailed test is -2.326, we use the Z critical value to decide whether to reject or fail to reject the null hypothesis.

a. The critical value procedure

Step 3: Use the sample information to calculate the test statistic value

The entries for the Z-test statistic formula as presented in Eq. (9.4) are already provided. The collected data showed that the mean concentration of mercury in water of Juru River (μ_1) is 0.13, the standard deviation is 0.02, and the sample size is 33; the mean concentration of mercury in water of Jejawi River (μ_2) is 0.20, the standard deviation is 0.07, and the sample size is 33.

We apply the formula presented in Eq. (9.4) to test the hypothesis regarding the mean concentration of mercury in water of Juru and Jejawi Rivers.

$$Z = \frac{\overline{Y}_1 - \overline{Y}_2}{\sqrt{\frac{\sigma_1^2}{n_1} + \frac{\sigma_1^2}{n_2}}} = \frac{0.13 - 0.20}{\sqrt{\frac{(0.02)^2}{33} + \frac{(0.07)^2}{33}}}$$

$$Z = -5.523535 = -5.52$$

The test statistic value for the mean concentration of mercury in water for Juru and Jejawi Rivers is found to be -5.52.

Step 4: Identify the critical and noncritical regions for the study

We can easily identify the rejection and nonrejection regions for a left-tailed test employing the Z critical value found in step 2. The rejection and nonrejection regions for the difference between the mean concentration of mercury in water for Juru and Jejawi Rivers are presented in Fig. 9.2 for the standard normal curve (orange shaded area).

Step 5: Make a decision and interpret the results

One can observe that the null hypothesis should be rejected because the test statistic value of the Z-test calculated from the sample data is -5.52, which is smaller than the Z critical value of -2.326 for a left-tailed test with $\alpha = 0.01$. Moreover, one can use the standard normal curve to decide, the same decision (reject the null hypothesis) can be reached using the normal curve and it can be seen that the test statistic value falls in the rejection region on the left tail as shown in Fig. 9.2. The null hypothesis is rejected in favor of the alternative hypothesis and we believe that the mean concentration of mercury in water of Juru River μ_1 is less than the mean concentration of mercury in water of Jejawi River μ_2.

We can conclude at a 1% significance level that there is sufficient evidence provided by the collected data to believe that the mean concentration of mercury in water of Juru River is less than the mean concentration of mercury in water of Jejawi River.

b. The *P*-value procedure

The first three steps of the general procedure are the same for critical value and *P*-value procedures. Thus we should start from step 4 to calculate the *P*-value.

Step 4: Calculate the P-value and specify the critical and noncritical regions for the study

Consider the null hypothesis is correct, calculating the *P*-value for a left-tailed test requires computing the probability of observing a value of Z of -5.52 or less than -5.52. This probability represents the area to the left of Z value of -5.52. The probability to the left of Z value (-5.52) is 0.0001 ($1 - 0.9999 = 0.0001$). The exact *P*-value (blue shaded area) using technology is very small and is equal to 0 ($1.661235e - 08 = 0.0$) as presented in Fig. 9.2.

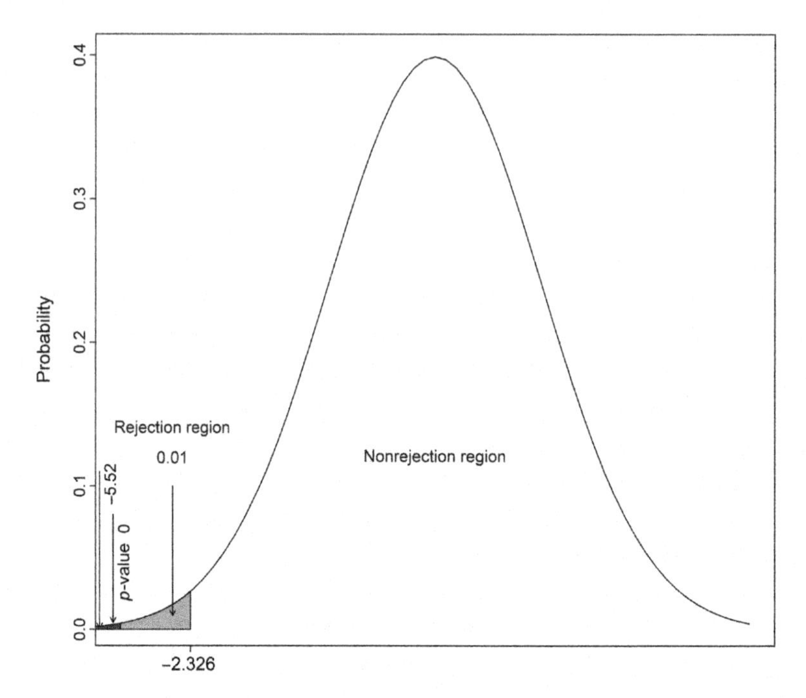

Figure 9.2 The rejection and nonrejection regions for the mean concentration of mercury in water of Juru and Jejawi Rivers.

Step 5: Make a decision using P-value and interpret the results

Small *P*-value leads to rejecting the null hypothesis in favor of the alternative hypothesis. The decision is made by comparing the *P*-value with the significance level (α), one can see that the calculated *P*-value is almost zero (0) which is less than the significance level of 0.01, $0.0 < 0.01$.

The same decision is reached by using the *P*-value and critical value procedures: reject the null hypothesis and one should believe that the mean concentration of mercury (Hg) in water of Juru River is less than the mean concentration of mercury in water of Jejawi River.

Example 9.6: The concentration of PM_1 in the air of palm oil mills: The concentration of PM_1 was investigated for two palm oil mills by a research team to verify the claim that the mean concentration of PM_1 μgm^{-3} in the air of a palm oil mill located in Kedah state is more than the mean concentration of PM_1 in the air of a palm oil mill located in Penang state of Malaysia. Thirty-four days were selected, and the concentration of particulate matter was measured. The collected data showed that the mean concentration of PM_1 in air of the Penang palm oil mill is 19.30 and the standard deviation is 16.07, while the collected data for the palm oil in the Kedah palm oil mill showed that the mean concentration of PM_1 is 74.87 and the standard deviation is 26.89. A significance level of $\alpha = 0.01$ is chosen to test the claim. Assume that the two populations are normally distributed.

The Z-test for two samples can be used to make a decision regarding the mean concentration of PM_1 in the air of a palm oil mill located in Kedah state (μ_1) and the mean concentration of PM_1 in the air of a palm oil mill located in Penang state of Malaysia (μ_2). Two procedures will be used to test the hypothesis regarding the mean concentration of PM_1, the two procedures are critical value (traditional) and P-value.

Step 1: Specify the null and alternative hypotheses

The mean concentration of PM_1 in the air of a palm oil mill located in Kedah state (μ_1) is more than the mean concentration of PM_1 in the air of a palm oil mill located in Penang state (μ_2), this claim should be under the alternative hypothesis because the claim represents only one direction which is $\mu_1 > \mu_2$. If the mean concentration of the PM_1 in the air of a palm oil mill located in Kedah state is not more than the mean concentration of PM_1 in the air of a palm oil mill located in Penang state, then two directions should be considered, the two directions are $\mu_1 < \mu_2$ and $\mu_1 = \mu_2$. The two directions (less than and equal to) can be represented mathematically as $\leq$. Thus we can write the two hypotheses (null and alternative) as presented in Eq. (9.7).

$$H_0: \mu_1 \leq \mu_2 \quad \text{vs} \quad H_1: \mu_1 > \mu_2 \tag{9.7}$$

We should make a decision regarding the null hypothesis as to whether the mean concentration of the PM_1 in the air of a palm oil mill located in Kedah state μ_1 is less than or equal to the mean concentration of PM_1 in the air of a palm oil mill located in Penang state μ_2 or $\mu_1 > \mu_2$.

Step 2: Select the significance level (α) for the study

The significance level of 0.01 ($\alpha = 0.01$) is selected to test the hypothesis. Because the test is a one-tailed test as presented by the alternative hypothesis (greater than), we should represent the rejection region on the right tail of the standard normal curve. The Z critical value for a one-tailed test with $\alpha = 0.01$ is 2.326 as appeared in the standard normal table (Table A in the Appendix). Thus the Z critical value for a right-tailed test is 2.326, we use the Z critical value to decide whether to reject or fail to reject the null hypothesis.

a. The critical value procedure

Step 3: Use the sample information to calculate the test statistic value

The entries for the Z-test statistic formula as presented in Eq. (9.4) are already provided. The collected data showed that the mean concentration of PM_1 in the air of a palm oil mill located in Kedah state μ_1 is 74.87, the standard deviation is 26.89, and the sample size is 34; the mean concentration of PM_1 in the air of a palm oil mill located in Penang state μ_2 is 19.30, the standard deviation is 16.07, and the sample size is 34.

We apply the formula presented in Eq. (9.4) to test the hypothesis regarding the mean concentration of PM_1 in the air of a palm oil mill located in Kedah and Penang states.

$$Z = \frac{\overline{Y}_1 - \overline{Y}_2}{\sqrt{\frac{\sigma_1^2}{n_1} + \frac{\sigma_1^2}{n_2}}} = \frac{74.87 - 19.30}{\sqrt{\frac{(26.89)^2}{34} + \frac{(16.07)^2}{34}}}$$

$$Z = 10.34368 = 10.34.$$

The test statistic value for the mean concentration of PM_1 in the air of a palm oil mill is found to be 10.34.

Step 4: Identify the critical and noncritical regions for the study

We can easily identify the rejection and nonrejection regions for a left-tailed test employing the Z critical value found in step 2. The rejection and nonrejection regions for the difference between the mean concentration of PM_1 in the air of a palm oil mill located in Kedah and Penang states are presented in Fig. 9.3 for the standard normal curve (orange shaded area).

Step 5: Make a decision and interpret the results

One can observe that the null hypothesis should be rejected because the test statistic value of the Z-test calculated from the sample data is 10.34, which is larger than the Z critical value of 2.326 for a right-tailed test with $\alpha = 0.01$. Moreover, one can use the standard normal curve to decide, the same decision (reject the null hypothesis) can be reached using the normal curve and it can be seen that the test statistic value falls in the rejection region as shown in Fig. 9.3. The null hypothesis is rejected, and we should believe that the mean concentration of PM_1 in the air of a palm oil mill located in Kedah state is more than the mean concentration of PM_1 in the air of a palm oil mill located in Penang state.

We can conclude at a 1% significance level that there is sufficient evidence provided by the collected data to believe that the mean concentration of PM_1 in the air of a palm oil mill located in Kedah state μ_1 is more than the mean concentration of PM_1 in the air of a palm oil mill located in Penang state μ_2.

b. The P-value procedure

The first three steps of the general procedure are the same for critical value and P-value procedures. Thus we should start from step 4 to calculate the P-value.

Step 4: Calculate the P-value and specify the critical and noncritical regions for the study

Consider the null hypothesis is correct, calculating the P-value for a right-tailed test requires computing the probability of observing a value of Z of 10.34 or greater than 10.34. This probability represents the area to the right of Z value of 10.34. The probability to the right of Z value (10.34) is 0.0. The exact P-value (blue shaded area) using technology is very small and is equal to 0.0 as presented in Fig. 9.3.

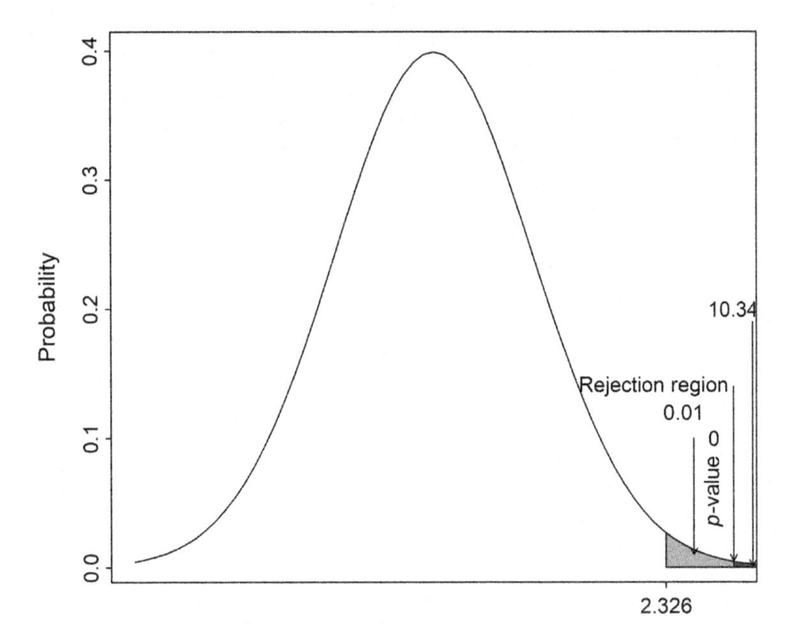

Figure 9.3 The rejection and nonrejection regions for the mean concentration of particulate matter in the air of a palm oil mill.

Step 5: Make a decision using P-value and interpret the results

A small P-value leads to rejecting the null hypothesis. The decision is made by comparing the P-value with the significance level (α), one can see that the calculated P-value is 0 which is smaller than the significance level of 0.01, $0 < 0.01$.

The same decision is reached by using the P-value and critical value procedures: reject the null hypothesis and we believe that the mean concentration of PM_1 in the air of palm oil mill in Kedah state is more than the mean concentration of PM_1 in the air of palm oil mill in Penang state.

9.4 Testing the difference between two means when the sample size is small

Consider two small independent random samples ($n < 30$) are selected, each sample is selected from a normally distributed population. Consider Y_1 and Y_2 are two variables of interest, n_1 and n_2 represent the sample sizes selected from population 1 and population 2, respectively, with $\overline{Y}_1$ and $\overline{Y}_2$ to represent the average values and σ_1^2 and σ_2^2 to represent the variances for populations 1 and 2, respectively. A claim regarding the difference between two population means of the variable of interest when the sample is small and the variance of the population is unknown can be tested employing a t-test for two small independent samples to make a decision regarding the hypothesis. The mathematical formula for computing the test statistic value for a two-sample t-test is presented in Eqs. (9.8) and (9.9) for unequal and equal variances, respectively.

a. Unequal variances: the t-test statistic for the difference between two samples when the variances are unequal is given in Eq. (9.8).

$$t = \frac{\overline{Y}_1 - \overline{Y}_2 - (\mu_1 - \mu_2)}{\sqrt{\frac{S_1^2}{n_1} + \frac{S_2^2}{n_2}}} \quad (9.8)$$

which has t distribution with degrees of freedom $(d \cdot f)$ equal to the smaller of $(n_1 - 1)$ and $(n_2 - 1)$.

Or we can use the following formula to calculate the degrees of freedom.

$$d \cdot f = \frac{\left[\frac{S_1^2}{n_1} + \frac{S_2^2}{n_2}\right]^2}{\frac{1}{(n_1 - 1)}\left(\frac{S_1^2}{n_1}\right)^2 + \frac{1}{(n_2 - 1)}\left(\frac{S_2^2}{n_2}\right)^2}$$

b. Equal variances: the t-test statistic for the difference between two samples when the variances are equal is given in Eq. (9.9).

$$t = \frac{\overline{Y}_1 - \overline{Y}_2 - (\mu_1 - \mu_2)}{\sqrt{\frac{S_P^2}{n_1} + \frac{S_P^2}{n_2}}} \quad (9.9)$$

which has t distribution with degrees of freedom $(d \cdot f)$ equal to $n_1 + n_2 - 2$ where $S_P^2 = \frac{(n_1 - 1)S_1^2 + (n_2 - 1)S_2^2}{n_1 + n_2 - 2}$ is called pooled variance. $\overline{Y}_1 - \overline{Y}_2$ represents the observed difference between the two-sample means, $\mu_1 - \mu_2$ represents the expected difference which equals to zero $(\mu_1 - \mu_2 = 0)$. S_1^2 and S_2^2 represent the sample variances.

The computed test statistic value obtained from the sample data (9.8 or 9.9) is used to make a decision to reject or fail to reject the null hypothesis regarding the difference between the two-sample means. The procedure to make a decision is to compare the absolute value of the test statistic with the theoretical value (critical value) of the t distribution or using a t distribution curve, a large test statistic value leads to rejecting the null hypothesis $(t_{\alpha,d.f} < |Computed\ t|)$.

Example 9.7: The concentration of arsenic in cockles: A professor at an environmental section wanted to verify the claim that the mean concentration of arsenic (As) in cockles (mg/L) obtained from region A is less than the concentration of arsenic in cockles obtained from region B. He selected 13 samples from each region and tested for the concentration of arsenic. The collected data showed that the mean concentration of arsenic obtained from region A is 2.66 and the standard deviation is 0.15, while the collected data for region B showed that the mean concentration of arsenic is 2.69 and the standard deviation is 0.13. A significance level of $\alpha = 0.01$ is chosen to test the claim. Assume that the two populations are normally distributed, and the variances of the populations are unequal ($\sigma_1^2 \neq \sigma_2^2$).

The t-test for two independent samples can be used to make a decision regarding the difference between the mean concentration of arsenic in cockles obtained from region A (μ_1) and the mean concentration of arsenic in cockles obtained from region B (μ_2). Two procedures will be used to test the hypothesis regarding the concentration of arsenic in cockles, the two procedures are critical value (traditional) and P-value.

Step 1: Specify the null and alternative hypotheses

The mean concentration of arsenic obtained from region A (μ_1) is less than mean concentration of arsenic obtained from region B (μ_2), this claim should be under the alternative hypothesis because the claim represents only one direction which is less than. If the mean concentration of arsenic obtained from region A is not less than the mean concentration of arsenic obtained from region B, then two directions should be considered, the first direction could be $\mu_1 > \mu_2$ and the second direction is $\mu_1 = \mu_2$. The two directions (greater than or equal to) can be represented mathematically as $\geq$ (represents the null hypothesis). Thus we can write the two hypotheses (null and alternative) as presented in Eq. (9.10).

$$H_0: \mu_1 \geq \mu_2 \quad \text{vs} \quad H_1: \mu_1 < \mu_2 \tag{9.10}$$

We should make a decision regarding the null hypothesis as to whether the mean concentration of arsenic obtained from region A (μ_1) is greater than or equal to the mean concentration of arsenic obtained from region B (μ_2), or the mean concentration of arsenic obtained from region A is less than that from region B.

Step 2: Select the significance level (α) for the study

The significance level of 0.01 ($\alpha = 0.01$) is selected to test the hypothesis. Because the test is a one-tailed test as presented by the alternative hypothesis (less than), we should represent the rejection region on the left tail of the t distribution curve. The t critical value for a one-tailed test with $\alpha = 0.01$ and $d \cdot f = 12$ ($d \cdot f =$ smaller sample size $= 13 - 1 = 12$) is 2.681 ($t_{(\alpha, d \cdot f)} = t_{(0.01, 12)} = 2.681$) as appeared in the t table for critical values (Table B in the Appendix). Thus the t critical value for the left-tailed test is -2.681, we use the t critical value to decide whether to reject or fail to reject the null hypothesis.

a The critical value procedure

Step 3: Use the sample information to calculate the test statistic value

The entries for the t-test statistic formula as presented in Eq. (9.8) are already provided, The collected data showed that the mean concentration of arsenic obtained from region A is 2.66, the standard deviation is 0.15, and the sample size is 13, while the collected data for region B showed that the mean concentration of arsenic is 2.69, the standard deviation is 0.13, and the sample size is 13.

We apply the formula presented in Eq. (9.8) to test the null hypothesis regarding the mean concentration of arsenic obtained from regions A and B.

$$t = \frac{\overline{Y}_1 - \overline{Y}_2}{\sqrt{\frac{S_1^2}{n_1} + \frac{S_1^2}{n_2}}} = \frac{2.66 - 2.69}{\sqrt{\frac{(0.15)^2}{13} + \frac{(0.13)^2}{13}}}$$

$$Z = -0.5449351 = -0.54$$

The test statistic value for the mean concentration of arsenic in cockles is found to be -0.54.

Step 4: Specify the critical and noncritical regions for the study

We can easily identify the rejection and nonrejection regions for a left-tailed test by employing the t critical value found in step 2. The rejection and nonrejection regions for the difference between the mean concentrations of arsenic in the two regions are shown in Fig. 9.4 for a t distribution curve (orange shaded area).

Step 5: Make a decision and interpret the results

One can observe that the null hypothesis should not be rejected because the test statistic value of the t-test calculated from the sample data is -0.54, which is greater than the t critical value of -2.681 for a left-tailed test with $\alpha = 0.01$ and $d \cdot f = 12$. Moreover, one can use a t distribution curve to decide, the same decision (do not reject the null hypothesis) can be reached using the t distribution curve and it can be seen that the test statistic value falls in the nonrejection region as shown in Fig. 9.4. The null hypothesis is not rejected and we believe that the mean concentration of arsenic in cockles obtained from region A is more than or equal to the mean concentration of arsenic in cockles obtained from region B.

We can conclude at a 1% significance level that there is sufficient evidence provided by the collected data to believe that the mean concentration of arsenic obtained from region A is greater than or equal to the mean concentration of arsenic obtained from region B.

b. The P-value procedure

The first three steps of the general procedure are the same for critical value and P-value procedures. Thus we should start from step 4 to calculate the P-value.

Step 4: Calculate the P-value and specify the critical and noncritical regions for the study

Consider that the null hypothesis is correct, calculating the P-value for the left-tailed test requires computing the probability of observing a value of t of -0.54 or less than -0.54. This probability represents the area to the left of the t value of -0.54. The probability to the left of the t value (-0.54) is more than 0.25 because the test statistic value position is less than the value of 0.695 in the t table for critical values at the row of $d \cdot f = 12$ and $\alpha = 0.25$. The exact P-value (blue shaded area) using technology is large and is equal to $0.2978904 = 0.30$ as presented in Fig. 9.4.

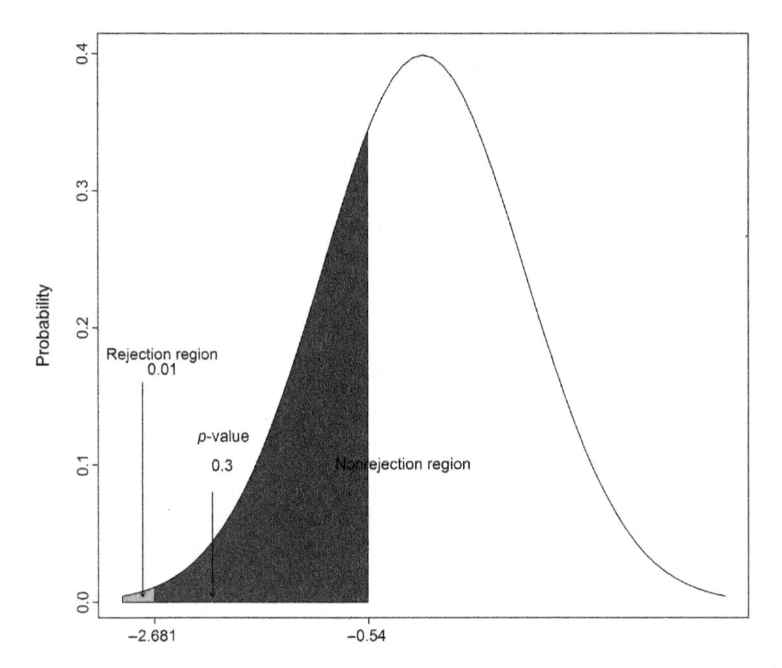

Figure 9.4 The rejection and nonrejection regions for the mean concentration of arsenic in cockles.

Step 5: Make a decision using P-value and interpret the results

A large P-value leads to not rejecting the null hypothesis. The decision is made by comparing the P-value with the significance level (α), one can see that the calculated P-value is greater than the significance level of 0.01 ($0.30 > 0.01$).

The same decision is reached by using the P-value and critical value procedures: fail to reject the null hypothesis and we believe that the mean concentration of arsenic obtained from region A is greater than or equal to the mean concentration of arsenic obtained from region B.

Example 9.8: The concentration of chromium in sediment: The concentration of chromium (Cr) (mg/L) in sediment was investigated to verify the claim that the mean concentration of chromium in Jejawi River is more than the mean concentration of chromium in Juru River. Ten samples were selected, and the concentration of chromium was measured. The collected data showed that the mean concentration of chromium of Jejawi River is 0.17 and the standard deviation is 0.034, while the collected data for Juru River showed that the mean concentration of chromium is 0.12 and the standard deviation is 0.029. A significance level of $\alpha = 0.01$ is chosen to test the claim. Assume that the two populations are normally distributed, and the variances are equal, $\sigma_1^2 = \sigma_2^2$.

The t-test for two independent samples can be used to make a decision regarding the difference between the mean concentration of chromium in sediment obtained from Jejawi River (μ_1) and the mean concentration of chromium in sediment obtained from Juru River (μ_2). Two procedures will be used to test the hypothesis regarding the concentration of chromium in sediment, the two procedures are critical value (traditional) and P-value.

Step 1: Specify the null and alternative hypotheses

The mean concentration of chromium in sediment obtained from Jejawi River (μ_1) is more than the mean concentration of chromium in sediment obtained from Juru River (μ_2), this claim should be under the alternative hypothesis because the claim represents only one direction which is more than. If the mean concentration of chromium in sediment obtained from Jejawi River is not more than the mean concentration of chromium in sediment obtained from Juru River, then two directions should be considered, the first direction could be $\mu_1 < \mu_2$ and the second direction is $\mu_1 = \mu_2$. The two directions (less than or equal to) can be represented mathematically as $\leq$ (represents the null hypothesis). Thus we can write the two hypotheses (null and alternative) as presented in Eq. (9.11).

$$H_0: \mu_1 \leq \mu_2 \quad \text{vs} \quad H_1: \mu_1 > \mu_2 \tag{9.11}$$

We should make a decision regarding the null hypothesis as to whether the concentration of chromium in sediment obtained from Jejawi River (μ_1) is less than or equal to the concentration of chromium in sediment obtained from Juru River (μ_2) or the concentration of chromium in sediment obtained from Jejawi River is more than the concentration of Jejawi River.

Step 2: Select the significance level (α) for the study

The significance level of 0.01 ($\alpha = 0.01$) is selected to test the hypothesis. Because the test is a one-tailed test as presented by the alternative hypothesis (greater than), we should represent the rejection region on the right tail of the t distribution curve. The t critical value for a one-tailed test with $\alpha = 0.01$ and $d \cdot f = (10 + 10 - 2 = 18)$ is 2.764 ($t_{(\alpha, d \cdot f)} = t_{(0.01,18)} = 2.552$) as appeared in the t table for critical values (Table B in the Appendix). Thus the t critical value for the right-tailed is 2.681, we use the t critical value to decide whether to reject or fail to reject the null hypothesis.

a. The critical value procedure

Step 3: Use the sample information to calculate the test statistic value

The entries for the t-test statistic formula as presented in Eq. (9.9) are already provided. The collected data showed that the mean concentration of chromium in sediment obtained from Jejawi River is 0.17 and the standard deviation is 0.034, while the collected data for Juru River showed that the mean concentration of chromium is 0.12 and the standard deviation is 0.029 and the sample size is 10 for each river.

We apply the formula presented in Eq. (9.9) to test the hypothesis regarding the mean concentration of chromium in sediment obtained from Jejawi and Juru Rivers.

$$S_P^2 = \frac{(n_1-1)S_1^2 + (n_2-1)S_2^2}{n_1+n_2-2} = \frac{(10-1)(0.034)^2 + (10-1)(0.029)^2}{10+10-2}$$

$$S_P^2 = 0.001156$$

$$t = \frac{\overline{Y}_1 - \overline{Y}_2 - (\mu_1 - \mu_2)}{\sqrt{\frac{S_P^2}{n_1} + \frac{S_P^2}{n_2}}} = \frac{0.17 - 0.12}{\sqrt{\frac{0.001936}{10} + \frac{0.001936}{10}}}$$

$$t = 3.288335 = 3.29$$

The test statistic value for the mean concentration of chromium in sediment for the two rivers is found to be 3.29.

Step 4: Identify the critical and noncritical regions for the study

We can easily identify the rejection and nonrejection regions for a right-tailed test employing the t critical value found in step 2. The rejection and nonrejection regions for the difference between the mean concentrations of chromium in sediment for the two rivers are shown in Fig. 9.5 for a t distribution curve (orange shaded area).

Step 5: Make a decision and interpret the results

One can observe that the null hypothesis should be rejected because the test statistic value of the t-test calculated from the sample data is 3.29, which is greater than the t critical value of 2.552 for a right-tailed test with $\alpha = 0.01$ and $d \cdot f = 18$. Moreover, one can use a t distribution curve to decide, the same decision (reject the null hypothesis) can be reached using the t distribution curve and it can be seen that the test statistic value falls in the rejection region as shown in Fig. 9.5. The null hypothesis is rejected and we believe that the mean concentration of chromium in sediment obtained from Jejawi River is more than the mean concentration of chromium in sediment obtained from Juru River.

We can conclude at a 1% significance level that there is sufficient evidence provided by the collected data to believe that the mean concentration of chromium in sediment obtained from Jejawi River is more than the mean concentration of chromium in sediment obtained from Juru River.

b. The P-value procedure

The first three steps of the general procedure are the same for critical value and P-value procedures. Thus we should start from step 4 to calculate the P-value.

Step 4: Calculate the P-value and specify the critical and noncritical regions for the study

Consider that the null hypothesis is correct, calculating the P-value for the right-tailed test requires computing the probability of observing a value of t of 3.29 or more than 3.29. This probability represents the area to the right of t value of 3.29. The probability to the right of t value 3.29 is very small and less than 0.01 because the test statistic value position is more than the critical value of 2.552 in the t table for critical values at the row of $d \cdot f = 18$ and $\alpha = 0.01$. The exact P-value (blue shaded area) using technology is very small and is equal to $0.002042563 = 0.002$ as presented in Fig. 9.5.

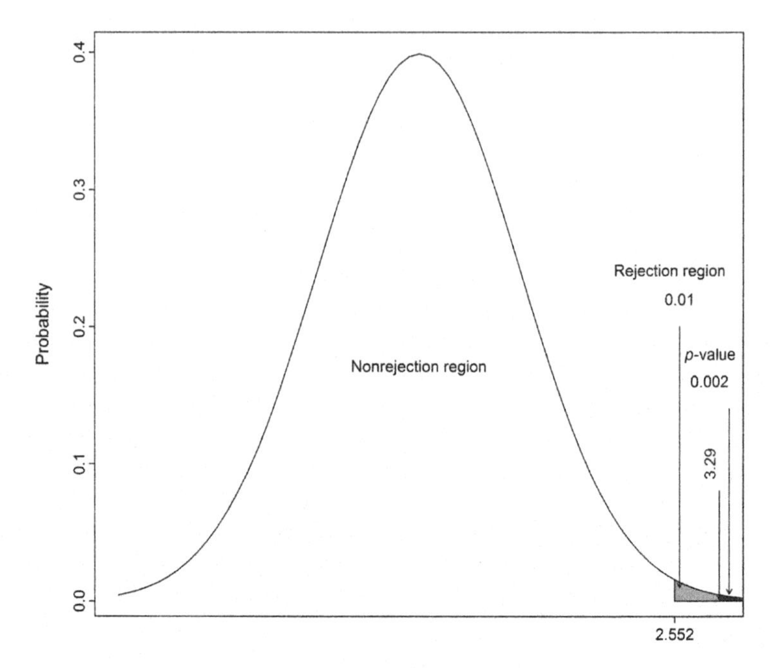

Figure 9.5 The rejection and nonrejection regions for the mean concentration of chromium in sediment.

Step 5: Make a decision using P-value and interpret the results

A small P-value leads to rejecting the null hypothesis. The decision is made by comparing the P-value with the significance level (α), one can see that the calculated P-value is smaller than the significance level of 0.01, $0.002042563 = 0.002 < 0.01$.

The same decision is reached by using the P-value and critical value procedures: reject the null hypothesis and we believe that the mean concentration of chromium in sediment obtained from Jejawi River is more than the mean concentration of chromium in sediment obtained from Juru River.

Example 9.9: The concentration of cobalt in two ponds: A professor wants to study the concentration of cobalt (Co) in two ponds of landfill leachate to verify the claim that the mean concentration of cobalt (mg/L) in the collection pond is equal to the mean concentration of cobalt in the aeration pond. Eleven samples were selected, and the concentration of cobalt was measured. The collected data showed that the mean concentration of cobalt of the collection pond is 53.75 and the standard deviation is 46.51, while the collected data from the aeration pond showed that the mean concentration of cobalt is 16.55 and the standard deviation is 14.21. A significance level of $\alpha = 0.01$ is chosen to test the claim. Assume that the two populations are normally distributed, and the variances are equal $\sigma_1^2 = \sigma_2^2$.

The t-test for two independent samples can be used to make a decision regarding the mean concentration of cobalt in the collection pond (μ_1) and the mean concentration of cobalt in the aeration pond (μ_2). Two procedures will be used to test the hypothesis regarding the concentration of cobalt of landfill leachate, the two procedures are critical value (traditional) and P-value.

Step 1: Specify the null and alternative hypotheses

The mean concentration of cobalt in the collection pond (μ_1) is equal to the mean concentration of cobalt in the aeration pond (μ_2), this claim should be under the null hypothesis because the claim represents equality (=). If the mean concentration of cobalt in the collection pond is not equal to the mean concentration of cobalt in the aeration pond, then two directions should be considered, the first direction could be $\mu_1 > \mu_2$ and the second direction could be $\mu_1 < \mu_2$. The two directions (greater than or less than) can be represented mathematically as $\neq$ (represents the alternative hypothesis). Thus we can write the two hypotheses (null and alternative) as presented in Eq. (9.12).

$$H_0: \mu_1 = \mu_2 \quad \text{vs} \quad H_1: \mu_1 \neq \mu_2 \tag{9.12}$$

We should make a decision regarding the null hypothesis as to whether the mean concentration of cobalt in the collection pond (μ_1) is equal to the mean concentration of cobalt in the aeration pond μ_2, or the mean concentration of cobalt in the collection pond differs from the aeration pond.

Step 2: Select the significance level (α) for the study

The significance level of 0.01 ($\alpha = 0.01$) is selected to test the hypothesis. We divide the value of the significance level (0.01) by 2 to represent the two-tailed test (more or less) of the alternative hypothesis, thus $\frac{\alpha}{2} = \frac{0.01}{2} = 0.005$ which represents the rejection region in each tail of the t distribution curve (the left and right tails). The t critical value for a two-tailed test with $\alpha = 0.01$ and $d \cdot f = 20$ is 2.845 as appeared in the t table (Table B in the Appendix). Thus the t critical values for both sides are ± 2.845, we use the two t critical values to make a decision whether to reject or fail to reject the null hypothesis.

a. The critical value procedure

Step 3: Use the sample information to calculate the test statistic value

The entries for the t-test statistic formula as presented in Eq. (9.9) are already provided. The collected data showed that the mean concentration of cobalt obtained from the collection pond is 53.75 and the standard deviation is 46.51, while the collected data for the aeration pond showed that the mean concentration of cobalt is 16.55 and the standard deviation is 14.21 and the sample size is 11 for each pond.

We apply the formula presented in Eq. (9.9) to test the hypothesis regarding the mean concentration of cobalt obtained from the collection and aeration ponds of landfill leachate.

$$S_P^2 = \frac{(n_1 - 1)S_1^2 + (n_2 - 1)S_2^2}{n_1 + n_2 - 2} = \frac{(11 - 1)(46.51)^2 + (11 - 1)(14.21)^2}{11 + 11 - 2}$$

$$S_P^2 = 2163.18$$

$$t = \frac{\overline{Y}_1 - \overline{Y}_2 - (\mu_1 - \mu_2)}{\sqrt{\frac{S_P^2}{n_1} + \frac{S_P^2}{n_2}}} = \frac{53.75 - 16.55}{\sqrt{\frac{2163.18}{11} + \frac{2163.18}{11}}}$$

$$t = 1.875763 = 1.88$$

The test statistic value for the mean concentration of cobalt of landfill leachate is found to be 1.88.

Step 4: Identify the critical and noncritical regions for the study

We can easily identify the rejection and nonrejection regions for a two-tailed test employing the t critical value found in step 2. The rejection and nonrejection regions for the difference between the mean concentrations of cobalt in two ponds of landfill leachate are shown in Fig. 9.6 for a t distribution curve (orange shaded area).

Step 5: Make a decision and interpret the results

One can observe that the null hypothesis should not be rejected because the test statistic value of the t-test calculated from the sample data is 1.88 which is less than the t critical value of 2.845 for a two-tailed test with $\alpha = 0.01$ and $d \cdot f = 11 + 11 - 2 = 20$. Moreover, one can use a t distribution curve to decide, the same decision (do not reject the null hypothesis) can be reached using the t distribution curve and it can be seen that the test statistic value falls in the nonrejection region as shown in Fig. 9.6. The null hypothesis is not rejected and we believe that the mean concentration of cobalt in the collection pond is equal to the mean concentration of cobalt in the aeration pond.

We can conclude at a 1% significance level that there is sufficient evidence provided by the collected data to believe that the mean concentration of cobalt in the collection pond is equal to the mean concentration of cobalt in the aeration pond.

b. The P-value procedure

The first three steps of the general procedure are the same for critical value and P-value procedures. Thus we should start from step 4 to calculate the P-value.

Step 4: Calculate the P-value and specify the critical and noncritical regions for the study

Consider that the null hypothesis is correct, calculating the P-value for the two-tailed test requires computing the probability of observing a value of t of 1.88 or greater or a value of t (-1.88) or less. The probability to the right of the t value (1.88) is $0.03768096 = 0.04$ and the area to the left of t (-1.88) is $0.03768096 = 0.04$. Because the problem is two-tailed the P-value should be multiplied by 2 to calculate for a two-sided test ($2 \times 0.03768096 = 0.07536192 = 0.08$). The exact P-value (blue shaded area) using technology is large and is equal to 0.08 as presented in Fig. 9.6.

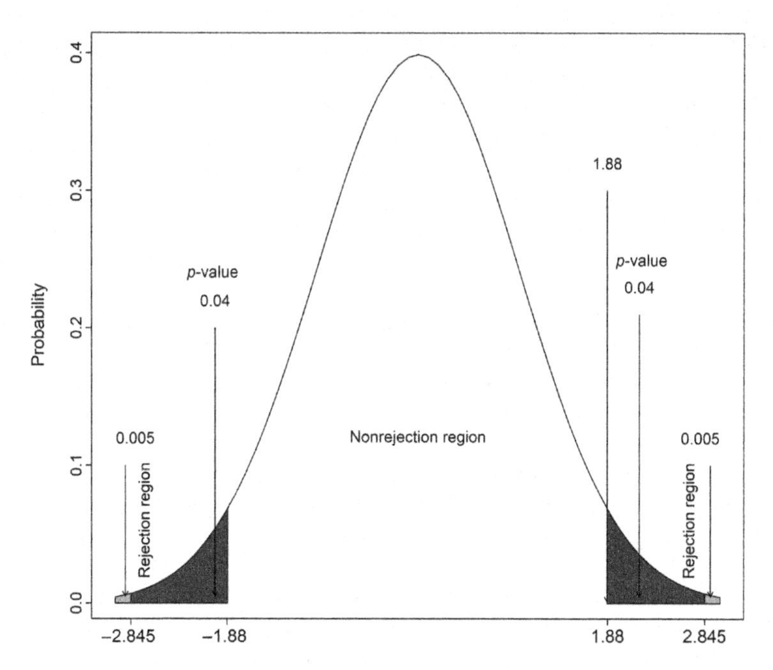

Figure 9.6 The rejection and nonrejection regions for the mean concentration of cobalt for landfill leachate.

Step 5: Make a decision using P-value and interpret the results

A large *P*-value leads to not rejecting the null hypothesis. The decision is made by comparing the *P*-value with the significance level (α), one can see that the calculated *P*-value is greater than the significance level of 0.01 ($0.08 > 0.01$).

The same decision is reached by using the *P*-value and critical value procedures: fail to reject the null hypothesis and we believe that the mean concentration of cobalt in the collection pond is equal to the mean concentration of cobalt in the aeration pond.

9.5 Testing two dependent samples

There are situations where the selected samples are dependent, for instance involving the same individuals for testing before and after a treatment employing the same procedure or employing two different treatments or techniques on the same individuals.

Consider that a random sample of size n is selected from a normally distributed population and tested/treated twice (different conditions) and the variable of interest is measured in both cases. Suppose that Y_1 and Y_2 represent two responses measured under the first and second conditions, respectively. A claim regarding the difference between the mean of the first condition and the second condition can be tested employing a t-test for two dependent (match, paired) samples to make a decision regarding the hypothesis. The mathematical formula for stating the two hypotheses for a two-tailed test is presented in Eq. (9.13).

$$H_0: \mu_{Y_1} = \mu_{Y_2} \quad \text{vs} \quad H_1: \mu_{Y_1} \neq \mu_{Y_2} \tag{9.13}$$

Or the two hypotheses can be expressed as the difference between the two averages.

$$H_0: \mu_{Y_1} - \mu_{Y_2} = 0 \quad \text{vs} \quad H_1: \mu_{Y_1} - \mu_{Y_2} \neq 0$$

The two hypotheses are usually expressed as given in Eq. (9.14).

$$H_0: \mu_d = 0 \quad \text{vs} \quad H_1: \mu_d \neq 0 \tag{9.14}$$

where,
μ_d represents the difference between μ_{Y_1} and μ_{Y_2}.
The hypothesis in Eq. (9.14) is two-tailed test. The one-tailed test is either right tailed or left tailed as shown below.

1. Right-tailed, the two hypotheses for a right-tailed test are given in Eq. (9.15).

$$H_0: \mu_d \leq 0 \quad \text{vs} \quad H_1: \mu_d > 0 \tag{9.15}$$

2. Left-tailed, the two hypotheses for a left-tailed test are given in Eq. (9.16).

$$H_0: \mu_d \geq 0 \quad \text{vs} \quad H_1: \mu_d < 0 \tag{9.16}$$

A claim regarding the difference between the two mean responses of the variable of interest can be tested employing a t-test for two dependent samples to make a decision regarding the hypothesis. The mathematical formula for computing the test statistic value for two dependent samples t-test is presented in Eq. (9.17).

$$t = \frac{\overline{d} - \mu_d}{S_d/\sqrt{n}} \tag{9.17}$$

which has t distribution with degrees of freedom ($d \cdot f = n - 1$).where $\overline{d}$ represents the mean of the differences between the two variables, μ_d represents the expected mean of the differences, and S_d represents the standard deviation of the differences.

$$\overline{d} = \frac{\sum d}{n}$$

$$S_d = \sqrt{\frac{\sum d^2 - \frac{(\sum d)^2}{n}}{n-1}}$$

The computed test statistic value obtained from the sample data (9.17) is used to make a decision to reject or not to reject the null hypothesis regarding the difference between the two-sample means. The procedure to make a decision is to compare the absolute value of the test statistic with the theoretical value (critical value) of the t distribution or using t distribution curve, large test statistic value leads to reject the null hypothesis ($t_{\alpha,d\cdot f} < |Computedt|$).

Example 9.10: The skills in measuring the concentration of zinc: A professor at an environmental department wanted to determine the difference between two students in his research team regarding the lab skills for measuring the concentration of zinc (Zn) in water. He believes that the students have different skills. He randomly selected 11 samples and half of each sample was given to each student to measure the concentration of zinc, the results are given in Table 9.1. A significance level of $\alpha = 0.01$ is chosen to test the claim. Assume that the population is normally distributed.

The t-test for dependent samples (match t-test) can be used to make a decision regarding the mean of differences for the skills between the two students. Two procedures will be used to test the hypothesis regarding lab skills of two students in measuring the concentration of Zn, the two procedures are critical value (traditional) and P-value.

Step 1: Specify the null and alternative hypotheses

The skills of the first student are equal to the skills of the second student, this direction should be under the null hypothesis because of equality (=). If the skills of the first student are different from the skills of the second student, then two directions should be considered, the first direction could be the first student has higher skills than the second student, and the second direction could be the first student has lower skills than the second student. The two directions (greater than or less than) can be represented mathematically as $\neq$ (represents the alternative hypothesis). Thus we can write the two hypotheses (null and alternative) as presented in Eq. (9.18).

$$H_0: \mu_d = 0 \quad \text{vs} \quad H_1: \mu_d \neq 0 \tag{9.18}$$

Table 9.1 The calculated data for a t-test including the concentration of zinc in water measured by two students, differences (D), and the squared valued of the differences D^2.

Y_1 = Student 1	Y_2 = Student 2	D = $Y_1 - Y_2$	D^2
0.05	0.03	0.02	0.0004
0.06	0.04	0.02	0.0004
0.06	0.03	0.03	0.0009
0.06	0.05	0.01	0.00004
0.05	0.03	0.02	0.0004
0.05	0.04	0.01	0.0001
0.05	0.05	0	0
0.06	0.05	0.01	1E-04
0.05	0.03	0.02	0.0004
0.06	0.03	0.03	0.0009
0.07	0.04	0.03	0.0009
		0.2	0.0046
	Mean	0.018182	
	SD	0.009816	

We should make a decision regarding the null hypothesis as to whether the two students have equal or different skills.

Step 2: Select the significance level (α) for the study

The significance level of 0.01 ($\alpha = 0.01$) is selected to test the hypothesis. We divide the value of the significance level (0.01) by 2 to represent the two-tailed test (more or less) of the alternative hypothesis, thus $\frac{\alpha}{2} = \frac{0.01}{2} = 0.005$ which represents the rejection region in each tail of the t distribution curve (the left and right tails). The t critical value for a two-tailed test with $\alpha = 0.01$ and $d \cdot f = 10$ is 3.169 as appeared in the t table (Table B in the Appendix). Thus the t critical values for both sides are $t_{(\frac{\alpha}{2}, d \cdot f)} = t_{(\frac{0.01}{2}, 10)} = \pm 3.169$, we use the two t critical values to make a decision as to whether to reject or fail to reject the null hypothesis.

a. The critical value procedure

Step 3: Use the sample information to calculate the test statistic value

The mean and standard deviation for the differences of the data presented in Table 9.1 are calculated below.

$$\bar{d} = \frac{\sum d}{n} = \frac{0.2}{11}$$

$$\bar{d} = 0.018181818 = 0.0182$$

$$S_d = \sqrt{\frac{\sum d^2 - \frac{(\sum d)^2}{n}}{n-1}} = \sqrt{\frac{0.0046 - \left[\frac{(0.2)^2}{11}\right]}{11}}$$

$$S_d = 0.009816498 = 0.0098$$

The entries for the matched (paired) t-test statistic formula as presented in Eq. (9.17) are already provided, the mean of the differences between the two students' records in measuring the Zn in water is 0.018, the standard deviation of the differences between the two students' records is 0.0098, and the sample size is 11.

We apply the formula presented in Eq. (9.17) to test the hypothesis regarding the lab skills of the two students.

$$t = \frac{\bar{d} - \mu_d}{S_d/\sqrt{n}} = \frac{0.0182 - 0}{\frac{0.0098}{\sqrt{11}}}$$

$$t = 6.142951 = 6.14$$

The test statistic value for the lab skills of the two students is found to be 6.14.

Step 4: Identify the critical and noncritical regions for the study

We can easily identify the rejection and nonrejection regions for a two-tailed test employing the t critical value found in step 2. The rejection and nonrejection regions for the mean differences between the two students' records are shown in Fig. 9.7 for a t distribution curve (orange shaded area).

Step 5: Make a decision and interpret the results

One can observe that the null hypothesis should be rejected because the test statistic value of the t-test calculated from the sample data is 6.14, which is greater than the t

critical value for a two-tailed test with $\alpha = 0.01$ and $d \cdot f = 10$ is 3.169. Moreover, one can use a t distribution curve to decide, the same decision (reject the null hypothesis) can be reached using the t distribution curve and it can be seen that the test statistic value falls in the rejection region as shown in Fig. 9.7. The null hypothesis is rejected and we believe that the two students have different skills.

We can conclude at a 1% significance level that there is sufficient evidence provided by the collected data to believe that the two students have different skills.

b. The P-value procedure

The first three steps of the general procedure are the same for critical value and P-value procedures. Thus we should start from step 4 to calculate the P-value.

Step 4: Calculate the P-value and specify the critical and noncritical regions for the study

Consider that the null hypothesis is correct, calculating the P-value for the two-tailed test requires computing the probability of observing a value of t of 6.14 or greater or a value of t (-6.14) or less. The t statistic value 6.14 falls beyond the value 3.169 with a significance level equal to 005, thus the P-value falls beyond 0.005. The exact P-value (blue) using technology is very small and is equal to $5.466498e - 05 = 0$ as presented in Fig. 9.7. The probability to the right of the t value is $5.466498e - 05 = 0$ and the area to the left of t (-6.14) is $5.466498e - 05 = 0.0$, because the problem is two-tailed the P-value should be multiplied by 2 to calculate for a two-sided test $(2 \times 5.466498e - 05 = 0)$.

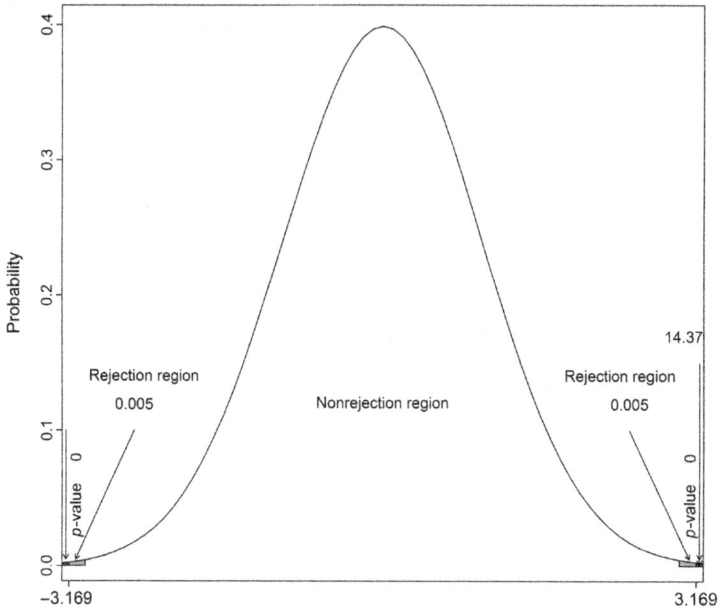

Figure 9.7 The rejection and nonrejection regions for the lab skills of two students.

Step 5: Make a decision using P-value and interpret the results

A small P-value leads to rejecting the null hypothesis in favor of the alternative hypothesis. The decision is made by comparing the P-value with the significance level (α), one can see that the calculated P-value is smaller than the significance level of 0.01 ($0 < 0.01$).

The same decision is reached by using the P-value and critical value procedures: reject the null hypothesis and we believe that the two students have different skills.

Example 9.11: The concentration of lead in cockles: A scientist wanted to determine the difference between two laboratories' equipment to measure the lead (Pb) concentration in cockles (mg/L). He believed that the mean concentration of lead measured by laboratory A is higher than the mean concentration of lead in cockles measured by laboratory B. He randomly selected 12 samples and half of each sample was sent to each laboratory, the data showed that the mean and standard deviation for the differences are 0.13 and 0.03, respectively. A significance level of $\alpha = 0.01$ is chosen to test the claim. Assume that the population is normally distributed.

The t-test for dependent samples (matched t-test) can be used to make a decision regarding the mean difference between the two laboratories for measuring the concentration of lead in cockles. Two procedures will be used to test the hypothesis regarding the accuracy of the two laboratories in measuring the concentration of lead, the two procedures are critical value (traditional) and P-value.

Step 1: Specify the null and alternative hypotheses

The mean records of laboratory A are higher than the mean records of laboratory B regarding the concentration of lead, this claim should be under the alternative hypothesis because the claim represents only one direction, which is more than. If the mean records of laboratory A are not more than the mean records of laboratory B, then two directions should be considered, the first direction could be the mean records of laboratory A are less than the mean records of laboratory B, and the second direction could be the mean records of laboratory A are equal to the mean records of laboratory B. The two directions (less than or equal to) can be represented mathematically as $\leq$ (represents the null hypothesis). Thus we can write the two hypotheses (null and alternative) as presented in Eq. (9.19).

$$H_0: \mu_d \leq 0 \quad \text{vs} \quad H_1: \mu_d > 0 \tag{9.19}$$

We should make a decision regarding the null hypothesis as to whether the records provided by laboratory A are less than or equal to the records provided by laboratory B regarding the lead concentration in cockles or the records provided by laboratory A are higher than the records provided by laboratory B.

Step 2: Select the significance level (α) for the study

The significance level of 0.01 ($\alpha = 0.01$) is selected to test the hypothesis. Because the test is a one-tailed test as presented by the alternative hypothesis (greater than), we should represent the rejection region on the right tail of the t distribution curve. The t critical value for a one-tailed test with $\alpha = 0.01$ and $d \cdot f = 12 - 1 = 11$ is 2.718 ($t_{(\alpha,d.f)} = t_{(0.01,11)} = 2.718$) as appeared in the t distribution table for critical values

(Table B in the Appendix). Thus the t critical value for the right-tailed is 2.718, we use the t critical value to decide whether to reject or fail to reject the null hypothesis.

a. The critical value procedure

Step 3: Use the sample information to calculate the test statistic value

The entries for the matched (paired) t-test statistic formula as presented in Eq. (9.17) are already provided, the mean of the differences between the two records in measuring the concentration of lead in cockles is 0.13, the standard deviation of the differences between the two records is 0.03, and the sample size is 12.

We apply the formula presented in Eq. (9.17) to test the hypothesis regarding the two laboratories in measuring the concentration of lead in cockles.

$$t = \frac{\overline{d} - \mu_d}{S_d/\sqrt{n}} = \frac{0.13 - 0}{\frac{0.03}{\sqrt{12}}}$$

$$t = 15.01111 = 15.01$$

The test statistic value for the two laboratories in measuring the concentration of lead in cockles is found to be 15.01.

Step 4: Identify the critical and noncritical regions for the study

We can easily identify the rejection and nonrejection regions for a right-tailed test employing the t critical value found in step 2. The rejection and nonrejection regions for the difference between the mean concentration of lead in cockles measured by two laboratories are shown in Fig. 9.8 for a t distribution curve (orange shaded area).

Step 5: Make a decision and interpret the results

One can observe that the null hypothesis should be rejected because the test statistic value of the t-test calculated from the sample data is 15.01, which is greater than the t critical value for a right-tailed test with $\alpha = 0.01$ and $d.f = 11$ is 2.718. Moreover, one can use a t distribution curve to decide, the same decision (reject the null hypothesis) can be reached using the t distribution curve and it can be seen that the test statistic value falls in the rejection region as shown in Fig. 9.8. The null hypothesis is rejected and we believe that the mean records of laboratory A are higher than the mean records of the laboratory B in measuring the concentration of lead in cockles.

We can conclude at a 1% significance level that there is sufficient evidence provided by the collected data to believe that the two laboratories provide different results (laboratory A provides higher records than laboratory B).

b. The P-value procedure

The first three steps of the general procedure are the same for critical value and P-value procedures. Thus we should start from step 4 to calculate the P-value.

Step 4: Calculate the P-value and specify the critical and noncritical regions for the study

Consider that the null hypothesis is correct, calculating the P-value for a right-tailed test requires computing the probability of observing a value of t of 15.01 or more than 15.01. This probability represents the area to the right of the t value of 15.01. The probability to the right of the t value (15.01) is less than 0.01 because the test statistic value position is more than the critical value of 2.178 in the t table for critical values at the row of $d \cdot f = 11$ and $\alpha = 0.01$. The exact P-value (blue shaded area) using technology is very small and is equal to $5.654066e - 09 = 0$ as presented in Fig. 9.8.

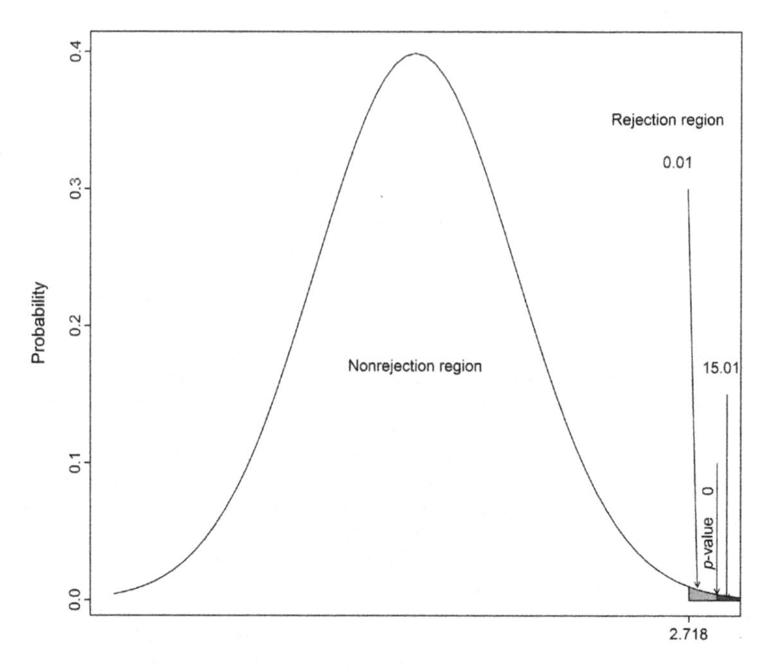

Figure 9.8 The rejection and nonrejection regions for the mean difference of the concentration of lead in cockles measured by two laboratories.

Step 5: Make a decision using P-value and interpret the results

A small *P*-value leads to rejecting the null hypothesis in favor of the alternative hypothesis. The decision is made by comparing the *P*-value with the significance level (α), one can see that the calculated *P*-value is smaller than the significance level of 0.01 ($0 < 0.01$).

The same decision is reached by using the *P*-value and critical value procedures: reject the null hypothesis and we believe that the two laboratories provide different results (laboratory A provides higher records than laboratory B regarding the concentration of lead in cockles).

Example 9.12: The sales plan for a new coagulant: An environmental company provided a new coagulant for treating wastewater. The company decided to start an intensive campaign to promote the new product. The sales before and after the intensive campaign covering 11 markets were recorded in thousands of dollars for a 1-month period, the data showed that the mean and standard deviation for the differences are -1.80 and 0.92, respectively. Can it be concluded that the sales before the advertising campaign are less than the sales after the campaign? A significance level of $\alpha = 0.01$ is chosen to test the claim. Assume that the population is normally distributed.

The *t*-test for dependent samples (match *t*-test) can be used to make a decision regarding the mean difference of sales of coagulants before and after the campaign covering 11 markets. Two procedures will be used to test the hypothesis regarding sales before and after the campaign, the two procedures are critical value (traditional) and *P*-value.

Step 1: Specify the null and alternative hypotheses

The mean records of sales before the campaign are less than the mean records of the sales after the campaign. This direction (less) can be represented mathematically as $<$ and placed under the alternative hypothesis. If the mean records of sales before the campaign are not less than the mean records of sales after the campaign, which means that the sales are more than or equal, this direction (more than or equal to) can be represented mathematically as $\geq$, this direction should be under the null hypothesis because of equality ($=$). Thus we can write the two hypotheses (null and alternative) as presented in Eq. (9.20).

$$H_0: \mu_d \geq 0 \quad \text{vs} \quad H_1: \mu_d < 0 \tag{9.20}$$

We should make a decision regarding the null hypothesis as to whether the sales before the campaign are more than or equal to the sales after the campaign regarding the new coagulant or the sales before the campaign are less than after the campaign.

Step 2: Select the significance level (α) for the study

The significance level of 0.01 ($\alpha = 0.01$) is selected to test the hypothesis. Because the test is a one-tailed test as presented by the alternative hypothesis (less than), we should represent the rejection region on the left tail of the t distribution curve. The t critical value for a one-tailed test with $\alpha = 0.01$ and $d \cdot f = 11 - 1 = 10$ is 2.764 ($t_{(\alpha,d.f)} = t_{(0.01,10)} = 2.764$) as appeared in the t table for critical values (Table B in the Appendix). Thus the t critical value for the left-tailed test is -1.812, we use the t critical value to decide whether to reject or fail to reject the null hypothesis.

a. The critical value procedure

Step 3: Use the sample information to calculate the test statistic value

The entries for the matched (paired) t-test statistic formula as presented in Eq. (9.17) are already provided, the mean of the differences between before and after the campaign is -1.80, the standard deviation of the differences between before and after the campaign is 0.92, and the sample size is 11.

We apply the formula presented in Eq. (9.17) to test the hypothesis regarding the sales of new coagulant before and after the advertising campaign.

$$t = \frac{\bar{d} - \mu_d}{S_d/\sqrt{n}} = \frac{-1.80 - 0}{\frac{0.92}{\sqrt{11}}}$$

$$t = -6.489049 = -6.49$$

The test statistic value for the sales before and after the campaign is found to be -6.49.

Step 4: Identify the critical and noncritical regions for the study

We can easily identify the rejection and nonrejection regions for a left-tailed test employing the t critical value found in step 2. The rejection and nonrejection regions for the difference between the mean sales for before and after the campaign are shown in Fig. 9.9 for a t distribution curve (orange shaded area).

Step 5: Make a decision and interpret the results

One can observe that the null hypothesis should be rejected because the test statistic value of the t-test calculated from the sample data is -6.49, which is smaller than the t critical value for a left-tailed test with $\alpha = 0.01$ and $d \cdot f = 11$ is -2.718. Moreover, one can use a t distribution curve to decide, the same decision (reject the null hypothesis) can be reached using the t distribution curve and it can be seen that the test statistic value falls in the rejection region as shown in Fig. 9.9. The null hypothesis is rejected and we believe that the mean of sales before the campaign is less than the mean of sales after the campaign regarding the new coagulant.

b. The P-value procedure

The first three steps of the general procedure are the same for critical value and P-value procedures. Thus we should start from step 4 to calculate the P-value.

Step 4: Calculate the P-value and specify the critical and noncritical regions for the study

Consider that the null hypothesis is correct, calculating the P-value for the left-tailed test requires computing the probability of observing a value of t of -6.49 or less than -6.49. This probability represents the area to the left of the t value of -6.49. The probability to the left of the t value (-6.49) is less than 0.01 because the test statistic value position is more than the value of 2.764 in the t table for critical values at the row of $d \cdot f = 10$ and $\alpha = 0.01$. The exact P-value (blue) using technology is very small and is equal to $3.495938e - 05 = 0.0$ as presented in Fig. 9.9.

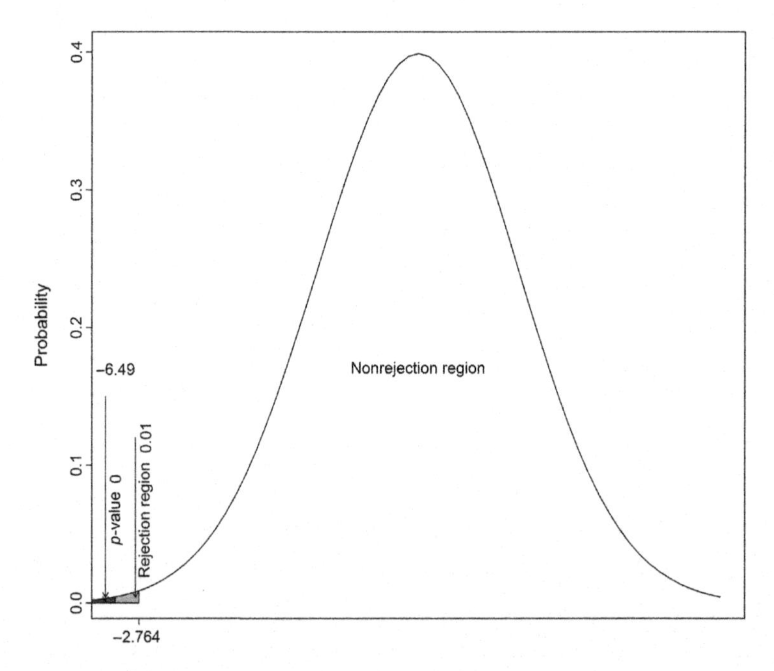

Figure 9.9 The rejection and nonrejection regions for the mean sales before and after the campaign.

Step 5: Make a decision using P-value and interpret the results

A small *P*-value leads to rejecting the null hypothesis in favor of the alternative hypothesis. The decision is made by comparing the *P*-value with the significance level (α), one can see that the calculated *P*-value is smaller than the significance level of 0.01 ($0 < 0.01$).

The same decision is reached by using the *P*-value and critical value procedures: reject the null hypothesis and we believe that the difference between before and after the campaign is less than zero, which indicates that the sales were lower before the campaign.

9.6 Testing the difference between two proportions

Consider that two independent random samples are selected, each sample is selected from a normally distributed population. Consider that Y_1 and Y_2 are two variables of interest to represent the number of units that carry a particular feature (property) for populations 1 and 2, respectively, $\hat{p}_2$ represents the proportion of the variables of interest, n_1 and n_2 represent the sample sizes selected from populations 1 and 2, respectively. A claim regarding the difference between population proportions of the variable of interest can be tested employing a *Z*-test for two independent samples to make a decision regarding the hypothesis. The mathematical formula for stating the null and alternative hypotheses for a two-tailed test is presented in Eq. (9.21).

$$H_0: p_1 = p_2 \quad \text{vs} \quad H_1: p_1 \neq p_2 \tag{9.21}$$

Or the two hypotheses can be expressed as a difference between the two proportions.

$$H_0: p_1 - p_2 = 0 \quad \text{vs} \quad H_1: p_1 - p_2 \neq 0$$

where p represents the population proportion.

The hypothesis in Eq. (9.21) is a two-tailed test. The one-tailed test is either right tailed or left tailed as shown below.

1. Right-tailed, the two hypotheses for a right-tailed test are given in Eq. (9.22).

$$H_0: p_1 \leq p_2 \quad \text{vs} \quad H_1: p_1 > p_2$$

or

$$H_0: p_1 - p_2 \leq 0 \quad \text{vs} \quad H_1: p_1 - p_2 > 0 \tag{9.22}$$

2. Left-tailed, the two hypotheses for a left-tailed test are given in Eq. (9.23).

$$H_0: p_1 \geq p_2 \quad \text{vs} \quad H_1: p_1 < p_2$$

or

$$H_0: p_1 - p_2 \geq 0 \quad \text{vs} \quad H_1: p_1 - p_2 < 0 \tag{9.23}$$

A claim regarding the difference between two population proportions can be tested employing a *Z*-test for two samples to make a decision regarding the hypothesis. The mathematical formula for computing the test statistic value for two sample proportions Z-test is presented in Eq. (9.24).

$$Z = \frac{\hat{p}_1 - \hat{p}_2 - (p_1 - p_2)}{\sqrt{\overline{p}\overline{q}(\frac{1}{n_1} + \frac{1}{n_2})}} \tag{9.24}$$

$\hat{p}$ is the sample proportion, and can be calculated as:

$$\hat{p}_1 = \frac{X_1}{n_1} \quad \hat{p}_2 = \frac{X_2}{n_2} \quad \overline{p} = \frac{X_1 + X_2}{n_1 + n_2} \quad \overline{q} = 1 - \overline{p}$$

where,
$\hat{p}_1 - \hat{p}_2$ represents the observed difference.
$p_1 - p_2$ represents the expected difference which equals zero ($p_1 - p_2 = 0$).
The computed test statistic value obtained from the sample data (9.24) is used to make a decision to reject or fail to reject the null hypothesis regarding the difference between two sample proportions. The procedure to make a decision is to compare the absolute value of the test statistic with the theoretical value (critical value) of the normal distribution or using a normal distribution curve, the large test statistic value leads to reject the null hypothesis ($Z_\alpha < |ComputedZ|$).

Example 9.13: The concentration of PM_{10} in the air (dry and wet seasons): A research team wants to investigate the claim regarding the proportion of PM_{10} in the air which says that PM_{10} is higher during the dry season than the wet season. The collected data during the dry season showed that 14 samples out of 100 showed high PM_{10} and the collected data during the wet season showed that two out of 100 showed high PM_{10}. A significance level of $\alpha = 0.01$ is chosen to test the claim. Assume that the two populations are normally distributed.

The Z-test for proportion can be used to make a decision regarding the difference in the proportion of PM_{10} in the air during the dry and wet seasons. Two procedures will be used to test the hypothesis regarding the concentration of PM_{10} in the air during the dry and wet seasons, the two procedures are critical value (traditional) and *P*-value.

Step 1: Specify the null and alternative hypotheses

The proportion of samples that show high PM_{10} during the dry season ($\hat{p}_1$) is more than the proportion of samples during the wet season ($\hat{p}_2$), this direction should be under the alternative hypothesis. This direction (greater than) can be represented mathematically as $>$. If the proportion of samples of PM_{10} obtained the during dry season is not more than the proportion of samples during the wet season, then two directions should be considered, the first direction could be $\hat{p}_1 < \hat{p}_2$ and the second direction could be $\hat{p}_1 = \hat{p}_2$. This direction (less than or equal to) can be represented mathematically as $\leq$ (represents the null hypothesis). Thus we can write the two hypotheses (null and alternative) as presented in Eq. (9.25).

$$H_0: \hat{p}_1 \leq \hat{p}_2 \quad \text{vs} \quad H_1: \hat{p}_1 > \hat{p}_2 \tag{9.25}$$

We should make a decision regarding the null hypothesis as to whether the proportion of PM_{10} in the air during the dry season $\hat{p}_1$ is less than or equal to the proportion of PM_{10} in the air during the wet season $\hat{p}_2$ or the proportion of PM_{10} in the air during the dry season is higher than that in the wet season.

Step 2: Select the significance level (α) for the study

The significance level of 0.01 ($\alpha = 0.01$) is selected to test the hypothesis. Because the test is a one-tailed test as presented by the alternative hypothesis (greater than), we should represent the rejection region on the right tail of the standard normal curve. The Z critical value for a one-tailed test with $\alpha = 0.01$ is 2.326 as appeared in the standard normal table (Table A in the Appendix). Thus the Z critical value for the right-tailed test is 2.326, we use the Z critical value to decide whether to reject or fail to reject the null hypothesis.

a. The critical value procedure

Step 3: Use the sample information to calculate the test statistic value

The entries for the Z-test statistic formula as presented in Eq. (9.24) are already provided, The collected data for PM_{10} during the dry season recorded 14 samples that exceeded the permissible limits out of 100, while the collected data calculated for PM_{10} during the wet season recorded two samples that exceeded the permissible limits out of 100.

The proportion for each season can be calculated using the available information.

$$\hat{p}_1 = \frac{X_1}{n_1} = \frac{14}{100} = 0.14$$

$$\hat{p}_2 = \frac{X_2}{n_2} = \frac{2}{100} = 0.02$$

$$\overline{p} = \frac{X_1 + X_2}{n_1 + n_2} = \frac{14 + 2}{100 + 100} = 0.08$$

$$\overline{q} = 1 - \overline{p} = 1 - 0.08 = 0.92$$

We apply the formula presented in Eq. (9.24) to test the hypothesis regarding the proportion of PM_{10} in the air during the dry ($\hat{p}_1$) and wet ($\hat{p}_2$) seasons.

$$Z = \frac{\hat{p}_1 - \hat{p}_2 - (p_1 - p_2)}{\sqrt{\overline{p}\overline{q}\left(\frac{1}{n_1} + \frac{1}{n_2}\right)}} = \frac{0.14 - 0.02}{\sqrt{(0.08)(0.92)\left(\frac{1}{100} + \frac{1}{100}\right)}}$$

$$Z = 3.127716 = 3.13$$

The test statistic value for the concentration of PM_{10} in the air during the dry and wet seasons is found to be 3.13.

Step 4: Identify the critical and noncritical regions for the study

We can easily identify the rejection and nonrejection regions for a right-tailed test employing the Z critical value found in step 2. The rejection and nonrejection regions for the proportion of samples that exceed the permissible limits regarding the concentration of PM_{10} during the dry and wet seasons are shown in Fig. 9.10 for the standard normal curve (orange shaded area).

Step 5: Make a decision and interpret the results

One can observe that the null hypothesis should be rejected because the test statistic value of the Z-test calculated from the sample data is 3.13, which is larger than the Z critical value of 2.326 for a right-tailed test with $\alpha = 0.01$. Moreover, one can use the standard normal curve to decide, the same decision (reject the null hypothesis) can be reached using the standard normal curve and it can be seen that the test statistic value falls in the rejection region as shown in Fig. 9.10. The null hypothesis is rejected and we believe that the proportion of samples that exceed the permissible limits of PM_{10} in the air during the dry season is higher than during the wet season.

b. The *P*-value procedure

The first three steps of the general procedure are the same for critical value and *P*-value procedures. Thus we should start from step 4 to calculate the *P*-value.

Step 4: Calculate the P-value and specify the critical and noncritical regions for the study

Consider that the null hypothesis is correct, calculating the *P*-value for a right-tailed test requires computing the probability of observing a value of *Z* of 3.13 or more than 3.13. This probability represents the area to the right of the *Z* value of 3.13. The probability to the right of the *Z* value 3.13 is 0.0001 ($1 - 0.9999 = 0.0001$). The exact *P*-value (blue shaded area) using technology is very small and is equal to $0.000880851 = 0$ as presented in Fig. 9.10.

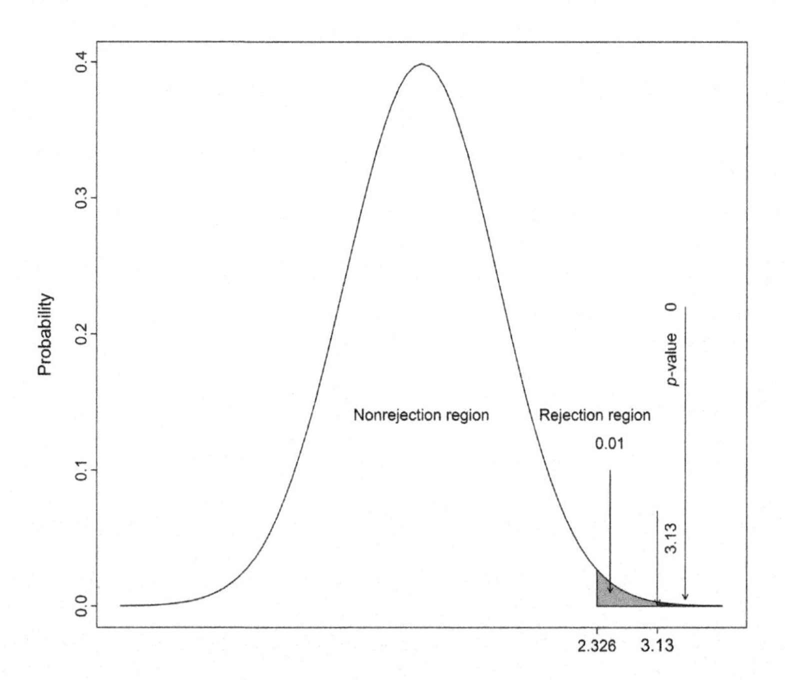

Figure 9.10 The rejection and nonrejection regions for the proportion of PM_{10} in the air during the dry and wet seasons.

Step 5: Make a decision using P-value and interpret the results

A small *P*-value leads to rejecting the null hypothesis in favor of the alternative hypothesis. The decision is made by comparing the *P*-value with the significance level (α), one can see that the calculated *P*-value is smaller than the significance level of 0.01 ($0 < 0.01$).

The same decision is reached by using the *P*-value and critical value procedures: reject the null hypothesis in favor of the alternative hypothesis and we believe that the proportion of samples that exceed the permissible limits of PM_{10} in the air during the dry season is higher than during the wet season.

Example 9.14: The level of turbidity in two locations: A research team wants to investigate the claim regarding the equality of the proportion of high turbidity in water of a dam and before the dam. The collected data for the water samples before the dam showed that 11 samples out of 80 showed high turbidity and collected data for the water samples from the dam showed that 10 out of 80 showed high turbidity. A significance level of $\alpha = 0.01$ is chosen to test the claim. Assume that the two populations are normally distributed.

The *Z*-test for proportion can be used to make a decision regarding the difference in the proportion of samples that showed high turbidity in the water of the dam and before the dam. Two procedures will be used to test the hypothesis regarding the proportion of high turbidity of water for two locations (before and at the dam), the two procedures are critical value (traditional) and *P*-value.

Step 1: Specify the null and alternative hypotheses

The proportion of water samples that showed high turbidity before the dam ($\hat{p}_1$) is equal to the proportion of water samples that showed high turbidity at the dam ($\hat{p}_2$), this claim should be under the null hypothesis because the claim represents equality (=). If the proportion $\hat{p}_1$ is not equal to $\hat{p}_2$, then two directions should be considered, the first direction could be $\hat{p}_1 > \hat{p}_2$ and the second direction could be $\hat{p}_1 < \hat{p}_2$. The two directions (greater than and less than) can be represented mathematically as $\neq$ (represents the alternative hypothesis). Thus we can write the two hypotheses (null and alternative) as presented in Eq. (9.26).

$$H_0: \hat{p}_1 = \hat{p}_2 \quad \text{vs} \quad H_1: \hat{p}_1 \neq \hat{p}_2 \tag{9.26}$$

We should make a decision regarding the null hypothesis as to whether the proportion of high turbidity before the dam $\hat{p}_1$ is equal to the proportion of high turbidity at the dam $\hat{p}_2$, or $\hat{p}_1$ differs from $\hat{p}_2$.

Step 2: Select the significance level (α) for the study

The significance level of 0.01 ($\alpha = 0.01$) is selected to test the hypothesis. We divide the value of the significance level (0.01) by 2 to represent the two-tailed test (more or less) of the alternative hypothesis, thus $\frac{\alpha}{2} = \frac{0.01}{2} = 0.005$ which represents the rejection region in each tail of the standard normal curve (the left and right tails). The *Z* critical value for a two-tailed test with $\alpha = 0.01$ is 2.58 as appeared in the standard normal table (Table A in the Appendix). Thus the *Z* critical values for both sides are ± 2.58, we use the two Z critical values to make a decision whether to reject or fail to reject the null hypothesis.

a. The critical value procedure

Step 3: Use the sample information to calculate the test statistic value

The entries for the Z-test statistic formula as presented in Eq. (9.24) are already provided. The collected data for the proportion of high turbidity before the dam recorded 11 samples showed high turbidity and the sample size is 80, while the collected data for the proportion of high turbidity at the dam recorded 10 samples that showed high turbidity out of 80.

The proportion for each region can be calculated using the available information.

$$\hat{p}_1 = \frac{X_1}{n_1} = \frac{11}{80} = 0.14$$

$$\hat{p}_2 = \frac{X_2}{n_2} = \frac{10}{80} = 0.12$$

$$\overline{p} = \frac{X_1 + X_2}{n_1 + n_2} = \frac{11 + 10}{80 + 80} = 0.13125 = 0.13$$

$$\overline{q} = 1 - \overline{p} = 1 - 0.13125 = 0.86875 = 0.87$$

We apply the formula presented in Eq. (9.24) to test the hypothesis regarding the proportion of high turbidity before the dam ($\hat{p}_1$) and at the dam ($\hat{p}_2$).

$$Z = \frac{\hat{p}_1 - \hat{p}_2 - (p_1 - p_2)}{\sqrt{\overline{p}\overline{q}\left(\frac{1}{n_1} + \frac{1}{n_2}\right)}} = \frac{0.14 - 0.12}{\sqrt{(0.13)(0.87)\left(\frac{1}{80} + \frac{1}{80}\right)}}$$

$$Z = 0.3745958 = 0.37$$

The test statistic value for the proportion of high turbidity in the water before and at the dam is found to be 0.37.

Step 4: Identify the critical and noncritical regions for the study

We can easily identify the rejection and nonrejection regions for right-tailed test employing the Z critical values found in step 2. The rejection and nonrejection regions for the difference between the proportions of high turbidity before and at the dam are shown in Fig. 9.11 for the standard normal curve (orange shaded area).

Step 5: Make a decision and interpret the results

One can observe that the null hypothesis should not be rejected because the test statistic value of the Z-test calculated from the sample data is 0.37, which is smaller than the Z critical value of 2.58 for two-tailed test with $\alpha = 0.01$. Moreover, one can use the standard normal curve to decide, the same decision (fail to reject the null hypothesis) can be reached using the standard normal curve and it can be seen that the test statistic value falls in the nonrejection region as shown in Fig. 9.11. The null hypothesis is not rejected and we believe that the proportion of high turbidity at the dam and before the dam is the same.

b. The P-value procedure

The first three steps of the general procedure are the same for critical value and P-value procedures. Thus we should start from step 4 to calculate the P-value.

Step 4: Calculate the P-value and specify the critical and noncritical regions for the study

Consider that the null hypothesis is correct, calculating the P-value for the two-tailed test requires computing the probability of observing a value of Z of 0.37 or greater or a value of Z (-0.37) or less. The probability to the right of the Z value (0.37) is 0.3557 ($1-0.6443=0.3557$) and the area to the left of Z (-0.37) is 0.3557. Because the problem is two-tailed the P-value should be multiplied by 2 to calculate for a two-sided test ($2\times 0.3557=0.7114$). The exact P-value (blue shaded area) using technology is very large and is equal to 0.71 for both sides ($0.7079611=0.71$) as presented in Fig. 9.11.

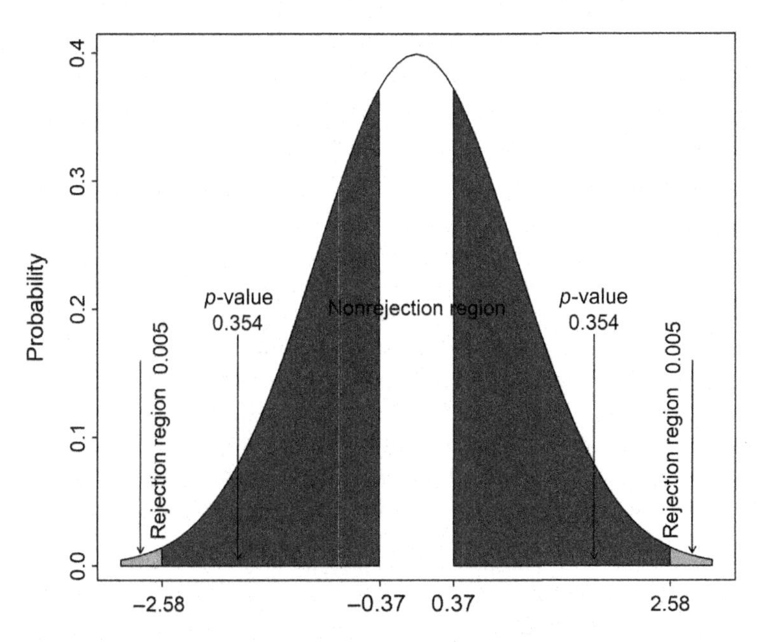

Figure 9.11 The rejection and nonrejection regions for the proportion of turbidity at the dam and before the dam.

Step 5: Make a decision using P-value and interpret the results

A large P-value leads to not rejecting the null hypothesis. The decision is made by comparing the P-value with the significance level (α), one can see that the calculated P-value is more than the significance level of 0.01 ($0.71 > 0.01$).

The same decision is reached by using the P-value and critical value procedures: fail to reject thc null hypothesis and we believe that the proportion of high turbidity at the dam and before the dam is the same.

Example 9.15: The concentration of manganese in sediment: A research team wants to investigate a claim regarding the proportion of samples that exceed the permissible limit of manganese (Mn) in the sediment of two rivers. The collected data for sediment samples obtained from Juru River showed that two samples out of 100 exceeded the permissible limits, and the collected data for sediment samples obtained from Jejawi River showed that 15 out of 100 exceeded the permissible limits. A significance level of $\alpha = 0.01$ is chosen to test the claim that Juru River showed a lower proportion than Jejawi River. Assume that the two populations are normally distributed.

The Z-test for proportion can be used to make a decision regarding the difference in the proportion of samples that exceeded the permissible limit regarding the concentration of manganese in sediment for Juru and Jejawi Rivers. Two procedures will be used to test the hypothesis regarding the proportion of manganese in sediment, the two procedures are critical value (traditional) and P-value.

Step 1: Specify the null and alternative hypotheses

The proportion of samples that exceeded the permissible limit regarding the concentration of manganese in sediment for Juru River ($\hat{p}_1$) is less than the proportion of samples that exceeded the permissible limit for Jejawi River ($\hat{p}_2$), this direction should be under the alternative hypothesis. This direction (less than) can be represented mathematically as $<$. If the proportion of samples that exceeded the permissible limit for Juru River is not less than the proportion of samples for Jejawi River; then two directions should be considered, the first direction could be $\hat{p}_1 > \hat{p}_2$ and the second direction could be $\hat{p}_1 = \hat{p}_2$. This direction (greater than or equal to) can be represented mathematically as $\geq$ (represents the null hypothesis). Thus we can write the two hypotheses (null and alternative) as presented in Eq. (9.27).

$$H_0: \hat{p}_1 \geq \hat{p}_2 \quad \text{vs} \quad H_1: \hat{p}_1 < \hat{p}_2 \tag{9.27}$$

We should make a decision regarding the null hypothesis as to whether the proportion of samples that exceeded the permissible limit regarding the concentration of manganese in sediment for Juru River ($\hat{p}_1$) is more than or equal to the proportion of samples that exceeded the permissible limit for Jejawi River ($\hat{p}_2$), or the proportion of samples that exceeded the permissible limit regarding the concentration of manganese in sediment for Juru River is less than of Jejawi River.

Step 2: Select the significance level (α) for the study

The significance level of 0.01 ($\alpha = 0.01$) is selected to test the hypothesis. Because the test is a one-tailed test as presented by the alternative hypothesis (less than), we should represent the rejection region on the left tail of the standard normal curve. The Z critical value for a one-tailed test with $\alpha = 0.01$ is 2.326 as appeared in the standard normal table (Table A in the Appendix). Thus the Z critical value for the left-tailed test is -2.326, we use the Z critical value to decide whether to reject or fail to reject the null hypothesis.

a. The critical value procedure

Step 3: Use the sample information to calculate the test statistic value

The entries for the Z-test statistic formula as presented in Eq. (9.24) are already provided, The collected data for the concentration of manganese in sediment showed only two samples that exceeded the permissible limit out of 100 samples for Juru River, while the collected data for Jejawi River showed 15 samples out of 100 samples that exceeded the permissible limit.

The proportion for each river can be calculated using the available information.

$$\hat{p}_1 = \frac{X_1}{n_1} = \frac{2}{100} = 0.02$$

$$\hat{p}_2 = \frac{X_2}{n_2} = \frac{15}{100} = 0.15$$

$$\overline{p} = \frac{X_1 + X_2}{n_1 + n_2} = \frac{2 + 15}{100 + 100} = 0.085$$

$$\overline{q} = 1 - \overline{p} = 1 - 0.085 = 0.915$$

We apply the formula presented in Eq. (9.24) to test the hypothesis regarding the proportion of samples that exceeded the permissible limit regarding the concentration of manganese in sediment for Juru River ($\hat{p}_1$) and Jejawi River ($\hat{p}_2$).

$$Z = \frac{\hat{p}_1 - \hat{p}_2 - (p_1 - p_2)}{\sqrt{\overline{p}\overline{q}\left(\frac{1}{n_1} + \frac{1}{n_2}\right)}} = \frac{0.02 - 0.15}{\sqrt{(0.085)(0.915)\left(\frac{1}{100} + \frac{1}{100}\right)}}$$

$$Z = -3.29616 = -3.30$$

The test statistic value for the concentration of manganese in sediment for Juru and Jejawi Rivers is found to be -3.30.

Step 4: Identify the critical and noncritical regions for the study

We can easily identify the rejection and nonrejection regions for a left-tailed test employing the Z critical value found in step 2. The rejection and nonrejection regions for the difference between the proportions of manganese that exceeded the permissible limit in water of two rivers are shown in Fig. 9.12 for the standard normal curve (orange shaded area).

Step 5: Make a decision and interpret the results

One can observe that the null hypothesis should be rejected because the test statistic value of the Z-test calculated from the sample data is -3.30, which is smaller than the Z critical value of -2.326 for a left-tailed test with $\alpha = 0.01$. Moreover, one can use the standard normal curve to decide, the same decision (reject the null hypothesis) can be reached using the standard normal curve and it can be seen that the test statistic value falls in the rejection region as shown in Fig. 9.12. The null hypothesis is rejected and we believe that the proportion of samples that exceeded the permissible limit regarding the concentration of manganese in sediment for Juru River ($\hat{p}_1$) is lower than the proportion of samples that exceeded the permissible limit for Jejawi River ($\hat{p}_2$).

b. The P-value procedure

The first three steps of the general procedure are the same for critical value and P-value procedures. Thus we should start from step 4 to calculate the P-value.

Step 4: Calculate the P-value and specify the critical and noncritical regions for the study

Consider that the null hypothesis is correct, calculating the P-value for a left-tailed test requires computing the probability of observing a value of Z of -3.30 or less than -3.30. This probability represents the area to the left of the Z value of -3.30. The probability to the left of the Z value -3.30 is 0.0001 ($1 - 0.9999 = 0.0001$). The exact P-value (blue shaded area) using technology is very small and is equal to $0.0004900801 = 0$ as presented in Fig. 9.12.

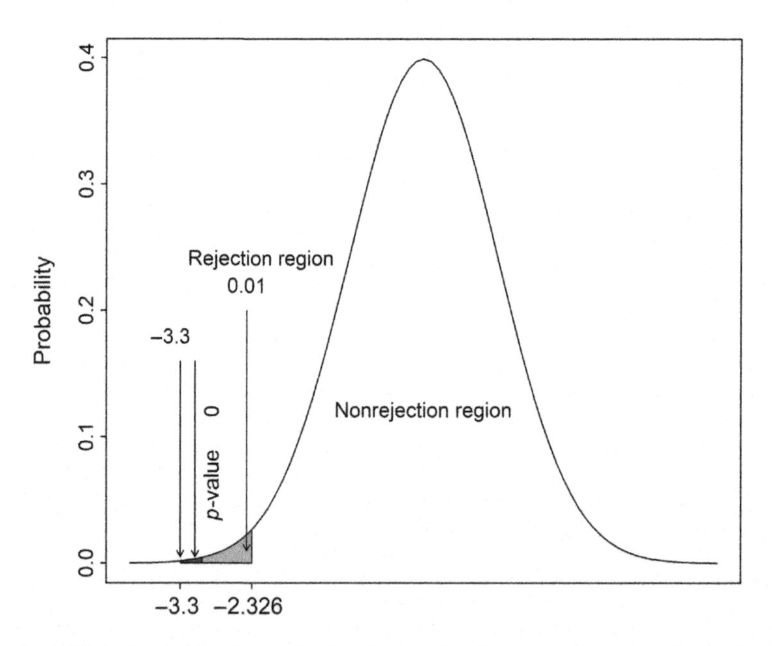

Figure 9.12 The rejection and nonrejection regions for the manganese in sediment for Juru and Jejawi Rivers.

Step 5: Make a decision using P-value and interpret the results

A small P-value leads to rejecting the null hypothesis in favor of the alternative hypothesis. The decision is made by comparing the P-value with the significance level (α), one can see that the calculated P-value is smaller than the significance level of 0.01 ($0 < 0.01$).

The same decision is reached by using the P-value and critical value procedures: reject the null hypothesis and believe that the proportion of samples that exceeded the permissible limit regarding the concentration of manganese in sediment for Juru River ($\hat{p}_1$) is less than the proportion of samples that exceeded the permissible limit for Jejawi River ($\hat{p}_2$).

9.7 Testing the ratio of two variances

Consider that two random samples are selected, each sample is selected from a normally distributed population, n_1 and n_2 represent the sample sizes selected from population 1 and population 2, respectively, and σ_1^2 and σ_2^2 to represent the variances for population 1 and 2, respectively. A claim regarding the ratio of two population variances can be tested employing an F-test for two sample variances to make a decision regarding the hypothesis. The mathematical formula for computing the test statistic value for a two-sample F-test is presented in Eq. (9.28).

$$F = \frac{S_1^2}{S_2^2} \tag{9.28}$$

Follow F distribution with $(n_1 - 1)$ and $(n_2 - 1)$ degrees of freedom for the numerator and denominator, respectively.

We can summarize the type of hypothesis into two types, the first type is called a two-sided test and the second type is called a one-sided test as shown below.

a. Two-sided or two-tailed test for the two population variances.

The two hypotheses (null and alternative) take the form given in Eq. (9.29).

$$H_0: \sigma_1^2 = \sigma_2^2$$

$$H_1: \sigma_1^2 \neq \sigma_2^2 \tag{9.29}$$

b. One-sided or one-tailed test.

The two hypotheses take the form given in Eqs. (9.30) and (9.31) for the right-tailed and left-tailed tests.

1. Right-tailed test

$$H_0: \sigma_1^2 \leq \sigma_2^2$$

$$H_1: \sigma_1^2 > \sigma_2^2 \tag{9.30}$$

2. Left-tailed test

$$H_0: \sigma_1^2 \geq \sigma_2^2$$

$$H_1: \sigma_1^2 < \sigma_2^2 \tag{9.31}$$

The computed test statistic value obtained from the sample data (9.28) is used to make a decision to reject or not to reject the null hypothesis regarding the two sample variances. The procedure to make a decision is to compare the value of the test statistic with the theoretical value (critical value) of the F distribution or using an F distribution curve. The null hypothesis should be rejected if

1. The calculated F-test statistic value is equal to or greater than the F critical value $F_{(\frac{\alpha}{2}, n_1-1, n_2-1)}$,

$$F \geq F_{(\frac{\alpha}{2}, n_1-1, n_2-1)}.$$

2. The calculated F-test statistic value is equal to or less than the F critical value $F_{(1-\frac{\alpha}{2}, n_1-1, n_2-1)}$

$$F \leq F_{(1-\frac{\alpha}{2}, n_1-1, n_2-1)} = \frac{1}{F_{(\frac{\alpha}{2}, n_2-1, n_1-1)}}$$

- α should be used instead of $\frac{\alpha}{2}$ if the test is a one-tailed test.

Example 9.16: The pH value of surface water before and after the dam: A professor at an environmental section wanted to verify the claim that the variance of pH of surface water before Beris dam is equal to the pH value of surface water after the dam. He selected 11 samples from each region and tested for the pH value. The collected data showed that the variance of pH before the dam is 0.26, while the collected data after the dam showed that the variance of pH is 0.71. A significance level of $\alpha = 0.01$ is chosen to test the claim. Assume that the two populations are normally distributed.

The general procedure for conducting hypothesis testing can be used to make the decision regarding the equality of variance of pH values of surface water before and after the dam. Two procedures will be used to test the hypothesis regarding the variance of pH values before and after the dam, the two procedures are critical value (traditional) and P-value.

Step 1: Specify the null and alternative hypotheses

The variance of pH values of surface water before the dam (σ_1^2) is equal to the variance of pH values after the dam (σ_2^2), this claim should be under the null hypothesis because the claim represents equality (=). If the variance of pH values of surface water before the dam is not equal to variance of pH values after the dam, then two directions should be considered, the first direction is could the variance of pH before the dam be greater than the variance of pH values after the dam or the second direction that it could be less. The two directions (greater than and less than) can be represented mathematically as $\neq$ (represents the null hypothesis). Thus we can write the two hypotheses (null and alternative) as presented in Eq. (9.32).

$$H_0: \sigma_1^2 = \sigma_2^2 \quad \text{vs} \quad H_1: \sigma_1^2 \neq \sigma_2^2 \tag{9.32}$$

We should make a decision regarding the null hypothesis as to whether the variance of pH values of surface water before the dam is equal to the variance of pH values after the dam, or the variance of pH before and after the dam are different.

Step 2: Select the significance level (α) for the study

The significance level of 0.01 ($\alpha = 0.01$) is selected to test the hypothesis. We divide the value of the significance level (0.01) by 2 to represent the two-tailed test (more or less) of the alternative hypothesis, thus $\frac{\alpha}{2} = \frac{0.01}{2} = 0.005$, which represents the rejection region in each tail of the F distribution curve (the left and right tails). The F critical value for a two-tailed test with $\alpha = 0.01$ and $d \cdot f_1 = (n_1 - 1) = 11 - 1 = 10$ for the numerator and $d \cdot f_2 = 10$ for the denominator can be computed using software as follows:

The left-tailed value: The F critical value for a left-tailed test with $d \cdot f_1 = 10$, $d \cdot f_2 = 10$, and $\alpha = 0.01$ is $F_{(1-\frac{\alpha}{2}, n_1-1, n_2-1)}$, the value that represents the point of intersection between $d \cdot f_1 = 10$, $d \cdot f_2 = 10$, and $1 - \frac{\alpha}{2} = 0.995$ is 0.1710, $F_{(1-\frac{\alpha}{2}, n_1-1, n_2-1)} = F_{(0.995, 11-1, 11-1)} = 0.1710$ as shown in Table D in the Appendix or using software and presented in Fig. 9.13.

The right-tailed value: The F critical value for a right-tailed test with $d \cdot f_1 = 10$, $d \cdot f_2 = 10$, and $\alpha = 0.01$ is $F_{(\frac{\alpha}{2}, n_1-1, n_2-1)}$, the value that represents the point of intersection between $d \cdot f_1 = 10$, $d \cdot f_2 = 10$, and $\frac{\alpha}{2} = 0.005$ is 5.8467, $F_{(\frac{\alpha}{2}, n_1-1, n_2-1)} = F_{(0.005, 11-1, 11-1)} = 5.8467$ as shown in Table D in the Appendix and presented in Fig. 9.13.

a. The critical value procedure

Step 3: Use the sample information to calculate the test statistic value

The entries for the F-test statistic formula as presented in Eq. (9.28) are already provided. The variance of pH values of surface water before the dam (S_1^2) is 0.26 and the sample size (n_1) is 11, while the variance of pH values of surface water after the dam (S_2^2) is 0.71 and the sample size (n_2) is 11.

We apply the formula presented in Eq. (9.28) to test the hypothesis regarding the variance of pH of surface water before and after the dam.

$$F = \frac{S_1^2}{S_2^2} = \frac{0.26}{0.71}$$

$$F = 0.3661972 = 0.37$$

The test statistic values for the variance of pH values of surface water before and after the dam is found to be 0.37.

Step 4: Identify the critical and noncritical regions for the study

We can easily identify the rejection and nonrejection regions for a two-tailed test employing the F critical values found in step 2. The rejection and nonrejection regions for the variance of pH before and after the dam are shown in Fig. 9.13 for the F distribution curve (orange shaded area).

Step 5: Make a decision and interpret the results

One can observe that we fail to reject the null hypothesis because the test statistic value of the F-test calculated from the sample data is 0.37, which is greater than the left critical value of 0.1710 and less than the right critical value of 5.8467. Moreover, one can use an F distribution curve to decide, the same decision (fail to reject the null hypothesis) can be reached using the F distribution curve and it can be seen that the test statistic value falls in the nonrejection region as shown in Fig. 9.13. The null hypothesis is not rejected and we believe that the variance of pH values before the dam (σ_1^2) is the same as the variance of pH values after the dam (σ_2^2).

b. The P-value procedure

The first three steps of the general procedure are the same for critical value and P-value procedures. Thus we should start from step 4 to calculate the P-value.

Step 4: Calculate the P-value and specify the critical and noncritical regions for the study

Consider that the null hypothesis is correct, calculating the P-value for a two-tailed test requires computing the probability of observing a value of F of 3.44 (the critical value corresponds with the P-value for the right side when the test statistic is 0.37) or greater (right tail) or a value of F of 0.29 (the critical value corresponds with the P-value for the left side when the test statistic is 0.37) or less (left tail). The probability to the right side is $0.0321787 = 0.03$ and the area to the left side is $0.0321787 = 0.03$. Because the problem is a two-tailed test, the P-value should be multiplied by 2 to calculate for the two sides ($2 \times 0.0321787 = 0.0643574 = 0.064$). The exact P-value (blue shaded area) using technology is equal to 0.064 and distributed on both sides as presented in Fig. 9.13.

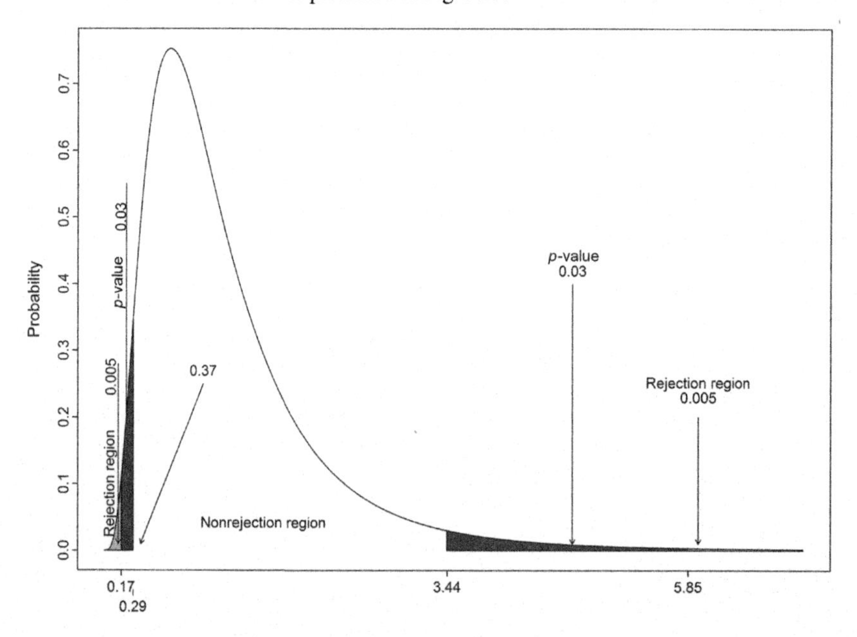

Figure 9.13 The rejection and nonrejection regions for the variance of pH values of surface water before and after the dam.

Step 5: Make a decision using P-value and interpret the results

A large P-value leads to not rejecting the null hypothesis. The decision is made by comparing the P-value with the significance level (α), one can see that the calculated P-value is greater than the significance level of 0.01 ($0.064 > 0.01$).

The same decision is reached by using the P-value and critical value procedures: fail to reject the null hypothesis and we believe that the variance of pH values before the dam (σ_1^2) is equal to the variance of pH values after the dam (σ_2^2).

Example 9.17: The concentration of iron in sediment of two rivers: A research group at a research center wishes to verify the claim that the variance of iron (Fe) concentration in sediment for the first river is less than the variance of iron in the second river. They selected 15 samples from each river and tested for iron concentration. The collected data showed that the variance of iron values for the first river is 0.29, while the collected data for the second river showed that the variance of iron is 1.42. A significance level of $\alpha = 0.01$ is chosen to test the claim. Assume that the two populations are normally distributed.

The general procedure for conducting hypothesis testing can be used to make the decision regarding the variance of iron in the sediment of two rivers. Two procedures will be used to test the hypothesis regarding the variance of iron values in the sediment of two rivers, the two procedures are critical value (traditional) and P-value.

Step 1: Specify the null and alternative hypotheses

The variance of iron values for the first river (σ_1^2) is less than the variance of iron values for the second river (σ_2^2), this claim should be under the alternative hypothesis because the claim does not include equality (=). If the variance of iron values for the first river is not less than the variance of iron values for the second river, then two directions should be considered, the first direction is whether the variance of iron for the first river is greater than the variance of iron for the second river or could they be equal. The two directions (greater than or equal to) can be represented mathematically as $\geq$ (represents the null hypothesis). Thus we can write the two hypotheses (null and alternative) as presented in Eq. (9.33).

$$H_0: \sigma_1^2 \geq \sigma_2^2 \quad \text{vs} \quad H_1: \sigma_1^2 < \sigma_2^2 \tag{9.33}$$

We should make a decision regarding the null hypothesis as to whether the variance of iron values for the first river is greater than or equal to the variance of iron values for the second river, or the variance of iron values for the first river is less than the variance of iron values for the second river.

Step 2: Select the significance level (α) for the study

The significance level of 0.01 ($\alpha = 0.01$) is selected to test the hypothesis. Because the test is a one-tailed test as presented by the alternative hypothesis (less than), we should represent the rejection region on the left tail of the F distribution curve. The F critical value for a one-tailed test with $d \cdot f_1 = 14$ and $d \cdot f_2 = 14$,

$\alpha = 0.01$ is $F_{(1-\alpha, n_1-1, n_2-1)} = F_{(0.995, 10, 10)} = 0.27045$ as appeared in the F distribution table for critical values (using software or Table D in the Appendix). Thus the F critical value for the left-tailed test is 0.27045, we use the F critical value to decide whether to reject or fail to reject the null hypothesis.

a. The critical value procedure

Step 3: Use the sample information to calculate the test statistic value

The entries for the F-test statistic formula as presented in Eq. (9.28) are already provided. The variance of iron values in sediment for the first river (S_1^2) is 0.29 and the sample size (n_1) is 15, while the variance of iron values in sediment for the second river (S_2^2) is 1.42 and the sample size (n_2) is 15.

We apply the formula presented in Eq. (9.28) to test the hypothesis regarding the variance of iron values in the sediment of two rivers.

$$F = \frac{S_1^2}{S_2^2} = \frac{0.29}{1.42}$$

$$F = 0.2042254 = 0.20$$

The test statistic value for the variance of iron values in sediment for the two rivers is found to be 0.20.

Step 4: Identify the critical and noncritical regions for the study

We can easily identify the rejection and nonrejection regions for the left-tailed test employing the F critical values found in step 2. The rejection and nonrejection regions for the variance of iron values in sediment for two rivers are shown in Fig. 9.14 for an F distribution curve (orange shaded area).

Step 5: Make a decision and interpret the results

One can observe that we rejected the null hypothesis because the test statistic value of the F-test calculated from the sample data is 0.20, which is less than the left critical value of $0.27045 = 0.27$. Moreover, one can use the F distribution curve to decide, the same decision (reject the null hypothesis) can be reached using the F distribution curve and it can be seen that the test statistic value falls in the rejection region as shown in Fig. 9.14. The null hypothesis is rejected and we believe that the variance of iron values in the sediment of the first river (σ_1^2) is less than the variance of iron values of the second river (σ_2^2).

b. The P-value procedure

The first three steps of the general procedure are the same for critical value and P-value procedures. Thus we should start from step 4 to calculate the P-value.

Step 4: Calculate the P-value and specify the critical and noncritical regions for the study

Consider that the null hypothesis is correct, calculating the P-value for the left-tailed test requires computing the probability of observing a value of F of 0.20 or less. The probability to the left side is $0.002660906 = 0.003$. The exact P-value (blue shaded area) using technology equals 0.003 as presented in Fig. 9.14.

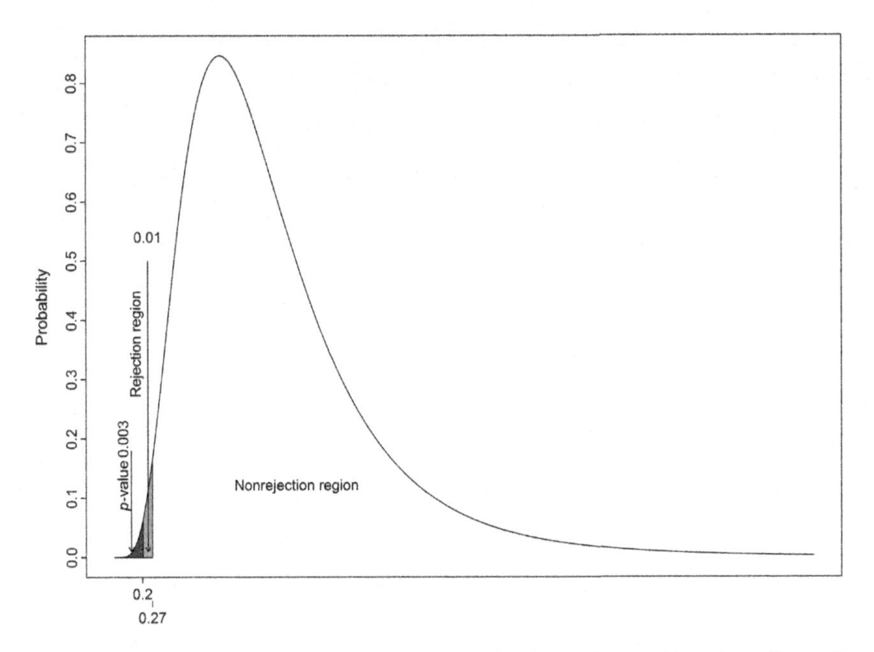

Figure 9.14 The rejection and nonrejection regions for the variance of iron in sediment for two rivers.

Step 5: Make a decision using P-value and interpret the results

A small *P*-value leads to rejecting the null hypothesis. The decision is made by comparing the P-value with the significance level (α), one can see that the calculated *P*-value is smaller than the significance level of 0.01 ($0.003 < 0.01$).

The same decision is reached by using the *P*-value and critical value procedures: reject the null hypothesis and we believe that the variance of iron values in sediment of the first river (σ_1^2) is less than the variance of iron values in sediment of the second river (σ_2^2).

Example 9.18: The amount of solid waste for palm oil mill: An environmentalist wishes to investigate the claim about the amount of potash as solid waste from a palm oil mill. Two palm oil mills (A and B) were selected and the amount of potash was recorded (tons) for 12 months. The collected data showed that the variance of potash values for mill A is 5.54, while the collected data for mill B showed that the variance of potash values is 2.68. A significance level of $\alpha = 0.01$ is chosen to test the claim that the variance of mill A is higher than the variance of mill B. Assume that the two populations are normally distributed.

The general procedure for conducting hypothesis testing can be used to make the decision regarding the variance of potash of the two mills. Two procedures will be used to test the hypothesis regarding the potash of the two mills A and B, the two procedures are critical value (traditional) and *P*-value.

Step 1: Specify the null and alternative hypotheses

The variance of potash for mill A is higher than the variance of potash for mill B, this claim should be under the alternative hypothesis because the claim does not include equality (=). If the variance of potash for mill A is not higher than the variance of potash for mill B, then two directions should be considered, the first direction is whether the variance of potash for mill A could be less than the variance of potash for mill B or could they be equal. The two directions (less than or equal) can be represented mathematically as ≤ (represents the null hypothesis). Thus we can write the two hypotheses (null and alternative) as presented in Eq. (9.34).

$$H_0: \sigma_1^2 \leq \sigma_2^2 \quad \text{vs} \quad H_1: \sigma_1^2 > \sigma_2^2 \tag{9.34}$$

We should make a decision regarding the null hypothesis as to whether the variance of potash for mill A is less than or equal to the variance of potash for mill B, or the variance of potash for mill A is greater than the variance of potash for mill B.

Step 2: Select the significance level (α) for the study

The significance level of 0.01 ($\alpha = 0.01$) is selected to test the hypothesis. Because the test is a one-tailed test as presented by the alternative hypothesis (greater than), we should represent the rejection region on the right tail of the F distribution curve. The F critical value for a one-tailed test with $\alpha = 0.01$, $d \cdot f_1 = 11$, and $d \cdot f_2 = 11$ is $F_{(\alpha, n_1 - 1, n_2 - 1)} = F_{(0.01, 11, 11)} = 4.4624$ as appeared in the F distribution table for critical values (using software or Table D in the Appendix). Thus the F critical value for a right-tailed test is 4.4624, we use the F critical value to decide whether to reject or fail to reject the null hypothesis.

a. The critical value procedure

Step 3: Use the sample information to calculate the test statistic value

The entries for the F-test statistic formula as presented in Eq. (9.28) are already provided. The variance of potash for mill A (S_1^2) is 5.54 and the sample size (n_1) is 12, while the variance of potash for mill B (S_2^2) is 2.68 and the sample size (n_2) is 12.

We apply the formula presented in Eq. (9.28) to test the hypothesis regarding the variance of potash for the two mills A and B.

$$F = \frac{S_1^2}{S_2^2} = \frac{5.54}{2.68}$$

$$F = 2.067164 = 2.067$$

The test statistic value for the variance of potash for the two mills A and B is found to be 2.067.

Step 4: Identify the critical and noncritical regions for the study

We can easily identify the rejection and nonrejection regions for the right-tailed test employing the F critical values found in step 2. The rejection and nonrejection regions for the variance of potash for the two mills A and B are shown in Fig. 9.15 for an F distribution curve (orange shaded area).

Step 5: Make a decision and interpret the results

One can observe that we failed to reject the null hypothesis because the test statistic value of the F-test calculated from the sample data is 2.067, which is less than the right critical value of 4.4624. Moreover, one can use the F distribution curve to decide, the same decision (fail to reject the null hypothesis) can be reached using the F distribution curve and it can be seen that the test statistic value falls in the nonrejection region as shown in Fig. 9.15. The null hypothesis is not rejected and we believe that the variance of potash for mill A (σ_1^2) is less than or equal to the variance of potash for mill B (σ_2^2).

b. The P-value procedure

The first three steps of the general procedure are the same for critical value and P-value procedures. Thus we should start from step 4 to calculate the P-value.

Step 4: Calculate the P-value and specify the critical and noncritical regions for the study

Consider the null hypothesis is correct, calculating the P-value for the right-tailed test requires computing the probability of observing a value of F of 2.067 or greater. The probability to the right side is $0.8779782 = 0.88$. The exact P-value (blue shaded area) using technology equals 0.88 as presented in Fig. 9.15.

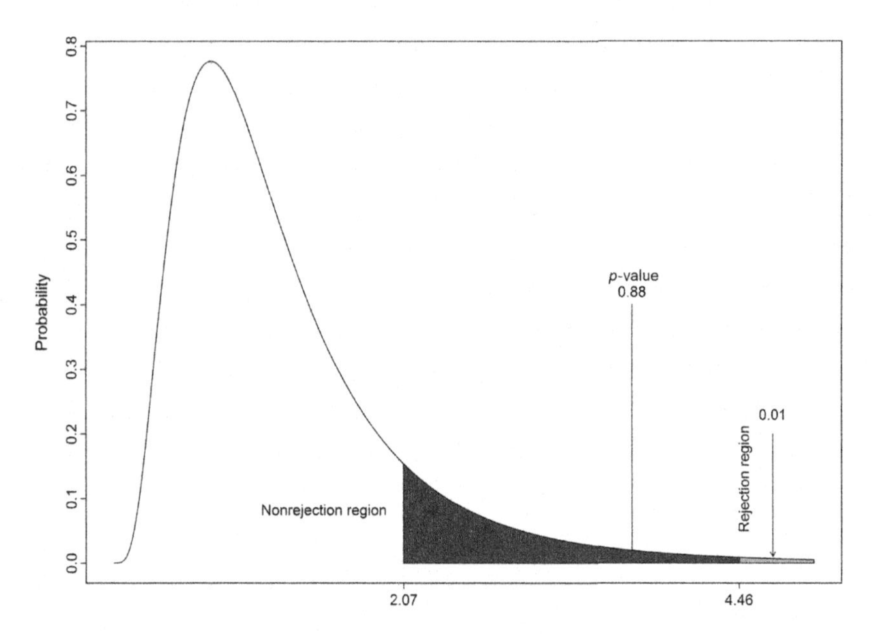

Figure 9.15 The rejection and nonrejection regions for the variance of potash of two mills.

Step 5: Make a decision using P-value and interpret the results

A large P-value leads to not rejecting the null hypothesis. The decision is made by comparing the P-value with the significance level (α), one can see that the calculated P-value is greater than the significance level of 0.01 ($0.88 > 0.01$).

The same decision is reached by using the P-value and critical value procedures: fail to reject the null hypothesis and we believe that the variance of potash for mill A (σ_1^2) is less than or equal to the variance of potash for mill B (σ_2^2).

Further reading

Alkarkhi, A. F. M., & ALqaraghuli, W. A. A. (2020). *Applied statistics for environmental science with R* (1st ed.). Elsevier.

Alkarkhi, A. F. M., & Chin, L. H. (2012). *Elementary statistics for technologist* (1st ed.). Malaysia: Universiti Sains Malaysia.

Alkarkhi, A. F. M., Ahmad, A., Ismail, N., & Easa, A. M. (2008). Multivariate analysis of heavy metals concentrations in river estuary. *Environmental Monitoring and Assessment, 143*, 179–186.

Alkarkhi, A. F. M., Ismail, N., Ahmed, A., & Easa, A. M. (2009). Analysis of heavy metal concentrations in sediments of selected estuaries of Malaysia—a statistical assessment. *Environmental Monitoring and Assessment, 153*, 179–185.

Alkarkhi, F. M. A., Ismail, N., & Easa, A. M. (2008). Assessment of arsenic and heavy metal contents in cockles (*Anadara granosa*) using multivariate statistical techniques. *Journal of Hazardous Material, 150*, 783–789.

Banch, T. J. H., Hanafiah, M. M., Alkarkhi, A. F. M., & Abu Amr, S. S. (2019). Statistical evaluation of landfill leachate system and its impact on groundwater and surface water in Malaysia. *Sains Malaysiana, 48*(11), 2391–2403.

Bluman, A. G. (1998). *Elementary statistics: A step by step approach* (3rd ed.). Boston, MA: WCB, McGraw-Hill.

Eberly College of Science-Department of Statistics. (2018). Two variances. In vol. 2020. Pennstate The Pennsylvania State University.

Hossain, M. A., AlKarkhi, A. F. M., & A., N. N. N. (2008). Statistical and trend analysis of water quality data for the Baris Dam of Darul Aman river in Kedah, Malaysia. In *Proceedings of the international conference on environmental research and technology*, (pp. 568–572). Penang-Malaysia: Universiti Sains Malaysia.

Ngu, C. C., Abdul Rahman, N. N. N., Alkarkhi, A. F. M., Ahmad, A., & Kadir, M. O. A. (2010). Assessment of major solid wastes generated in palm oil mills. *International Journal of Environmental Technology and Management, 13*(3/4), 245–252.

Rahman, N. N. N. A., Chard, N. C., Al-Karkhi, A. F. M., Rafatullah, M., & Kadir, M. O. A. (2017). Analysis of particulate matters in air of palm oil mills—A statistical assessment. *Environmental Engineering and Management Journal, 16*(11), 2537–2543.

Weiss, N. A. (2012). *Introductory statistics* (9th ed.). Pearson.

Interval estimation for the difference between two populations

Abstract

This chapter addresses the confidence interval for the difference between two population parameters covering the concept of confidence interval and the general procedure for calculating confidence interval. The confidence interval for the difference between two population means cove ring both large and small samples and for independent and dependent samples, two population proportions, and for two population variances are given. Furthermore, examples from the fields of environmental science and engineering are used to explain the general procedure for calculating confidence interval for each case with sufficient demonstration for each step and matched to the field of study.

Keywords: Confidence interval; confidence level; mean value; proportion value; variance

Learning outcomes

After completing this chapter, readers should be able to:

- Understand the meaning and importance of interval estimation (confidence interval) for the difference between two population parameters;
- Know the general procedure for computing the confidence interval for two populations;
- Describe the steps for computing the confidence interval for two population means;
- Describe the steps for computing the confidence interval for two population proportions;
- Describe the steps for computing the confidence interval for two population variances;
- Understand the steps used to interpret the confidence interval for the difference between two population parameters;
- Choose a suitable formula to compute the confidence interval for the difference between two selected population parameters;
- Know two-tailed and one-tailed confidence intervals for the difference between two populations;
- Write a report for the results.

10.1 Introduction

Interval estimation for one population parameter was studied and investigated in terms of calculating the confidence interval and how to use it in hypothesis

Applications of Hypothesis Testing for Environmental Science. DOI: https://doi.org/10.1016/B978-0-12-824301-5.00002-2

testing covering the confidence interval for the mean, proportion, and variance. The idea of the confidence interval for one parameter can be extended to include two parameters and to study the difference between two population parameters. Many examples are available in the environmental field, for instance, comparing two rivers based on the concentration of turbidity or other parameters.

The confidence interval for the difference between two population parameters is delivered in this chapter including the confidence interval for two means, two proportions, and two variances.

10.2 The steps for the confidence interval procedure for the difference between two populations

The general procedure for calculating the confidence interval for one population parameter can be used to calculate the confidence interval for the difference between two population parameters. The general procedure for calculating the confidence interval for the difference between two parameters includes four steps.

Step 1: Use the sample data to compute the sample statistics for sample 1 and sample 2

Step 2: Select the significance level (α) for the study and find the critical values

Step 3: Compute the confidence interval

Step 4: Interpret the results

The meaning and interpretation of the confidence interval for the difference between two population parameters are the same as for one population parameter taking into account the two populations as population 1 and population 2 and the main concern is the difference.

The general procedure can be used to calculate the difference between two means, two proportions, and two variances using different distribution.

10.3 Confidence interval for the difference between two means when the sample size is large

Consider that two large random samples ($n \geq 30$) are selected, each sample is selected from a normally distributed population. Consider that Y_1 and Y_2 are two variables of interest, n_1 and n_2 represent the sample sizes selected from population 1 and population 2, respectively, with $\overline{Y}_1$ and $\overline{Y}_2$ to represent the average value, and σ_1^2 and σ_2^2 to represent the variance for populations 1 and 2, respectively. A confidence interval for the difference between two population means and specific confidence level $(1 - \alpha)$ can be computed employing the sample data. The mathematical formula for computing the confidence interval for the difference between two means is presented in Eq. (10.1).

$$(\overline{Y}_1 - \overline{Y}_2) \pm Z_{\frac{\alpha}{2}}\sqrt{\frac{\sigma_1^2}{n_1} + \frac{\sigma_2^2}{n_2}} \tag{10.1}$$

The confidence interval can be written using another form:

$$(\overline{Y}_1 - \overline{Y}_2) - Z_{\frac{\alpha}{2}}\sqrt{\frac{\sigma_1^2}{n_1} + \frac{\sigma_2^2}{n_2}} \leq \mu_1 - \mu_2 \leq (\overline{Y}_1 - \overline{Y}_2) + Z_{\frac{\alpha}{2}}\sqrt{\frac{\sigma_1^2}{n_1} + \frac{\sigma_2^2}{n_2}}$$

where

$\overline{Y}_1 - \overline{Y}_2$ represents the observed difference between the two-sample means;
$\mu_1 - \mu_2$ represents the expected difference;
μ represents the population mean;
$Z_{\frac{\alpha}{2}}$ represents the Z critical value; and
$Z_{\frac{\alpha}{2}}\sqrt{\frac{\sigma_1^2}{n_1} + \frac{\sigma_2^2}{n_2}}$ represents the margin of error (E).

Note

- When the sample size is large and the population standard deviation (σ) is not provided, then we can use the sample standard deviation (S).
- The margin of error (E) can be used to write the confidence interval.

$$(\overline{Y}_1 - \overline{Y}_2) - E \leq \mu_1 - \mu_2 \leq (\overline{Y}_1 - \overline{Y}_2) + E$$

- The lower one-tailed confidence interval is given in Eq. (10.2):

$$(\overline{Y}_1 - Y_2) - Z_{\frac{\alpha}{2}}\sqrt{\frac{\sigma_1^2}{n_1} + \frac{\sigma_2^2}{n_2}} \leq \mu_1 - \mu_2 \tag{10.2}$$

or it can be written as the interval $\left[(\overline{Y}_1 - \overline{Y}_2) - Z_{\frac{\alpha}{2}}\sqrt{\frac{\sigma_1^2}{n_1} + \frac{\sigma_2^2}{n_2}},\ \infty\right)$

- The upper one-tailed confidence interval is given in Eq. (10.3):

$$\mu_1 - \mu_2 \leq (\overline{Y}_1 - \overline{Y}_2) + Z_{\frac{\alpha}{2}}\sqrt{\frac{\sigma_1^2}{n_1} + \frac{\sigma_2^2}{n_2}} \tag{10.3}$$

or it can be written as the interval $\left(-\infty,\ (\overline{Y}_1 - \overline{Y}_2) + Z_{\frac{\alpha}{2}}\sqrt{\frac{\sigma_1^2}{n_1} + \frac{\sigma_2^2}{n_2}}\right]$

Example 10.1: The amount of solid waste in two palm oil mills: An environmentalist wants to build a range for the difference in the amount of solid waste generated in two palm oil mills. She selected 45 days and the amount of solid waste in each palm oil mill was recorded in tons. The collected data showed that the mean of the solid waste generated in mill A is 2267 and the standard deviation is 268.31, the collected data for mill B resulted in an average of 5101 and the standard deviation is 208.79. A significance level of $\alpha = 0.01$ is chosen to construct the confidence interval for the difference in the solid waste generated in mill A and mill B. Assume that the two populations are normally distributed.

The general procedure can be employed to calculate the confidence interval for the difference in the amount of solid waste generated in the two mills.

Step 1: Use the sample data to compute the sample statistics for sample 1 and sample 2

The mean value of solid waste generated in palm mill A is already calculated to be 2267 and the mean value for mill B was calculated to be 5101.

Step 2: Select the significance level (α) for the study and find the critical values

The level of significance is chosen to be 0.01 ($\alpha = 0.01$) and $\frac{\alpha}{2} = \frac{0.01}{2} = 0.005$. The Z critical value is $Z_{\frac{\alpha}{2}} = 2.58$ (Table A in the Appendix).

Step 3: Compute the confidence interval

The entries for the confidence interval formula as presented in Eq. (10.1) (two-sided confidence interval) are already provided, the mean value of the solid waste for mill A is 2267 and the standard deviation is 268.31, while the mean value for mill B is 5101, standard deviation equals 208.79, and the sample size is 45 for each mill.

We apply the formula presented in Eq. (10.1) to calculate the confidence interval for the difference in the amount of solid waste in the two palm oil mills.

$$(\overline{Y}_1 - \overline{Y}_2) - Z_{\frac{\alpha}{2}}\sqrt{\frac{\sigma_1^2}{n_1} + \frac{\sigma_2^2}{n_2}} \leq \mu_1 - \mu_2 \leq (\overline{Y}_1 - \overline{Y}_2) + Z_{\frac{\alpha}{2}}\sqrt{\frac{\sigma_1^2}{n_1} + \frac{\sigma_2^2}{n_2}}$$

$$(2267 - 5101) - 2.58\sqrt{\frac{(268.31)^2}{45} + \frac{(208.79)^2}{45}}$$

$$\leq \mu_1 - \mu_2 \leq (2267 - 5101) + 2.58\sqrt{\frac{(268.31)^2}{45} + \frac{(208.79)^2}{45}}$$

$$-2834 - 130.5446 \leq \mu_1 - \mu_2 \leq -2834 + 130.5446$$

$$-2964.545 \leq \mu_1 - \mu_2 \leq -2703.455$$

Step 4: Interpret the results

It can be said that we are 99% confident that the mean difference in the amount of solid waste for the two mills is somewhere between -2964.545 and -2703.455. One can conclude that mill B generates more solid waste than mill A.

Example 10.2: The concentration of trace metal cadmium in the air: A researcher at an environmental section wishes to build a range for the difference in the mean concentration of trace metal cadmium (Cd) ($\mu g/m^3$) on coarse particulate matter (PM_{10}) in the air during the dry and wet seasons of an equatorial urban coastal location. The collected data for 42 days during the dry season showed that the mean concentration of cadmium is 0.02 and the standard deviation is 0.019, and the collected data for 33 days during the wet season showed that the mean concentration of cadmium is 0.007 and the standard deviation is 0.015. A significance level of $\alpha = 0.01$ is chosen to construct the confidence interval for the difference in cadmium during the dry and wet seasons. Assume that the two populations are normally distributed.

The general procedure can be employed to calculate the confidence interval for the difference in the mean concentration of cadmium on coarse particulate matter (PM_{10}) in the air during the dry and wet seasons.

Step 1: Use the sample data to compute the sample statistics for sample 1 and sample 2

The mean value of cadmium during the dry season is already calculated to be 0.02 and the mean value for the wet season was calculated to be 0.007.

Step 2: Select the significance level (α) for the study and find the critical values

The level of significance is chosen to be 0.01 ($\alpha = 0.01$) and $\frac{\alpha}{2} = \frac{0.01}{2} = 0.005$. The Z critical value is $Z_{\frac{\alpha}{2}} = 2.58$ (Table A in the Appendix).

Step 3: Compute the confidence interval

The entries for the confidence interval formula as presented in Eq. (10.1) (two-sided confidence interval) are already provided, the mean value of cadmium during the dry season is 0.02, the standard deviation is 0.019, and the sample size is 42, while the mean concentration of cadmium during the wet season is 0.007, standard deviation is equal to 0.015, and the sample size is 33.

We apply the formula presented in Eq. (10.1) to calculate the confidence interval for the difference in the concentration of trace metal cadmium in the air.

$$(\overline{Y}_1 - \overline{Y}_2) - Z_{\frac{\alpha}{2}}\sqrt{\frac{\sigma_1^2}{n_1} + \frac{\sigma_2^2}{n_2}} \le \mu_1 - \mu_2 \le (\overline{Y}_1 - \overline{Y}_2) + Z_{\frac{\alpha}{2}}\sqrt{\frac{\sigma_1^2}{n_1} + \frac{\sigma_2^2}{n_2}}$$

$$(0.02 - 0.007) - 2.58\sqrt{\frac{(0.019)^2}{42} + \frac{(0.015)^2}{33}}$$

$$\le \mu_1 - \mu_2 \le (0.02 - 0.007) + 2.58\sqrt{\frac{(0.019)^2}{42} + \frac{(0.015)^2}{33}}$$

$$0.013 - 0.01011269 \le \mu_1 - \mu_2 \le 0.013 - 0.01011269$$

$$0.003 \le \mu_1 - \mu_2 \le 0.023$$

Step 4: Interpret the results

It can be said that we are 99% confident that the mean difference of cadmium for the dry and wet seasons is somewhere between 0.003 and 0.023. One can conclude that the concentration of cadmium during the dry season is higher than during the wet season.

10.4 Confidence interval for the difference between two means when the sample size is small

Consider that two small independent random samples ($n < 30$) are selected, each sample is selected from a normally distributed population. Consider that Y_1 and Y_2 are two variables of interest, n_1 and n_2 represent the sample sizes selected from population 1 and population 2, respectively, with $\overline{Y}_1$ and $\overline{Y}_2$ to represent the average value and S_1^2 and S_2^2 to represent the variances computed from sample 1 and sample 2, respectively. A confidence interval for the difference between two means and the specific confidence level $(1 - \alpha)$ can be computed employing the sample data. The mathematical formula for computing the confidence interval for two sample means when the sample is small are given in Eqs. (10.4) and (10.5) for unequal and equal variances, respectively.

1. Unequal variances

The mathematical formula for computing the confidence interval for the difference between two sample means when the variances are unequal is given in Eq. (10.4).

$$(\overline{Y}_1 - Y_2) \pm t_{\frac{\alpha}{2},d.f}\sqrt{\frac{S_1^2}{n_1} + \frac{S_2^2}{n_2}} \tag{10.4}$$

The confidence interval in Eq. (10.4) can be written using another form:

$$(\overline{Y}_1 - \overline{Y}_2) - t_{\frac{\alpha}{2},d.f}\sqrt{\frac{S_1^2}{n_1} + \frac{S_2^2}{n_2}} \le \mu_1 - \mu_2 \le (\overline{Y}_1 - \overline{Y}_2) + t_{\frac{\alpha}{2},d.f}\sqrt{\frac{S_1^2}{n_1} + \frac{S_2^2}{n_2}}$$

has t distribution with degrees of freedom $(d \cdot f)$ equals to the smaller of $(n_1 - 1)$ and $(n_2 - 1)$.

Or we can use the following formula to calculate the degrees of freedom.

$$d{\cdot}f = \frac{\left[\frac{S_1^2}{n_1} + \frac{S_2^2}{n_2}\right]^2}{\frac{1}{(n_1 - 1)}\left(\frac{S_1^2}{n_1}\right)^2 + \frac{1}{(n_2 - 1)}\left(\frac{S_2^2}{n_2}\right)^2}$$

where

$\overline{Y}_1 - \overline{Y}_2$ represents the observed difference between the two sample means;
$\mu_1 - \mu_2$ represents the expected difference;
μ represents the population mean;
$t_{\frac{\alpha}{2}}$ represents the t critical value; and
$t_{\frac{\alpha}{2},d\cdot f}\sqrt{\frac{S_1^2}{n_1} + \frac{S_2^2}{n_2}}$ represents the margin of error (E).

- The margin of error (E) can be used to write the confidence interval.

$$(\overline{Y}_1 - \overline{Y}_2) - E \leq \mu_1 - \mu_2 \leq (\overline{Y}_1 - \overline{Y}_2) + E$$

- The lower one-tailed confidence interval is given in Eq. (10.5):

$$(\overline{Y}_1 - Y_2) - t_{\frac{\alpha}{2},d\cdot f}\sqrt{\frac{S_1^2}{n_1} + \frac{S_2^2}{n_2}} \leq \mu_1 - \mu_2 \tag{10.5}$$

or it can be written as the interval $\left[\left(\overline{Y}_1 - \overline{Y}_2\right) - t_{\frac{\alpha}{2},d\cdot f}\sqrt{\frac{S_1^2}{n_1} + \frac{S_2^2}{n_2}},\ \infty\right)$

- The upper one-tailed confidence interval is given in Eq. (10.6):

$$\mu_1 - \mu_2 \leq (\overline{Y}_1 - \overline{Y}_2) + t_{\frac{\alpha}{2},d\cdot f}\sqrt{\frac{S_1^2}{n_1} + \frac{S_2^2}{n_2}} \tag{10.6}$$

or it can be written as the interval $\left(-\infty,\ \left(\overline{Y}_1 - \overline{Y}_2\right) + t_{\frac{\alpha}{2},d\cdot f}\sqrt{\frac{S_1^2}{n_1} + \frac{S_2^2}{n_2}}\right]$

2. Equal variances

The mathematical formula for computing the confidence interval for the difference between two samples when the variances are equal is given in Eq. (10.7).

$$(\overline{X}_1 - \overline{X}_2) \pm t_{\frac{\alpha}{2},d\cdot f}\sqrt{\frac{S_P^2}{n_1} + \frac{S_P^2}{n_2}} \tag{10.7}$$

The confidence interval in Eq. (10.7) can be written using another form:

$$(\overline{Y}_1 - \overline{Y}_2) - t_{\frac{\alpha}{2},d\cdot f}\sqrt{\frac{S_P^2}{n_1} + \frac{S_P^2}{n_2}} \leq \mu_1 - \mu_2 \leq (\overline{Y}_1 - \overline{Y}_2) + t_{\frac{\alpha}{2},d\cdot f}\sqrt{\frac{S_P^2}{n_1} + \frac{S_P^2}{n_2}}$$

has t distribution with degrees of freedom $(d \cdot f)$ equal to $(n_1 + n_2 - 2)$,

where

$S_P^2 = \frac{(n_1 - 1)S_1^2 + (n_2 - 1)S_2^2}{n_1 + n_2 - 2}$ is called pooled variance, and

$$t_{\frac{\alpha}{2},d\cdot f}\sqrt{\frac{S_P^2}{n_1}+\frac{S_P^2}{n_2}}:\ \text{the margin of error }(E).$$

- The margin of error (E) can be used to write the confidence interval.

$$(\overline{Y}_1-\overline{Y}_2)-E\le\mu_1-\mu_2\le(\overline{Y}_1-\overline{Y}_2)+E$$

- The lower one-tailed confidence interval is given in Eq. (10.8):

$$(\overline{Y}_1-Y_2)-t_{\frac{\alpha}{2},d\cdot f}\sqrt{\frac{S_P^2}{n_1}+\frac{S_P^2}{n_2}}\le\mu_1-\mu_2 \tag{10.8}$$

or it can be written as the interval $\left[(\overline{Y}_1-\overline{Y}_2)-t_{\frac{\alpha}{2},d\cdot f}\sqrt{\frac{S_P^2}{n_1}+\frac{S_P^2}{n_2}},\ \infty\right)$

- The upper one-tailed confidence interval is given in Eq. (10.9):

$$\mu_1-\mu_2\le(\overline{Y}_1-\overline{Y}_2)+t_{\frac{\alpha}{2},d\cdot f}\sqrt{\frac{S_P^2}{n_1}+\frac{S_P^2}{n_2}} \tag{10.9}$$

or it can be written as the interval $\left(-\infty,\ (\overline{Y}_1-\overline{Y}_2)+t_{\frac{\alpha}{2},d\cdot f}\sqrt{\frac{S_P^2}{n_1}+\frac{S_P^2}{n_2}}\right]$

Example 10.3: The electrical conductivity in collection and aeration ponds: A researcher at an environmental section wishes to build a range for the difference in the mean of electrical conductivity (EC) ($\mu S/cm$) of the collection and aeration ponds for a leachate landfill system. The collected data for 13 samples obtained from a collection pond showed that the mean of EC is 6565 and the standard deviation is 324, and the collected data for 13 samples obtained from an aeration pond showed that the mean of EC is 2055 and the standard deviation is 424. A significance level of $\alpha=0.01$ is chosen to construct the confidence interval for the difference in the mean of EC of collection and aeration ponds. Assume that the two populations are normally distributed, and the variances are unequal ($\sigma_1^2\neq\sigma_2^2$).

The general procedure can be employed to calculate the confidence interval for the difference in the mean of EC of collection and aeration ponds for a leachate landfill system.

Step 1: Use the sample data to compute the sample statistics for sample 1 and sample 2

The mean value of EC for the collection pond is already calculated to be 6565 and for aeration pond the mean value of EC was calculated to be 2055.

Step 2: Select the significance level (α) for the study and find the critical values

The level of significance is chosen to be 0.01 ($\alpha = 0.01$) and $\frac{\alpha}{2} = \frac{0.01}{2} = 0.005$. The t critical value is $t_{\frac{\alpha}{2},d\cdot f} = t_{(0.005,12)} = 3.055$ (Table B in the Appendix).

Step 3: Compute the confidence interval

The entries for the confidence interval formula as presented in Eq. (10.4) (two-sided confidence interval) are already provided, the mean value of EC for the collection pond is 6565, the standard deviation is 324, and the sample size is 13, while the mean value of EC for the aeration pond is 2055, the standard deviation is equal to 424, and the sample size is 13.

We apply the formula presented in Eq. (10.4) to calculate the confidence interval for EC in the collection and aeration ponds.

$$(\overline{Y}_1 - \overline{Y}_2) - t_{\frac{\alpha}{2},d\cdot f}\sqrt{\frac{S_1^2}{n_1} + \frac{S_2^2}{n_2}} \leq \mu_1 - \mu_2 \leq (\overline{Y}_1 - \overline{Y}_2) + t_{\frac{\alpha}{2},d\cdot f}\sqrt{\frac{S_1^2}{n_1} + \frac{S_2^2}{n_2}}$$

$$(6565 - 2055) - 3.055\sqrt{\frac{(324)^2}{13} + \frac{(424)^2}{13}}$$

$$\leq \mu_1 - \mu_2 \leq (6565 - 2055) + 3.055\sqrt{\frac{(324)^2}{13} + \frac{(424)^2}{13}}$$

$$4510 - 469.0524 \leq \mu_1 - \mu_2 \leq 4510 + 469.0524]$$

$$4040.948 \leq \mu_1 - \mu_2 \leq 4979.052$$

Step 4: Interpret the results

It can be said that we are 99% confident that the mean difference of EC for the collection and aeration ponds is somewhere between 4040.948 and 4979.052. One can conclude that EC for the collection pond is higher than for the aeration pond.

Example 10.4: The concentration of biochemical oxygen demand for two rivers: An environmentalist wishes to find a range of values for the difference in the mean concentration of biochemical oxygen demand (BOD) (mg/L) for the surface water of two rivers. The collected data for 10 samples from river 1 showed that the concentration of BOD is 7.53 and the standard deviation is 3.22, and the collected data for 10 samples from river 2 showed that the mean concentration of BOD of is 7.64 and the standard deviation is 4.18. A significance level of $\alpha = 0.01$ is chosen to construct the confidence interval for the difference in the mean concentration of BOD for the two rivers. Assume that the two populations are normally distributed, and the variances are equal ($\sigma_1^2 = \sigma_2^2$).

The general procedure can be employed to calculate the confidence interval for the difference in the mean concentration of BOD for the surface waters of the two rivers.

Step 1: Use the sample data to compute the sample statistics for sample 1 and sample 2

The mean concentration of BOD for river 1 is already calculated to be 7.53, and for the second river the mean concentration of BOD was calculated to be 7.64.

Step 2: Select the significance level (α) for the study and find the critical values

The level of significance is chosen to be 0.01 ($\alpha = 0.01$) and $\frac{\alpha}{2} = \frac{0.01}{2} = 0.005$. The t critical value is $t_{\frac{\alpha}{2},d\cdot f} = t_{0.005,10+10-2=18} = 2.878$ (Table B in the Appendix).

Step 3: Compute the confidence interval

The entries for the confidence interval formula as presented in Eq. (10.7) (two-sided confidence interval) are already provided, the mean concentration of BOD for the first river is 7.53, the standard deviation is 3.22, and the sample size is 10, while the mean concentration of BOD for the second river is 7.64, the standard deviation is 4.18, and the sample size is 10.

We apply the formula presented in Eq. (10.7) to calculate the confidence interval for the difference in the concentration of BOD for the two rivers.

$$(\overline{Y}_1 - \overline{Y}_2) - t_{\frac{\alpha}{2},d\cdot f}\sqrt{\frac{S_P^2}{n_1} + \frac{S_P^2}{n_2}} \leq \mu_1 - \mu_2 \leq (\overline{Y}_1 - \overline{Y}_2) + t_{\frac{\alpha}{2},d\cdot f}\sqrt{\frac{S_P^2}{n_1} + \frac{S_P^2}{n_2}}$$

$$S_P^2 = \frac{(n_1 - 1)S_1^2 + (n_2 - 1)S_2^2}{n_1 + n_2 - 2} = \frac{(10 - 1)(3.22)^2 + (10 - 1)(4.18)^2}{10 + 10 - 2} = 13.9204$$

$$(7.53 - 7.64) - 2.552\sqrt{\frac{13.9204}{10} + \frac{13.9204}{10}}$$

$$\leq \mu_1 - \mu_2 \leq (7.53 - 7.64) - 2.552\sqrt{\frac{13.9204}{10} + \frac{13.9204}{10}}$$

$$-0.11 - 10.73948 \leq \mu_1 - \mu_2 \leq -0.11 + 10.73948$$

$$-10.849 \leq \mu_1 - \mu_2 \leq 10.629$$

Step 4: Interpret the results

It can be said that we are 99% confident that the mean difference of BOD for the two rivers is somewhere between -10.849 and 10.629. One can conclude that the mean concentration of BOD of river 1 sometimes is higher than the concentration of river 2 and sometimes lower than that of river 2.

10.5 Confidence interval for dependent samples

Consider that a random sample of size n is selected from a normally distributed population and tested/treated twice (different conditions) and the variable of interest is measured in both cases. Suppose that the response measured under the first condition is represented by Y_1, and Y_2 represents the result of the second condition. A confidence interval for the difference between the two responses and specific confidence level $(1-\alpha)$ can be computed employing the sample data. The mathematical formula for computing the confidence interval for two dependent samples is given in Eq. (10.10).

$$\overline{d} \pm t_{(\frac{\alpha}{2},n-1)}\left(\frac{S_d}{\sqrt{n}}\right) \tag{10.10}$$

The confidence interval in Eq. (10.10) can be written using another form:

$$\overline{d} - t_{(\frac{\alpha}{2},n-1)}\left(\frac{S_d}{\sqrt{n}}\right) \le \mu_d \le \overline{d} + t_{(\frac{\alpha}{2},n-1)}\left(\frac{S_d}{\sqrt{n}}\right)$$

- $t_{\frac{\alpha}{2},(n-1)}\left(\frac{S_d}{\sqrt{n}}\right)$ represents the margin of error (E).
- The margin of error (E) can be used to write the confidence interval.

$$\overline{d} - E \le \mu_d \le \overline{d} + E$$

- The lower one-tailed confidence interval is given in Eq. (10.11):

$$\overline{d} - t_{(\frac{\alpha}{2},n-1)}\left(\frac{S_d}{\sqrt{n}}\right) \le \mu_d \tag{10.11}$$

or it can be written as the interval $\left[\overline{d} - t_{(\frac{\alpha}{2},n-1)}\left(\frac{S_d}{\sqrt{n}}\right),\ \infty\right)$

- The upper one-tailed confidence interval is given in Eq. (10.12):

$$\mu_d \le \overline{d} + t_{(\frac{\alpha}{2},n-1)}\left(\frac{S_d}{\sqrt{n}}\right) \tag{10.12}$$

or it can be written as the interval $\left(-\infty,\ \overline{d} + t_{(\frac{\alpha}{2},n-1)}\left(\frac{S_d}{\sqrt{n}}\right)\right]$

Example 10.5: The concentration of chemical oxygen demand in a collection pond: A professor at a research center wishes to investigate the difference between his laboratory and outside laboratory results in measuring the mean concentration of chemical oxygen demand (COD) for the collection pond in the leachate landfill system. He randomly selected 10 samples and half of each sample was given to each laboratory, the data showed that the mean and standard deviation for the differences are 0.53 and 0.05, respectively. A significance level of $\alpha = 0.01$ is chosen to construct the confidence interval for the difference in the mean concentration of COD for the two laboratories. Assume that the population is normally distributed.

The general procedure can be employed to calculate the confidence interval for the difference in measuring the concentration of COD for the collection pond in the leachate landfill system produced by the two laboratories.

Step 1: Use the sample data to compute the sample statistics for sample 1 and sample 2

The mean of the differences is calculated to be 0.53 and the standard deviation is calculated to be 0.05.

Step 2: Select the significance level (α) for the study and find the critical values

The level of significance is chosen to be 0.01 ($\alpha = 0.01$) and $\frac{\alpha}{2} = 0.005$. The t critical value for $d \cdot f = 9$ and $\alpha = 0.01$ is $t_{(\frac{\alpha}{2}, d \cdot f)} = t_{(0.005, 9)} = 3.250$ (Table B in the Appendix).

Step 3: Compute the confidence interval

The entries for the confidence interval formula as presented in Eq. (10.10) (two-sided confidence interval) are already provided, the mean value of the differences of concentration of COD provided by the two laboratories is 0.53, the standard deviation is 0.05, and the sample size is 10.

We apply the formula presented in Eq. (10.10) to calculate the confidence interval for the difference between the records provided by the two laboratories regarding the concentration of COD for the collection pond in the leachate landfill system.

$$\bar{d} - t_{(\frac{\alpha}{2}, n-1)}\left(\frac{S_d}{\sqrt{n}}\right) \le \mu_d \le \bar{d} + t_{(\frac{\alpha}{2}, n-1)}\left(\frac{S_d}{\sqrt{n}}\right)$$

$$0.53 - 3.250\left(\frac{0.05}{\sqrt{10}}\right) \le \mu_d \le 0.53 + 3.250\left(\frac{0.05}{\sqrt{10}}\right)$$

$$0.53 - 0.05138441 \le \mu_d \le 0.53 + 0.05138441$$

$$0.479 \le \mu_d \le 0.581$$

Step 4: Interpret the results

It can be said that we are 99% confident that the mean differences of COD records between the two laboratories are somewhere between 0.479 and 0.581. One can conclude that the researcher's laboratory gives higher results than the outside laboratory.

Example 10.6: The competency assessment of a new device: An environmental company produced a new machine to measure the concentration of arsenic (AS) (mg/L) and wanted to measure the competency of the new device by comparing the results with the existing device. They randomly selected 11 samples of water and tested for the concentration of arsenic, the data showed that the mean and standard deviation for the differences between the two devices are 1.35 and 0.63, respectively. A significance level of $\alpha = 0.01$ is chosen to construct the confidence interval for the difference in the mean concentration of arsenic for the two devices. Assume that the population is normally distributed.

The general procedure can be employed to calculate the confidence interval for the difference in measuring the concentration of arsenic for the two devices to measure the competency of the new device.

Step 1: Use the sample data to compute the sample statistics for sample 1 and sample 2

The mean of the differences is calculated to be 1.35 and the standard deviation was calculated to be 0.63.

Step 2: Select the significance level (α) for the study and find the critical values

The level of significance is chosen to be 0.01 ($\alpha = 0.01$) and $\frac{\alpha}{2} = 0.005$. The t critical value for $d \cdot f = 10$ and $\alpha = 0.01$ is $t_{(\frac{\alpha}{2}, d \cdot f)} = t_{(0.005,10)} = 3.169$ (Table B in the Appendix).

Step 3: Compute the confidence interval

The entries for the confidence interval formula as presented in Eq. (10.10) (two-sided confidence interval) are already provided, the mean of the differences of the concentration of arsenic for the results provided by the two devices is 1.35, the standard deviation is 0.63, and the sample size is 11.

We apply the formula presented in Eq. (10.10) to calculate the confidence interval for the differences between the records provided by the two devices for the concentration of arsenic.

$$\overline{d} - t_{(\frac{\alpha}{2}, n-1)}\left(\frac{S_d}{\sqrt{n}}\right) \leq \mu_d \leq \overline{d} + t_{(\frac{\alpha}{2}, n-1)}\left(\frac{S_d}{\sqrt{n}}\right)$$

$$1.35 - 3.269\left(\frac{0.63}{\sqrt{11}}\right) \leq \mu_d \leq 1.35 + 3.269\left(\frac{0.63}{\sqrt{11}}\right)$$

$$1.35 - 0.6020101 \leq \mu_d \leq 1.35 + 0.6020101$$

$$0.748 \leq \mu_d \leq 1.952$$

Step 4: Interpret the results

It can be said that we are 99% confident that the mean differences of arsenic records produced by the two devices are somewhere between 0.748 and 1.952. One can conclude that the new device gives higher results than the existing one.

10.6 Confidence interval for the difference between two proportions

Consider that two independent random samples are selected, each sample is selected from a normally distributed population. Consider that Y_1 and Y_2 are two variables of interest to represent the number of units that carry a particular feature (property) for population 1 and population 2, respectively, $\hat{p}_1$ and $\hat{p}_2$ represent the proportion of the variables of interest for population 1 and population 2 respectively, and n_1 and n_2 represent the sample sizes selected from population 1 and population 2, respectively. A confidence interval for the difference between two population proportions and specific confidence level $1 - \alpha$ can be computed employing the sample data. The mathematical formula for computing the confidence interval for the difference between two sample proportions is given in Eq. (10.13).

$$(\hat{p}_1 - \hat{p}_2) \pm Z_{\frac{\alpha}{2}}\sqrt{\frac{\hat{p}_1(1-\hat{p}_1)}{n_1} + \frac{\hat{p}_2(1-\hat{p}_2)}{n_2}} \tag{10.13}$$

The confidence interval in Eq. (10.13) can be written using another form:

$$(\hat{p}_1 - \hat{p}_2) - Z_{\frac{\alpha}{2}}\sqrt{\frac{\hat{p}_1(1-\hat{p}_1)}{n_1} + \frac{\hat{p}_2(1-\hat{p}_2)}{n_2}}$$

$$\leq p_1 - p_2 \leq (\hat{p}_1 - \hat{p}_2) + Z_{\frac{\alpha}{2}}\sqrt{\frac{\hat{p}_1(1-\hat{p}_1)}{n_1} + \frac{\hat{p}_2(1-\hat{p}_2)}{n_2}}$$

where

$\hat{p}_1 - \hat{p}_2$ represents the observed difference between two sample proportions;
$p_1 - p_2$ represents the expected difference;
$Z_{\frac{\alpha}{2}}$ represents the Z critical value; and
$Z_{\frac{\alpha}{2}}\sqrt{\frac{\hat{p}_1\cdot(1-\hat{p}_1)}{n_1} + \frac{\hat{p}_2(1-.\hat{p}_2)}{n_2}}$ represents the margin of error (E).

- The margin of error (E) can be used to write the confidence interval.

$$(\hat{p}_1 - \hat{p}_2) - E \leq p_1 - p_2 \leq (\hat{p}_1 - \hat{p}_2) + E$$

- The lower one-tailed confidence interval is given in Eq. (10.14):

$$(\hat{p}_1 - \hat{p}_2) - Z_{\frac{\alpha}{2}}\sqrt{\frac{\hat{p}_1(1-\hat{p}_1)}{n_1} + \frac{\hat{p}_2(1-\hat{p}_2)}{n_2}} \leq p_1 - p_2 \tag{10.14}$$

or it can be written as the interval $\left[(\hat{p}_1 - \hat{p}_2) - Z_{\frac{\alpha}{2}}\sqrt{\frac{\hat{p}_1(1-\hat{p}_1)}{n_1} + \frac{\hat{p}_2(1-\hat{p}_2)}{n_2}},\ \infty\right)$

- The upper one-tailed confidence interval is given in Eq. (10.15):

$$p_1 - p_2 \leq (\hat{p}_1 - \hat{p}_2) + Z_{\frac{\alpha}{2}}\sqrt{\frac{\hat{p}_1(1-\hat{p}_1)}{n_1} + \frac{\hat{p}_2(1-\hat{p}_2)}{n_2}} \tag{10.15}$$

or it can be written as the interval $\left(-\infty,\ (\hat{p}_1 - \hat{p}_2) - Z_{\frac{\alpha}{2}}\sqrt{\frac{\hat{p}_1(1-\hat{p}_1)}{n_1} + \frac{\hat{p}_2(1-\hat{p}_2)}{n_2}}\right]$

Example 10.7: The high electrical conductivity in two rivers: A study was carried out to identify the range of values for the difference in the proportion of high electrical conductivity (EC) ($\mu S/cm$) of the surface water of two rivers. The collected data from river 1 showed that only two samples exhibited high conductivity out of 90 samples, and the collected data from river 2 showed that 25 samples exhibited high conductivity out of 90. A significance level of $\alpha = 0.01$ is chosen to construct the confidence interval for the difference in the proportion of high conductivity of water for the two rivers. Assume that the two populations are normally distributed.

The general procedure can be employed to calculate the confidence interval for the difference between the proportion of samples that exhibited high conductivity by water samples collected from two rivers.

Step 1: Use the sample data to compute the sample statistics for sample 1 and sample 2

The number of samples that exhibited high conductivity is already calculated to be two out of 90 from river 1 and the number of samples that exhibited high conductivity in river 2 was calculated to be 25 out of 90.

The proportion of high conductivity for each river can be calculated using the available information.

$$\hat{p}_1 = \frac{X_1}{n_1} = \frac{2}{90} = 0.0222 = 0.02$$

$$\hat{p}_2 = \frac{X_2}{n_2} = \frac{25}{90} = 0.2778 = 0.28$$

Step 2: Select the significance level (α) for the study and find the critical values

The level of significance is chosen to be 0.01 ($\alpha = 0.01$) and $\frac{\alpha}{2} = \frac{0.01}{2} = 0.005$. The Z critical value is $Z_{\frac{\alpha}{2}} = 2.58$ (Table A in the Appendix).

Step 3: Compute the confidence interval

The entries for the confidence interval formula as presented in Eq. (10.13) (two-sided confidence interval) are already provided. The collected data showed that the proportion of high conductivity for river 1 is 0.02, while the collected data from river 2 showed that the proportion of high conductivity is 0.28.

We apply the formula presented in Eq. (10.13) to calculate the confidence interval for the proportion of samples that showed high conductivity collected from two rivers.

$$(\hat{p}_1 - \hat{p}_2) - Z_{\frac{\alpha}{2}}\sqrt{\frac{\hat{p}_1(1-\hat{p}_1)}{n_1} + \frac{\hat{p}_2(1-\hat{p}_2)}{n_2}}$$

$$\leq p_1 - p_2 \leq (\hat{p}_1 - \hat{p}_2) + Z_{\frac{\alpha}{2}}\sqrt{\frac{\hat{p}_1(1-\hat{p}_1)}{n_1} + \frac{\hat{p}_2(1-\hat{p}_2)}{n_2}}(0.02 - 0.28)$$

$$-2.58\sqrt{\frac{0.02(1-0.02)}{90} + \frac{0.28(1-0.28)}{90}}$$

$$\leq p_1 - p_2 \leq (0.02 - 0.28) + 2.58\sqrt{\frac{0.02(1-0.02)}{90} + \frac{0.28(1-0.28)}{90}}$$

$$-0.26 - 0.128 \leq p_1 - p_2 \leq -0.26 + 0.128$$

$$-0.388 \leq p_1 - p_2 \leq -0.132$$

Step 4: Interpret the results

It can be said that we are 99% confident that the difference in the proportion of high conductivity between the two rivers is somewhere between -0.388 and -0.132. One can conclude that the conductivity of river 1 is lower than the conductivity of river 2.

Example 10.8: The permissible limit of PM_{10} in the air: A research center started collecting samples regarding the PM_{10} level in the air of two regions and compared the records with the permissible limits for 100 days. The collected data from station 1 showed that only four out of 100 samples were out of the permissible limits, while the collected data from station 2, of 100 samples, only two samples were out of the permissible limits. A significance level of $\alpha = 0.01$ is chosen to construct the confidence interval for the difference in the proportion of samples that are recorded out of permissible limits of PM_{10} in the air sampled from two stations. Assume that the two populations are normally distributed.

The general procedure can be employed to calculate the confidence interval for the proportion of samples that showed out of permissible limits regarding PM_{10} sampled from two stations.

Step 1: Use the sample data to compute the sample statistics for sample 1 and sample 2

The number of samples that showed out of permissible limits is already calculated to be four out of 100 from station 1, and the number of samples that showed out of permissible limits from station 2 is calculated to be two out of 100.

The proportion of samples that showed out of permissible limits for each station can be calculated using the available information.

$$\hat{p}_1 = \frac{X_1}{n_1} = \frac{4}{100} = 0.04$$

$$\hat{p}_2 = \frac{X_2}{n_2} = \frac{2}{100} = 0.02$$

Step 2: Select the significance level (α) for the study and find the critical values

The level of significance is chosen to be 0.01 ($\alpha = 0.01$) and $\frac{\alpha}{2} = \frac{0.01}{2} = 0.005$. The Z critical value is $Z_{\frac{\alpha}{2}} = 2.58$ (Table A in the Appendix).

Step 3: Compute the confidence interval

The entries for the confidence interval formula as presented in Eq. (10.13) (two-sided confidence interval) are already provided. The collected data showed that the proportion of samples that showed out of permissible limits for station 1 is 0.04, while the collected data from station 2 showed that the proportion of samples that showed out of permissible limits is 0.02.

We apply the formula presented in Eq. (10.13) to calculate the confidence interval for the proportion of samples that showed out of permissible limits collected from the two stations.

$$(\hat{p}_1 - \hat{p}_2) - Z_{\frac{\alpha}{2}}\sqrt{\frac{\hat{p}_1(1-\hat{p}_1)}{n_1} + \frac{\hat{p}_2(1-\hat{p}_2)}{n_2}}$$

$$\leq p_1 - p_2 \leq (\hat{p}_1 - \hat{p}_2) + Z_{\frac{\alpha}{2}}\sqrt{\frac{\hat{p}_1(1-\hat{p}_1)}{n_1} + \frac{\hat{p}_2(1-\hat{p}_2)}{n_2}}\ (0.04 - 0.02)$$

$$-\ 2.58\sqrt{\frac{0.04(1-0.04)}{100} + \frac{0.02(1-0.02)}{100}}$$

$$\leq p_1 - p_2 \leq (0.04 - 0.02) + 2.58\sqrt{\frac{0.04(1-0.04)}{100} + \frac{0.02(1-0.02)}{100}}$$

$$0.020 - 0.062 \leq p_1 - p_2 \leq 0.020 + 0.062$$

$$-0.042 \leq p_1 - p_2 \leq 0.082$$

Step 4: Interpret the results

It can be said that we are 99% confident that the difference in the proportion of samples that showed out of permissible limits between the two stations is somewhere between −0.042 and 0.082. One can conclude that the proportion of out of permissible limit samples fluctuates between the two stations, which means that sometimes station 1 showed lower and sometimes higher proportions of out of permissible limits samples regarding the PM_{10} level.

10.7 Confidence interval for the ratio of two variances

Consider that two random samples are selected, each sample is selected from a normally distributed population, n_1 and n_2 represent the sample sizes selected from population 1 and population 2, respectively, and σ_1^2 and σ_2^2 represent the variances for population 1 and population 2, respectively. A confidence interval for the ratio of two variances and specific confidence level $(1-\alpha)$ can be computed employing the sample data. The mathematical formula for computing the confidence interval for two sample variances is given in Eq. (10.16).

$$\frac{1}{F_{(\frac{\alpha}{2}, n_1-1, n_2-1)}}\frac{S_1^2}{S_2^2} \leq \frac{\sigma_1^2}{\sigma_2^2} \leq \frac{1}{F_{(1-\frac{\alpha}{2}, n_1-1, n_2-1)}}\frac{S_1^2}{S_2^2} \tag{10.16}$$

where

$F_{(\frac{\alpha}{2},n_1-1,n_2-1)}$ and $F_{(1-\frac{\alpha}{2},n_1-1,n_2-1)}$ represent the critical F values for the right and left tails.

S_1^2 and S_2^2 represent the sample variance for sample 1 and sample 2, respectively.

- The lower one-tailed confidence interval is given in Eq. (10.17):

$$\frac{1}{F_{(\frac{\alpha}{2},n_1-1,n_2-1)}}\frac{S_1^2}{S_2^2} \leq \frac{\sigma_1^2}{\sigma_2^2} \tag{10.17}$$

- The upper one-tailed confidence interval is given in Eq. (10.18):

$$\frac{\sigma_1^2}{\sigma_2^2} \leq \frac{1}{F_{(1-\frac{\alpha}{2},n_1-1,n_2-1)}}\frac{S_1^2}{S_2^2} \tag{10.18}$$

Example 10.9: The concentration of nitrate in two rivers: Samples from two rivers were selected and tested for nitrate concentration (NO_3^-) (mg/L) to investigate the range of differences between the variance of nitrate values in the two rivers (Jejawi and Juru Rivers). Fifteen samples were collected from each river and the concentration of nitrate was recorded, the collected data from Jejawi River showed that the standard deviation is 6.76, while the collected data from Juru River showed that the standard deviation is 4.11. A significance level of $\alpha = 0.01$ is chosen to construct the confidence interval for the variances of nitrate samples obtained from Jejawi and Juru Rivers. Assume that the two populations are normally distributed.

The general procedure can be employed to calculate the confidence interval for the variance of nitrate samples obtained from Jejawi and Juru Rivers.

Step 1: Use the sample data to compute the sample statistics for sample 1 and sample 2

The variance of nitrate values of surface water of Jejawi River is already calculated to be 6.76 and the variance of nitrate values of surface water of Juru River is calculated to be 4.11.

Step 2: Select the significance level (α) for the study and find the critical values

The level of significance is chosen to be 0.01 ($\alpha = 0.01$) and $\frac{\alpha}{2} = \frac{0.01}{2} = 0.005$. The F critical value for the two-tailed test with $\alpha = 0.01$ and $d \cdot f_1 = (n_1 - 1) = 15 - 1 = 14$ for numerator and $d \cdot f_2 = (n_2 - 1) = 14$ for denominator can be computed as follows:

The left-tailed value: The F critical value for the left-tailed test for $d \cdot f_1 = 14$, $d \cdot f_2 = 14$, and $1 - \frac{\alpha}{2} = 0.995$ is 0.2326, $F_{(1-\frac{\alpha}{2},m-1,m_2-1)} = F_{(0.995,14,14)} = 0.2326$ as shown in Table D in the Appendix or using software.

The right-tailed value: The F critical value for right-tailed test for $d \cdot f_1 = 14$, $d \cdot f_2 = 14$, and $\frac{\alpha}{2} = 0.005$ is 4.2993, $F_{(\frac{\alpha}{2}, n_1-1, n_2-1)} = F_{(0.005,14,14)} = 4.2993$ as shown in Table D in the Appendix or using software.

Step 3: Compute the confidence interval

The entries for the confidence interval formula as presented in Eq. (10.16) (two-sided confidence interval) are already provided. The collected data showed that the variance of nitrate samples obtained from Jejawi River is 6.76, while the collected data for nitrate samples obtained from Juru River showed that the variance is 4.11.

We apply the formula presented in Eq. (10.16) to calculate the confidence interval for the ratio of two variances of nitrate samples collected from Jejawi and Juru Rivers.

$$\frac{1}{F_{(\frac{\alpha}{2}, n_1-1, n_2-1)}} \frac{S_1^2}{S_2^2} \leq \frac{\sigma_1^2}{\sigma_2^2} \leq \frac{1}{F_{(1-\frac{\alpha}{2}, n_1-1, n_2-1)}} \frac{S_1^2}{S_2^2}$$

$$\frac{1}{4.2993} \frac{(6.76)^2}{(4.11)^2} \leq \frac{\sigma_1^2}{\sigma_2^2} \leq \frac{1}{0.2326} \frac{(6.76)^2}{(4.11)^2}$$

$$\frac{1}{4.2993} 2.705265 \leq \frac{\sigma_1^2}{\sigma_2^2} \leq \frac{1}{0.2326} 2.705265$$

$$0.629 \leq \frac{\sigma_1^2}{\sigma_2^2} \leq 11.631$$

Step 4: Interpret the results

It can be said that we are 99% confident that the difference in variances of the concentration of nitrate values of surface water of the two rivers is somewhere between 0.629 and 11.631. One can conclude that most of times the variance of nitrate samples obtained from Jejawi River is higher than the variance of nitrate samples obtained from Juru River.

Example 10.10: The comparison of variance of Zinc in sediment: Samples from two different rivers were selected and tested for zinc concentration (Zn) (mg/L) to investigate the range of differences between the variance of zinc values in the sediment of two rivers (river 1 and river 2). Ten samples were collected from each river and the concentration of zinc was recorded, the collected data from the first river showed that the standard deviation is 0.93, while the collected data from river 2 showed that the standard deviation is 0.55. A significance level of $\alpha = 0.01$ is chosen to construct the confidence interval for the variances of zinc samples obtained from the two rivers. Assume that the two populations are normally distributed.

The general procedure can be employed to calculate the confidence interval for the variance values of zinc concentration in the sediment samples obtained from river 1 and river 2.

Step 1: Use the sample data to compute the sample statistics for sample 1 and sample 2

The variance of zinc values in sediment for the first river is already calculated to be 0.93 and variance of zinc values in sediment for river 2 is calculated to be 0.55.

Step 2: Select the significance level (α) for the study and find the critical values

The level of significance is chosen to be 0.01 ($\alpha = 0.01$) and $\frac{\alpha}{2} = \frac{0.01}{2} = 0.005$. The F critical value for a two-tailed test with $\alpha = 0.01$ and $d \cdot f_1 = (n_1 - 1) = 10 - 1 = 9$ for numerator and $d \cdot f_2 = (n_2 - 1) = 9$ for denominator also can be computed as follows:

The right-tailed value: The F critical value for the left-tailed test for $d \cdot f_1 = 9$, $d \cdot f_2 = 9$, and $\frac{\alpha}{2} = 0.005$ is 6.5411, $F_{(\frac{\alpha}{2}, n_1 - 1, n_2 - 1)} = F_{(0.005, 9, 9)} = 6.5411$ as shown in Table D in the Appendix or using software.

The left-tailed value: The F value for the right-tailed test for $d \cdot f_1 = 9$, $d \cdot f_2 = 9$, and $1 - \frac{\alpha}{2} = 0.995$ is 0.1529, $F_{(1-\frac{\alpha}{2}, n_1 - 1, n_2 - 1)} = F_{(0.995, 9, 9)} = 0.1529$ as shown in Table D in the Appendix or using software.

Step 3: Compute the confidence interval

The entries for the confidence interval formula as presented in Eq. (10.16) (two-sided confidence interval) are already provided. The collected data showed that the variance of zinc samples selected from river 1 is 0.93, while the collected data for zinc samples from river 2 showed that the variance is 0.55.

We apply the formula presented in Eq. (10.16) to build the confidence interval for the variances of zinc samples in sediment collected from river 1 and river 2.

$$\frac{1}{F_{(\frac{\alpha}{2}, n_1 - 1, n_2 - 1)}} \frac{S_1^2}{S_2^2} \le \frac{\sigma_1^2}{\sigma_2^2} \le \frac{1}{F_{(1-\frac{\alpha}{2}, n_1 - 1, n_2 - 1)}} \frac{S_1^2}{S_2^2}$$

$$\frac{1}{6.5411} \frac{(0.93)^2}{(0.55)^2} \le \frac{\sigma_1^2}{\sigma_2^2} \le \frac{1}{0.1529} \frac{(0.93)^2}{(0.55)^2}$$

$$\frac{1}{4.2993} 2.8592 \le \frac{\sigma_1^2}{\sigma_2^2} \le \frac{1}{0.2326} 2.8592$$

$$0.437 \le \frac{\sigma_1^2}{\sigma_2^2} \le 18.702$$

Step 4: Interpret the results

It can be said that we are 99% confident that the difference in variances of the concentration of zinc values in the sediment of two rivers is somewhere between 0.437 and 18.702. One can conclude that most of the times the variance of zinc sample exhibited by the first river is higher than the variance of zinc samples obtained from the second river.

Further reading

Alkarkhi, A. F. M., Ahmad, A., & Easa, A. M. (2009). Assessment of surface water quality of selected estuaries of Malaysia: Multivariate statistical techniques. *The Environmentalist, 29*, 255–262.

Alkarkhi, A. F. M., & Chin, L. H. (2012). *Elementary statistics for technologist* (1st ed.). Malaysia: Universiti Sains Malaysia.

Alkarkhi, A. F. M., Ismail, N., Ahmed, A., & Easa, A. M. (2009). Analysis of heavy metal concentrations in sediments of selected estuaries of Malaysia—A statistical assessment. *Environmental Monitoring and Assessment, 153*, 179–185.

Banch, T. J. H., Hanafiah, M. M., Alkarkhi, A. F. M., & Abu Amr, S. S. (2019). Statistical evaluation of landfill leachate system and its impact on groundwater and surface water in Malaysia. *Sains Malaysiana, 48*(11), 2391–2403.

Bluman, A. G. (1998). *Elementary statistics: A step by step approach* (3rd ed.). Boston: WCB/McGraw-Hill.

Eberly College of Science-Department of Statistics (2018). Two variances. The Pennsylvania State University. <https://online.stat.psu.edu/stat414/node/206/>.

Rahman, N. N. N. A., Chard, N. C., Al-Karkhi, A. F. M., Rafatullah, M., & Kadir, A. F. M (2017). Analysis of particulate matters in air of palm oil mills – A statistical assessment. *Environmental Engineering and Management Journal, 16*(11), 2537–2543.

Weiss, N. A. (2012). *Introductory statistics* (9th ed.). Pearson.

Yusup, Y., & Alkarkhi, A. F. M. (2011). Cluster analysis of inorganic elements in particulate matter in the air environment of an equatorial urban coastal location. *Chemistry and Ecology, 27*(3), 273–286.

The interval estimation procedure: hypothesis testing for two populations

Abstract

This chapter addresses the application of the confidence interval procedure for testing various hypotheses including one-tailed and two-tailed tests for the difference between two population parameters. Moreover, the result of confidence interval in testing various hypotheses is compared with the results of two other procedures: critical value (traditional) and *P*-value. Three examples for each test are given including Z-test for two means, Z-test for two proportions, t-test for two population means, t-test for two paired means, and F-test for the ratio of two variances. The three procedures provide the same decision regarding one-tailed and two-tailed tests. The procedure is explained in an easy and enjoyable way to show clearly the individual steps.

Keywords: Confidence interval; *P*-value procedure; critical value procedure; mean; proportion; variance

Learning outcomes

After completing this chapter, readers should be able to:

- Know the steps on how to use confidence intervals for testing various hypotheses for the difference between two population parameters;
- Understand the procedure for testing a hypothesis concerning the difference between two sample means
- Understand the procedure for testing a hypothesis concerning the difference between two sample proportions;
- Understand the procedure for testing a hypothesis concerning the difference between two sample variances;
- Describe the procedure for using a confidence interval to make a decision regarding various types of hypothesis testing including one-tailed and two-tailed;
- Draw smart conclusions regarding the problem under investigation.

11.1 Introduction

Interval estimation (confidence interval) for one population parameter was used to test a hypothesis as the third procedure for performing hypothesis testing. The concept of one population parameter can be extended to test a hypothesis for the difference between two parameters using confidence intervals. We develop a procedure

Applications of Hypothesis Testing for Environmental Science. DOI: https://doi.org/10.1016/B978-0-12-824301-5.00005-8

for testing various hypotheses for two population parameters including the difference between two means, two proportions, and two variances. The results produced by employing a confidence interval will be compared with the results of other two procedures: the critical value and P-value procedures. Examples presented in Chapter 9, Hypothesis testing for the difference between two populations, will be reproduced to make the comparison easy and straightforward.

11.2 The steps for the confidence interval procedure for the difference between two populations

Similar steps to critical value and P-value procedures can be applied to perform hypothesis testing using the confidence intervals for the difference between two population parameters. We have studied hypothesis testing using an interval estimation procedure for one population parameter including four steps. The four steps for performing hypothesis testing using interval estimation (confidence interval) are:

Step 1: Specify the null and alternative hypotheses

Step 2: Select the significance level (α) for the study and find the critical values

Step 3: Use the sample information to calculate the confidence interval

Step 4: Make a decision using the confidence interval and interpret the results

- Do not reject the null hypothesis if the hypothesized (claimed) value falls in the confidence interval.
- Reject the null hypothesis if the hypothesized (claimed) value does not fall in the confidence interval, which indicates that the result is statistically significant.

11.3 Confidence interval for testing the difference between two means when sample size is large

Hypothesis testing for the difference between two population means when the sample size is large ($n \geq 30$) can be tested using the confidence interval including one-tailed and two-tailed tests. A step-by-step procedure will be employed to illustrate the steps of making a decision concerning each hypothesis, whether left-tailed, right-tailed, or two-tailed.

Let us start with the two-tailed and one-tailed confidence interval formulas before giving examples to illustrate the procedure.

- The two-tailed confidence interval for the difference between two population means is given in Eq. (11.1).

$$(\overline{Y}_1 - \overline{Y}_2) - Z_{\frac{\alpha}{2}}\sqrt{\frac{\sigma_1^2}{n_1} + \frac{\sigma_2^2}{n_2}} \leq \mu_1 - \mu_2 \leq (\overline{Y}_1 - \overline{Y}_2) + Z_{\frac{\alpha}{2}}\sqrt{\frac{\sigma_1^2}{n_1} + \frac{\sigma_2^2}{n_2}} \quad (11.1)$$

or it can be written as $\left[(\overline{Y}_1 - \overline{Y}_2) - Z_{\frac{\alpha}{2}}\sqrt{\frac{\sigma_1^2}{n_1} + \frac{\sigma_2^2}{n_2}}, (\overline{Y}_1 - \overline{Y}_2) + Z_{\frac{\alpha}{2}}\sqrt{\frac{\sigma_1^2}{n_1} + \frac{\sigma_2^2}{n_2}}\right]$.

Do not reject the null hypothesis if the hypothesized (claimed) value falls in the confidence interval and reject the null hypothesis if the hypothesized (claimed) value does not fall in the confidence interval, which indicates that the result is statistically significant.

- The one-tailed confidence interval for the difference between two population means.

The lower one-tailed confidence interval for the difference between two population means is given in Eq. (11.2).

$$\left(\overline{Y}_1 - \overline{Y}_2\right) - Z_\alpha\sqrt{\frac{\sigma_1^2}{n_1} + \frac{\sigma_2^2}{n_2}} \le \mu_1 - \mu_2 \tag{11.2}$$

or it can be written as $\left[\left(\overline{Y}_1 - \overline{Y}_2\right) - Z_\alpha\sqrt{\frac{\sigma_1^2}{n_1} + \frac{\sigma_2^2}{n_2}}, \infty\right]$.

Reject the null hypothesis if the hypothesized (claimed) value does not fall in the confidence interval $\left(\overline{Y}_1 - \overline{Y}_2\right) - Z_\alpha\sqrt{\frac{\sigma_1^2}{n_1} + \frac{\sigma_2^2}{n_2}} \le \mu_1 - \mu_2$, which means we reject the null hypothesis if $\mu_1 - \mu_2 < \left(\overline{Y}_1 - \overline{Y}_2\right) - Z_\alpha\sqrt{\frac{\sigma_1^2}{n_1} + \frac{\sigma_2^2}{n_2}}$.

The upper one-tailed confidence interval for the difference between two population means is given in Eq. (11.3).

$$\mu_1 - \mu_2 \le \left(\overline{Y}_1 - \overline{Y}_2\right) + Z_\alpha\sqrt{\frac{\sigma_1^2}{n_1} + \frac{\sigma_2^2}{n_2}} \tag{11.3}$$

or it can be written as $\left[-\infty, \left(\overline{Y}_1 - \overline{Y}_2\right) + Z_\alpha\sqrt{\frac{\sigma_1^2}{n_1} + \frac{\sigma_2^2}{n_2}}\right]$.

Reject the null hypothesis if the hypothesized value does not fall in the confidence interval $\mu_1 - \mu_2 \le \left(\overline{Y}_1 - \overline{Y}_2\right) + Z_\alpha\sqrt{\frac{\sigma_1^2}{n_1} + \frac{\sigma_2^2}{n_2}}$, which means we reject the null hypothesis if $\mu_1 - \mu_2 > \left(\overline{Y}_1 - \overline{Y}_2\right) + Z_\alpha\sqrt{\frac{\sigma_1^2}{n_1} + \frac{\sigma_2^2}{n_2}}$.

Example 11.1: The concentration of dissolved oxygen before and after the dam: Example 9.4 is reproduced "A professor at an environmental section wanted to verify the claim that the mean concentration of dissolved oxygen (DO) (mg/L) of water before Beris dam is the same as the mean concentration of dissolved oxygen of water after the dam. He selected 35 samples from each region and tested for dissolved oxygen concentration. The collected data showed that the mean concentration of dissolved oxygen before the dam is 5.20 and the standard deviation of the population is 0.41, while the collected data after the dam showed that the mean concentration of dissolved oxygen is 2.50 and the standard deviation of the population is 1.10. A significance level of $\alpha = 0.01$ is chosen to test the claim. Assume that the two populations are normally distributed."

The four steps for performing hypothesis testing employing the confidence interval procedure can be used to test the hypothesis regarding the mean concentration of dissolved oxygen of water before and after the dam. The result of the confidence interval procedure will be compared with the critical value (traditional) and *P*-value procedures.

The confidence interval procedure

Step 1: Specify the null and alternative hypotheses

The two hypotheses regarding the concentration of dissolved oxygen before and after the dam are presented in Eq. (11.4).

$$H_0{:}\mu_1 = \mu_2 \text{ vs } H_1{:}\mu_1 \neq \mu_2 \tag{11.4}$$

We should make a decision regarding the null hypothesis as to whether the mean concentration of dissolved oxygen of water before the dam μ_1 is equal to the mean concentration of dissolved oxygen of water after the dam μ_2, or μ_1 differs from μ_2.

Step 2: Select the significance level (α) for the study

The Z critical value for a two-tailed test with a significance level of 0.01 ($\alpha = 0.01$) and $\frac{\alpha}{2} = \frac{0.01}{2} = 0.005$ is 2.58 as appeared in the standard normal table (Table A in the Appendix).

Step 3: Use the sample information to calculate the confidence interval

The hypothesis in Eq. (11.4) represents a two-tailed test. Thus, the confidence interval formula presented in Eq. (11.1) should be used to calculate the confidence interval for the difference between the mean concentration of dissolved oxygen of water before and after the dam, this formula represents the alternative hypothesis ($\neq$) for a two-sided test.

The entries for the confidence interval formula as presented in Eq. (11.1) are already provided. The mean concentration of dissolved oxygen before the dam is 5.20 and the standard deviation is 0.41, while the mean concentration of dissolve oxygen after the dam is 2.50 and the standard deviation is 1.10. The sample size is 35 for each region.

$$(\overline{Y}_1 - \overline{Y}_2) - Z_{\frac{\alpha}{2}}\sqrt{\frac{\sigma_1^2}{n_1} + \frac{\sigma_2^2}{n_2}} \leq \mu_1 - \mu_2 \leq (\overline{Y}_1 - \overline{Y}_2) + Z_{\frac{\alpha}{2}}\sqrt{\frac{\sigma_1^2}{n_1} + \frac{\sigma_2^2}{n_2}}$$

$$(5.20 - 2.50) - 2.58\sqrt{\frac{(0.41)^2}{35} + \frac{(1.10)^2}{35}} \leq \mu_1 - \mu_2 \leq (5.20 - 2.50) + 2.58\sqrt{\frac{(0.41)^2}{35} + \frac{(1.10)^2}{35}}$$

$$2.7 - 0.511 \leq \mu_1 - \mu_2 \leq 2.7 + 0.511$$

$$2.189 \leq \mu_1\text{-}\mu_2 \leq 3.211$$

Step 4: Make a decision using the confidence interval and interpret the results

Making a decision using the confidence interval procedure requires checking the confidence interval as to whether it contains the hypothesized (claimed) value or not. It can be seen that the hypothesized value ($\mu_1 - \mu_2 = 0$) is not in the calculated confidence interval $2.189 \leq \mu_1 - \mu_2 \leq 3.211$.

1. The decision reached by the confidence interval procedure: reject the null hypothesis and we believe that the mean concentration of dissolved oxygen of water before and after the dam is different.
   The result of the confidence interval is compared with the decisions reached by the critical value and *P*-value procedures:
2. The critical value (traditional) procedure: reject the null hypothesis as presented in Example 9.4.
3. The *P*-value procedure: reject the null hypothesis as presented in Example 9.4.

We can say that the three procedures have reached to the same decision: Reject the null hypothesis at 1% and we believe that the mean concentration of dissolved oxygen of water before and after the dam is not equal as stated by the null hypothesis.

Example 11.2: The concentration of mercury in river water: Example 9.5 is reproduced "A researcher at a research institution wanted to verify the claim that the mean concentration of mercury (Hg) (mg/L) in water of Juru River is less than the mean concentration of mercury (Hg) in water of Jejawi River. He selected 33 samples from each river and tested for mercury concentration. The collected data showed that the mean concentration of mercury of Juru River is 0.13 and the standard deviation is 0.02, while the collected data for Jejawi River showed that the mean concentration of mercury is 0.20 and the standard deviation is 0.07. A significance level of $\alpha = 0.01$ is chosen to test the claim. Assume that the two populations are normally distributed."

The four steps for performing hypothesis testing employing the confidence interval procedure can be used to test the hypothesis regarding the mean concentration of mercury in water of Juru and Jejawi Rivers. The result of the confidence interval procedure will be compared with the critical value (traditional) and *P*-value procedures.

The confidence interval procedure

Step 1: Specify the null and alternative hypotheses

The two hypotheses regarding the mean concentration of mercury in water of Juru and Jejawi Rivers are presented in Eq. (11.5).

$$H_0{:}\mu_1 \geq \mu_2 \text{ vs } H_1{:}\mu_1 < \mu_2 \tag{11.5}$$

We should make a decision regarding the null hypothesis as to whether the mean concentration of mercury in water of Juru River μ_1 is greater than or equal to the mean concentration of mercury in water of Jejawi River μ_2 or μ_1 is less than μ_2.

Step 2: Select the significance level (α) for this study

The Z critical value for a one-tailed test with a significance level of 0.01 ($\alpha = 0.01$) and $\frac{\alpha}{2} = \frac{0.01}{2} = 0.005$ is 2.326 as appeared in the standard normal table (Table A in the Appendix).

Step 3: Use the sample information to calculate the test statistic value

The hypothesis in Eq. (11.5) represents a one-tailed test. Thus, the upper confidence interval formula presented in Eq. (11.3) should be used to calculate the confidence interval for the difference between the mean concentration of mercury in Juru and Jejawi Rivers because of the alternative hypothesis of the form less than (<), this formula represents the alternative hypothesis for a one-sided test.

The entries for the confidence interval formula as presented in Eq. (11.3) are already provided. The mean concentration of mercury in Juru River is 0.13 and the standard deviation is 0.02, while the mean concentration of mercury in Jejawi River is 0.20 and the standard deviation is 0.07. The sample size is 33 for each river.

$$\mu_1 - \mu_2 \leq (\overline{Y}_1 - \overline{Y}_2) + Z_\alpha \sqrt{\frac{\sigma_1^2}{n_1} + \frac{\sigma_2^2}{n_2}}$$

$$\mu_1 - \mu_2 \leq (0.13 - 0.20) - 2.326\sqrt{\frac{(0.02)^2}{33} + \frac{(0.07)^2}{33}}$$

$$\mu_1 - \mu_2 \leq (-0.07) + 0.033$$

$$\mu_1 - \mu_2 \leq -0.037$$

Step 4: Make a decision using the confidence interval and interpret the results

Making a decision using the confidence interval procedure requires checking the confidence interval as to whether it contains the hypothesized (claimed) value or not. It can be seen that the hypothesized value ($\mu_1 - \mu_2 = 0$) is not in the calculated confidence interval $\mu_1 - \mu_2 \leq -0.03$.

1. The decision reached by confidence interval procedure: reject the null hypothesis and we believe that the mean concentration of mercury in water of Juru River is less than the mean concentration of mercury in water of Jejawi River.

   The result of the confidence interval is compared with the decisions reached by the critical value and *P*-value procedures:
2. The critical value (traditional) procedure: reject the null hypothesis as presented in Example 9.5.
3. The *P*-value procedure: reject the null hypothesis as presented in Example 9.5.

We can say that the three procedures have reached the same decision: Reject the null hypothesis at 1% and believe that the mean concentration of mercury in water of Juru River is less than the mean concentration of mercury in water of Jejawi River.

Example 11.3: The concentration of PM_1 in the air of palm oil mills: Example 9.6 is reproduced "The concentration of PM_1 was investigated for two palm oil mills by a research team to verify the claim that the mean concentration of PM_1 $\mu g m^{-3}$ in the air of a palm oil mill located in Kedah state is more than the mean concentration of PM_1 in the air of a palm oil mill located in Penang state of Malaysia. Thirty-four days were selected, and the concentration of particulate matter was measured. The collected data showed that the mean concentration of PM_1 in air of the Penang palm oil mill is 19.30 and the standard deviation is 16.07, while the collected data for the palm oil in Kedah showed that the mean concentration of PM_1 is 74.87 and the standard deviation is 26.89. A significance level of $\alpha = 0.01$ is chosen to test the claim. Assume that the two populations are normally distributed."

The four steps for performing hypothesis testing employing the confidence interval procedure can be used to test the hypothesis regarding the mean concentration of PM_1 in the air of a palm oil mill located in Penang and Kedah states of Malaysia. The result of the confidence interval procedure will be compared with the critical value (traditional) and P-value procedures.

The confidence interval procedure

Step 1: Specify the null and alternative hypotheses

The two hypotheses regarding the mean concentration of PM_1 in the air of a palm oil mill located in Kedah and Penang states of Malaysia are presented in Eq. (11.6).

$$H_0: \mu_1 \leq \mu_2 \text{ vs } H_1: \mu_1 > \mu_2 \tag{11.6}$$

We should make a decision regarding the null hypothesis as to whether the mean concentration of PM_1 in the air of a palm oil mill located in Kedah state μ_1 is less than or equal the mean concentration of PM_1 in the air of a palm oil mill located in Penang state μ_2 or μ_1 is greater than μ_2.

Step 2: Select the significance level (α) for the study

The Z critical value for one-tailed test with a significance level of 0.01 ($\alpha = 0.01$) and $\frac{\alpha}{2} = \frac{0.01}{2} = 0.005$ is 2.326 as appeared in the standard normal table (Table A in the Appendix).

Step 3: Use the sample information to calculate the test statistic value

The hypothesis in Eq. (11.6) represents a one-tailed test. Thus, the lower confidence interval formula presented in Eq. (11.2) should be used to calculate the confidence interval for the difference between the mean concentration of PM_1 in the air of a palm oil mill located in Kedah and Penang states, this formula represents the alternative hypothesis > for a one-sided test.

The entries for the confidence interval formula as presented in Eq. (11.2) are already provided. The mean concentration of PM_1 in the air of a palm oil mill located in Penang state is 19.30 and the standard deviation is 16.07, while the mean concentration of PM_1 in the air of a palm oil mill located in Kedah state is 74.87 and the standard deviation is 26.89. The sample size is 34 for each location.

$$\left(\overline{Y}_1 - \overline{Y}_2\right) - Z_\alpha \sqrt{\frac{\sigma_1^2}{n_1} + \frac{\sigma_2^2}{n_2}} \leq \mu_1 - \mu_2$$

$$(74.87 - 19.30) - 2.326 \sqrt{\frac{(26.89)^2}{34} + \frac{(16.07)^2}{34}} \leq \mu_1 - \mu_2$$

$$55.57 - 13.838 \leq \mu_1 - \mu_2$$

$$41.732 \leq \mu_1 - \mu_2$$

Step 4: Make a decision using the confidence interval and interpret the results

Making a decision using the confidence interval procedure requires checking the confidence interval as to whether it contains the hypothesized (claimed) value or not. It can be seen that the hypothesized value ($\mu_1 - \mu_2 = 0$) is not in the calculated confidence interval $41.732 \leq \mu_1 - \mu_2$.

1. The decision reached by the confidence interval procedure: reject the null hypothesis and we believe that the mean concentration of PM_1 in the air of a palm oil mill located in Kedah state μ_1 is more than the mean concentration of PM_1 in the air of a palm oil mill located in Penang state μ_2.

   The result of the confidence interval is compared with the decisions reached by the critical value and *P*-value procedures:
2. The critical value (traditional) procedure: reject the null hypothesis as presented in Example 9.6.
3. The *P*-value procedure: reject the null hypothesis as presented in Example 9.6.

We can say that the three procedures have reached the same decision: Reject the null hypothesis at 1% and we believe that the mean concentration of PM_1 in the air of a palm oil mill located in Kedah state μ_1 is more than the mean concentration of PM_1 in the air of a palm oil mill located in Penang state μ_2.

11.4 Confidence interval for testing the difference between two means when the sample size is small

Hypothesis testing for the difference between two population means when the sample size is small ($n < 30$) can be tested using the confidence interval including one-tailed and two-tailed tests. A step-by-step procedure will be employed to illustrate the steps of making a decision concerning each hypothesis including left-tailed, right-tailed, and two-tailed. There are two cases for the confidence interval when the sample size is small, the two cases are based on the variances: unequal variances and equal variances.

Let us start with the two-tailed and one-tailed confidence interval formulas for *unequal variances* first.

- The two-tailed confidence interval for the difference between two population means when the variances are *unequal* is given in Eq. (11.7).

$$(\overline{Y}_1 - \overline{Y}_2) - t_{(\frac{\alpha}{2},d.f)}\sqrt{\frac{S_1^2}{n_1} + \frac{S_2^2}{n_2}} \leq \mu_1 - \mu_2 \leq (\overline{Y}_1 - \overline{Y}_2) + t_{(\frac{\alpha}{2},d.f)}\sqrt{\frac{S_1^2}{n_1} + \frac{S_2^2}{n_2}} \tag{11.7}$$

or it can be written as $\left[(\overline{Y}_1 - \overline{Y}_2) - t_{(\frac{\alpha}{2},d.f)}\sqrt{\frac{S_1^2}{n_1} + \frac{S_2^2}{n_2}}, (\overline{Y}_1 - \overline{Y}_2) + t_{(\frac{\alpha}{2},d.f)}\sqrt{\frac{S_1^2}{n_1} + \frac{S_2^2}{n_2}}\right]$.

Do not reject the null hypothesis if the hypothesized (claimed) value falls in the confidence interval and reject the null hypothesis if the hypothesized (claimed) value does not fall in the confidence interval, which indicates that the result is statistically significant.

- The one-tailed confidence intervals for *unequal variances* are given in Eqs. (11.8) and (11.9).

The lower one-tailed confidence interval for the difference between two population means is given in Eq. (11.8).

$$\left(\overline{Y}_1 - \overline{Y}_2\right) - t_{(\alpha,d.f)}\sqrt{\frac{S_1^2}{n_1} + \frac{S_2^2}{n_2}} \le \mu_1 - \mu_2 \tag{11.8}$$

or it can be written as $\left[\left(\overline{Y}_1 - \overline{Y}_2\right) - t_{(\alpha,d.f)}\sqrt{\frac{S_1^2}{n_1} + \frac{S_2^2}{n_2}},\ \infty\right]$.

Reject the null hypothesis if the hypothesized (claimed) value does not fall in the confidence interval $\left(\overline{Y}_1 - \overline{Y}_2\right) - t_{(\alpha,d.f)}\sqrt{\frac{S_1^2}{n_1} + \frac{S_2^2}{n_2}} \le \mu_1 - \mu_2$, which means we reject the null hypothesis if $\mu_1 - \mu_2 < \left(\overline{Y}_1 - \overline{Y}_2\right) - t_{(\alpha,d.f)}\sqrt{\frac{S_1^2}{n_1} + \frac{S_2^2}{n_2}}$.

The upper one-tailed confidence interval for the difference between two population means is given in Eq. (11.9).

$$\mu_1 - \mu_2 \le \left(\overline{Y}_1 - \overline{Y}_2\right) + t_{(\alpha,d.f)}\sqrt{\frac{S_1^2}{n_1} + \frac{S_2^2}{n_2}} \tag{11.9}$$

or it can be written as $\left[-\infty,\ \left(\overline{Y}_1 - \overline{Y}_2\right) + t_{(\alpha,d.f)}\sqrt{\frac{S_1^2}{n_1} + \frac{S_2^2}{n_2}}\right]$.

Reject the null hypothesis if the hypothesized value does not fall in the confidence interval $\mu_1 - \mu_2 \le \left(\overline{Y}_1 - \overline{Y}_2\right) + t_{(\alpha,d.f)}\sqrt{\frac{S_1^2}{n_1} + \frac{S_2^2}{n_2}}$, which means we reject the null hypothesis if $\mu_1 - \mu_2 > \left(\overline{Y}_1 - \overline{Y}_2\right) + t_{(\alpha,d.f)}\sqrt{\frac{S_1^2}{n_1} + \frac{S_2^2}{n_2}}$.

The confidence interval for *equal variances* including two-tailed and one-tailed are:

- The two-tailed confidence interval for the difference between two population means when the variances are *equal* is given in Eq. (11.10).

$$\left(\overline{Y}_1 - \overline{Y}_2\right) - t_{(\frac{\alpha}{2},d.f)}\sqrt{\frac{S_P^2}{n_1} + \frac{S_P^2}{n_2}} \le \mu_1 - \mu_2 \le \left(\overline{Y}_1 - \overline{Y}_2\right) + t_{(\frac{\alpha}{2},d.f)}\sqrt{\frac{S_P^2}{n_1} + \frac{S_P^2}{n_2}} \tag{11.10}$$

or it can be written as $\left[\left(\overline{Y}_1 - \overline{Y}_2\right) - t_{(\frac{\alpha}{2},d.f)}\sqrt{\frac{S_P^2}{n_1} + \frac{S_P^2}{n_2}},\ \left(\overline{Y}_1 - \overline{Y}_2\right) + t_{(\frac{\alpha}{2},d.f)}\sqrt{\frac{S_P^2}{n_1} + \frac{S_P^2}{n_2}}\right]$.

S_P^2 is called the pooled variance, we use Eq. (11.11) to compute the pooled variance.

$$S_P^2 = \frac{(n_1 - 1)S_1^2 + (n_2 - 1)S_2^2}{n_1 + n_2 - 2} \tag{11.11}$$

Do not reject the null hypothesis if the hypothesized (claimed) value falls in the confidence interval and reject the null hypothesis if the hypothesized (claimed) value does not fall in the confidence interval, which indicates that the result is statistically significant.

- The one-tailed confidence intervals for *equal variances* are given in Eqs. (11.12) 1nd (11.13).

  The lower one-tailed confidence interval for the difference between two population means is given in Eq. (11.12).

$$(\overline{Y}_1 - \overline{Y}_2) - t_{(\alpha,d.f)}\sqrt{\frac{S_P^2}{n_1} + \frac{S_P^2}{n_2}} \leq \mu_1 - \mu_2 \tag{11.12}$$

or it can be written as $\left[(\overline{Y}_1 - \overline{Y}_2) - t_{(\alpha,d.f)}\sqrt{\frac{S_P^2}{n_1} + \frac{S_P^2}{n_2}}, \infty\right]$.

Reject the null hypothesis if the hypothesized (claimed) value does not fall in the confidence interval $(\overline{Y}_1 - \overline{Y}_2) - t_{(\alpha,d.f)}\sqrt{\frac{S_P^2}{n_1} + \frac{S_P^2}{n_2}} \leq \mu_1 - \mu_2$, which means we reject the null hypothesis if $\mu_1 - \mu_2 < (\overline{Y}_1 - \overline{Y}_2) - t_{(\alpha,d.f)}\sqrt{\frac{S_P^2}{n_1} + \frac{S_P^2}{n_2}}$.

The upper one-tailed confidence interval for the difference between two population means is given in Eq. (11.13).

$$\mu_1 - \mu_2 \leq (\overline{Y}_1 - \overline{Y}_2) + t_{(\alpha,d.f)}\sqrt{\frac{S_P^2}{n_1} + \frac{S_P^2}{n_2}} \tag{11.13}$$

or it can be written as $\left[-\infty, (\overline{Y}_1 - \overline{Y}_2) + t_{(\alpha,d.f)}\sqrt{\frac{S_P^2}{n_1} + \frac{S_P^2}{n_2}}\right]$.

Reject the null hypothesis if the hypothesized value does not fall in the confidence interval $\mu_1 - \mu_2 \leq (\overline{Y}_1 - \overline{Y}_2) + t_{(\alpha,d.f)}\sqrt{\frac{S_P^2}{n_1} + \frac{S_P^2}{n_2}}$, which means we reject the null hypothesis if $\mu_1 - \mu_2 > (\overline{Y}_1 - \overline{Y}_2) + t_{(\alpha,d.f)}\sqrt{\frac{S_P^2}{n_1} + \frac{S_P^2}{n_2}}$.

Example 11.4: The concentration of arsenic in cockles: Example 9.7 is reproduced "A professor at an environmental section wanted to verify the claim that the mean concentration of arsenic (As) in cockles (mg/L) obtained from region A is less than the concentration of arsenic in cockles obtained from region B. He selected 13 samples from each region and tested for the concentration of arsenic. The collected data showed that the mean concentration of arsenic obtained from region A is 2.66 and the standard deviation is 0.15, while the collected data for region B showed that the mean concentration of arsenic is 2.69 and the standard deviation is 0.13. A significance level of $\alpha = 0.01$ is chosen to test the claim. Assume that the two populations are normally distributed, and the variances of the populations are unequal $(\sigma_1^2 \neq \sigma_2^2)$."

The four steps for performing hypothesis testing employing the confidence interval procedure can be used to test the hypothesis regarding the difference between the mean concentration of arsenic in cockles obtained from the two regions A and B. The result of the confidence interval procedure will be compared with the critical value (traditional) and *P*-value procedures.

The confidence interval procedure

Step 1: Specify the null and alternative hypotheses

The two hypotheses regarding the mean concentration of arsenic in cockles obtained from the two regions A and B are presented in Eq. (11.14).

$$H_0{:}\mu_1 \geq \mu_2 \text{ vs } H_1{:}\mu_1 < \mu_2 \tag{11.14}$$

We should make a decision regarding the null hypothesis as to whether the mean concentration of arsenic in cockles obtained from region A (μ_1) is greater than or equal to the mean concentration of arsenic in cockles obtained from region B (μ_2) or μ_1 is less than μ_2.

Step 2: Select the significance level (α) for the study

The t critical value for a one-tailed test with a significance level of 0.01 ($\alpha = 0.01$) is $t_{(\alpha,d.f)} = t_{(0.01,12)} = 2.681$ as appeared in the t table for critical values (Table B in the Appendix).

Step 3: Use the sample information to calculate the test statistic value

The hypothesis in Eq. (11.14) represents a one-tailed test. Thus, the upper confidence interval formula presented in Eq. (11.9) should be used to calculate the confidence interval for the mean concentration of arsenic in cockles obtained from regions A and B because of the alternative hypothesis of the form <, this formula represents the alternative hypothesis for a one-sided test.

The entries for the confidence interval formula as presented in Eq. (11.9) are already provided. The mean concentration of arsenic in cockles obtained from region A is 2.66 and the standard deviation is 0.15, while the mean concentration of arsenic in cockles obtained from region B is 2.69 and the standard deviation is 0.13. The sample size is 13 for each region.

$$\mu_1 - \mu_2 \leq \left(\overline{Y}_1 - \overline{Y}_2\right) + t_{(\alpha,d.f)}\sqrt{\frac{S_1^2}{n_1} + \frac{S_2^2}{n_2}}$$

$$\mu_1 - \mu_2 \leq (2.66 - 2.69) + (-2.681)\sqrt{\frac{(0.15)^2}{13} + \frac{(0.13)^2}{13}}$$

$$\mu_1 - \mu_2 \leq (-0.03) + 0.148$$

$$\mu_1 - \mu_2 \leq 0.118$$

Step 4: Make a decision using the confidence interval and interpret the results

Making a decision using the confidence interval procedure requires checking the confidence interval as to whether it contains the hypothesized (claimed) value or not. It can be seen that the hypothesized value ($\mu_1 - \mu_2 = 0$) is in the calculated confidence interval $\mu_1 - \mu_2 \leq 0.118$.

1. The decision reached by the confidence interval procedure: fail to reject the null hypothesis and we believe that the mean concentration of arsenic in cockles obtained from region A (μ_1) is greater than or equal to the mean concentration of arsenic in cockles obtained from region B (μ_2).

   The result of the confidence interval is compared with the decisions reached by the critical and *P*-value procedures:
2. The critical value (traditional) procedure: fail to reject the null hypothesis as presented in Example 9.7.
3. The *P*-value procedure: fail to reject the null hypothesis as presented in Example 9.7.

We can say that the three procedures have reached the same decision: Fail to reject the null hypothesis at 1% and we believe that the mean concentration of arsenic in cockles obtained from region A (μ_1) is greater than or equal to the mean concentration of arsenic in cockles obtained from region B (μ_2).

Example 11.5: The concentration of chromium in sediment: Example 9.8 is reproduced "The concentration of chromium (Cr) (mg/L) in sediment was investigated to verify the claim that the mean concentration of chromium in Jejawi River is more than the mean concentration of chromium in Juru River. Ten samples were selected, and the concentration of chromium was measured. The collected data showed that the mean concentration of chromium of Jejawi River is 0.17 and the standard deviation is 0.034, while the collected data for Juru River showed that the mean concentration of chromium is 0.12 and the standard deviation is 0.029. A significance level of $\alpha = 0.01$ is chosen to test the claim. Assume that the two populations are normally distributed, and the variances are equal $\sigma_1^2 = \sigma_2^2$."

The four steps for performing hypothesis testing employing the confidence interval procedure can be used to test the hypothesis regarding the difference between the mean concentration of chromium in sediment obtained from Jejawi and Juru Rivers. The result of the confidence interval procedure will be compared with the critical value (traditional) and *P*-value procedures.

The confidence interval procedure

Step 1: Specify the null and alternative hypotheses

The two hypotheses regarding the mean concentration of chromium in sediment obtained from Jejawi and Juru Rivers are presented in Eq. (11.15).

$$H_0 : \mu_1 \leq \mu_2 \text{ vs } H_1 : \mu_1 > \mu_2 \tag{11.15}$$

We should make a decision regarding the null hypothesis as to whether the mean concentration of chromium in sediment obtained from Jejawi River (μ_1) is less than or equal to the mean concentration of chromium in sediment obtained from Juru River (μ_2) or μ_1 is greater than μ_2.

Step 2: Select the significance level (α) for the study

The t critical value for a one-tailed test with a significance level of 0.01 ($\alpha = 0.01$) is $t_{(\alpha,d.f)} = t_{(0.01,18)} = 2.552$ as appeared in the t table for critical values (Table B in the Appendix).

Step 3: Use the sample information to calculate the test statistic value

The hypothesis in Eq. (11.15) represents a one-tailed t-test. Thus, the lower confidence interval formula presented in Eq. (11.12) should be used to calculate the confidence interval for the mean concentration of chromium in sediment obtained from Jejawi and Juru Rivers because of the alternative hypothesis of the form $>$, this formula represents the alternative hypothesis for a one-sided test.

The entries for the confidence interval formula as presented in Eq. (11.12) are already provided. The mean concentration of chromium in sediment obtained from Jejawi River is 0.17 and the standard deviation is 0.034, while the mean concentration of chromium in sediment obtained from Juru River is 0.12 and the standard deviation is 0.029. The sample size is 10 for each river.

$$(\overline{Y}_1 - \overline{Y}_2) - t_{(\alpha,d.f)}\sqrt{\frac{S_P^2}{n_1} + \frac{S_P^2}{n_2}} \leq \mu_1 - \mu_2$$

First, we should calculate the pooled variance S_P^2 using the formula presented in Eq. (11.11).

$$S_P^2 = \frac{(n_1 - 1)S_1^2 + (n_2 - 1)S_2^2}{n_1 + n_2 - 2}$$

$$S_P^2 = \frac{(10 - 1)(0.034)^2 + (10 - 1)(0.029)^2}{10 + 10 - 2}$$

$$S_P^2 = 0.001156$$

$$(0.17 - 0.12) - 2.552\sqrt{\frac{0.001156}{10} + \frac{0.001156}{10}} \leq \mu_1 - \mu_2$$

$$0.05 - 0.04090891 \leq \mu_1 - \mu_2$$

$$0.009 \leq \mu_1 - \mu_2$$

Step 4: Make a decision using the confidence interval and interpret the results

Making a decision using the confidence interval procedure requires checking the confidence interval as to whether it contains the hypothesized (claimed) value or not. It can be seen that the hypothesized value ($\mu_1 - \mu_2 = 0$) is not in the calculated confidence interval $0.009 \leq \mu_1 - \mu_2$.

1. The decision reached by the confidence interval procedure: reject the null hypothesis and we believe that the mean concentration of chromium in sediment obtained from Jejawi River (μ_1) is greater than the mean concentration of chromium in sediment obtained from Juru River (μ_2).

   The result of the confidence interval is compared with the decisions reached by the critical value and *P*-value procedures:
2. The critical value (traditional) procedure: reject the null hypothesis as presented in Example 9.8.
3. The *P*-value procedure: reject the null hypothesis as presented in Example 9.8.

We can say that the three procedures have reached the same decision: reject the null hypothesis at 1% and we believe that the mean concentration of chromium in sediment obtained from Jejawi River (μ_1) is greater than the mean concentration of chromium in sediment obtained from Juru River (μ_2).

Example 11.6: The concentration of cobalt in two ponds: Example 9.9 is reproduced "A professor wants to study the concentration of cobalt (Co) in two ponds of landfill leachate to verify the claim that the mean concentration of cobalt (mg/L) in the collection pond is equal to the mean concentration of cobalt in the aeration pond. Eleven samples were selected, and the concentration of cobalt was measured. The collected data showed that the mean concentration of cobalt of the collection pond is 53.75 and the standard deviation is 46.51, while the collected data from the aeration pond showed that the mean concentration of cobalt is 16.55 and the standard deviation is 14.21. A significance level of $\alpha = 0.01$ is chosen to test the claim. Assume that the two populations are normally distributed, and the variances are equal $\sigma_1^2 = \sigma_2^2$."

The four steps for performing hypothesis testing employing the confidence interval procedure can be used to test the hypothesis regarding the mean concentration of cobalt in the collection and aeration ponds. The result of the confidence interval procedure will be compared with the critical value (traditional) and *P*-value procedures.

The confidence interval procedure

Step 1: Specify the null and alternative hypotheses

The two hypotheses regarding the mean concentration of cobalt obtained from the collection and aeration ponds are presented in Eq. (11.16).

$$H_0{:}\mu_1 = \mu_2 \text{ vs } H_1{:}\mu_1 \neq \mu_2 \tag{11.16}$$

We should make a decision regarding the null hypothesis as to whether the mean concentration of cobalt obtained from the collection pond (μ_1) equals the mean concentration of cobalt obtained from the aeration pond (μ_2) or μ_1 differs from μ_2.

Step 2: Select the significance level (α) for the study

The t critical value for a two-tailed test with a significance level of 0.01 ($\alpha = 0.01$), $\frac{\alpha}{2} = \frac{0.01}{2} = 0.005$ and $d.f = 11 + 11 - 2 = 20$ is $t_{(\frac{\alpha}{2}, d.f)} = t_{(0.005,20)} = 2.845$ as appeared in the t table for critical values (Table B in the Appendix).

Step 3: Use the sample information to calculate the test statistic value

The hypothesis in Eq. (11.16) represents a two-tailed t-test. Thus, the two-tailed confidence interval formula presented in Eq. (11.10) should be used to calculate the confidence interval for the mean concentration of cobalt in the collection and aeration ponds because of the alternative hypothesis of the form $\neq$, this formula represents the alternative hypothesis for a one-sided test.

The entries for the confidence interval formula as presented in Eq. (11.10) are already provided. The mean concentration of cobalt in the collection pond is 53.75 and the standard deviation is 46.51, while the mean concentration of cobalt in the aeration pond is 16.55 and the standard deviation is 14.21. The sample size is 11 for each pond.

$$(\overline{Y}_1 - \overline{Y}_2) - t_{(\frac{\alpha}{2},d.f)}\sqrt{\frac{S_P^2}{n_1} + \frac{S_P^2}{n_2}} \le \mu_1 - \mu_2 \le (\overline{Y}_1 - \overline{Y}_2) + t_{(\frac{\alpha}{2},d.f)}\sqrt{\frac{S_P^2}{n_1} + \frac{S_P^2}{n_2}}$$

First, we should calculate the pooled variance S_P^2 using the formula presented in Eq. (11.11).

$$S_P^2 = \frac{(n_1 - 1)S_1^2 + (n_2 - 1)S_2^2}{n_1 + n_2 - 2} = \frac{(11 - 1)(46.51)^2 + (11 - 1)(14.21)^2}{11 + 11 - 2}$$

$$S_P^2 = 2163.18$$

$$(53.75 - 16.55) - 2.845\sqrt{\frac{2163.18}{11} + \frac{2163.18}{11}} \le \mu_1 - \mu_2 \le (53.75 - 16.55)$$

$$+ 2.845\sqrt{\frac{2163.18}{11} + \frac{2163.18}{11}}$$

$$37.2 - 52.58168 \le \mu_1 - \mu_2 \le 37.2 - 52.58168$$

$$-15.382 \le \mu_1 - \mu_2 \le 89.782$$

Step 4: Make a decision using the confidence interval and interpret the results

Making a decision using the confidence interval procedure requires checking the confidence interval as to whether it contains the hypothesized (claimed) value or not. It can be seen that the hypothesized value ($\mu_1 - \mu_2 = 0$) is in the calculated confidence interval $-15.382 \le \mu_1 - \mu_2 \le 89.782$.

1. The decision reached by the confidence interval procedure: fail to reject the null hypothesis and we believe that the mean concentration of cobalt in the collection pond (μ_1) is equal to the mean concentration of cobalt in the aeration pond (μ_2).

   The result of the confidence interval is compared with the decisions reached by the critical value and P-value:
2. The critical value (traditional) procedure: fail to reject the null hypothesis as presented in Example 9.9.
3. The P-value procedure: fail to reject the null hypothesis as presented in Example 9.9.

We can say that the three procedures have reached the same decision which is fail to reject the null hypothesis at 1% and we believe that the mean concentration of cobalt in the collection pond (μ_1) is equal to the mean concentration of cobalt in the aeration pond (μ_2).

11.5 Confidence interval for testing two dependent samples

Hypothesis testing for two dependent samples (the difference between the mean of the first condition and the second condition) can be tested employing the confidence interval for two dependent (matched, paired) samples to test various hypotheses including one-tailed and two-tailed tests. A step-by-step procedure will be employed to illustrate the steps of making a decision concerning each hypothesis, including left-tailed, right-tailed, and two-tailed.

The two-tailed and one-tailed confidence interval formulas for two dependent samples are:

- The two-tailed confidence interval for two dependent samples is given in Eq. (11.17).

$$\overline{d} - t_{(\frac{\alpha}{2},n-1)}\left(\frac{S_d}{\sqrt{n}}\right) \le \mu_d \le \overline{d} + t_{(\frac{\alpha}{2},n-1)}\left(\frac{S_d}{\sqrt{n}}\right) \tag{11.17}$$

  or it can be written as $\left[\overline{d} - t_{\frac{\alpha}{2},(n-1)}\left(\frac{S_d}{\sqrt{n}}\right) \le \mu_d, \overline{d} + t_{(\frac{\alpha}{2},n-1)}\left(\frac{S_d}{\sqrt{n}}\right)\right]$.

  Do not reject the null hypothesis if the hypothesized (claimed) value falls in the confidence interval and reject the null hypothesis if the hypothesized (claimed) value does not fall in the confidence interval which indicates that the result is statistically significant.

- The one-tailed confidence intervals for two dependent samples are given in Eqs. (11.18) and (11.19).

  The lower one-tailed confidence interval for two dependent samples is given in Eq. (11.18).

$$\overline{d} - t_{(\alpha,n-1)}\left(\frac{S_d}{\sqrt{n}}\right) \le \mu_d \tag{11.18}$$

  or it can be written as $\left[\overline{d} - t_{(\alpha,n-1)}\left(\frac{S_d}{\sqrt{n}}\right), \infty\right]$.

  Reject the null hypothesis if the hypothesized (claimed) value does not fall in the confidence interval $\overline{d} - t_{(\alpha,n-1)}\left(\frac{S_d}{\sqrt{n}}\right) \le \mu_d$, which means we reject the null hypothesis if $\overline{d} - t_{(\alpha,n-1)}\left(\frac{S_d}{\sqrt{n}}\right) > \mu_d$.

  The upper one-tailed confidence interval for two dependent samples is given in Eq. (11.19).

$$\mu_d \le \overline{d} - t_{(\alpha,n-1)}\left(\frac{S_d}{\sqrt{n}}\right) \tag{11.19}$$

  or it can be written as $[-\infty, \overline{d} - t_{(\alpha,n-1)}\left(\frac{S_d}{\sqrt{n}}\right)]$.

  Reject the null hypothesis if the hypothesized value does not fall in the confidence interval $\mu_d \le \overline{d} - t_{(\alpha,n-1)}\left(\frac{S_d}{\sqrt{n}}\right)$, which means we reject the null hypothesis if $\mu_d > \overline{d} - t_{(\alpha,n-1)}\left(\frac{S_d}{\sqrt{n}}\right)$.

Example 11.7: The skills in measuring the concentration of zinc: Example 9.10 is reproduced "A professor at an environmental department wanted to determine the difference between two students in his research team regarding the lab skills for measuring the concentration of zinc (Zn) in water. He believes that the students have different skills. He randomly selected 11 samples and half of each sample was given to each student to measure the concentration of zinc, the results are given in Table 9.1. A significance level of $\alpha = 0.01$ is chosen to test the claim. Assume that the population is normally distributed."

The four steps for performing hypothesis testing employing the confidence interval procedure can be used to test the hypothesis regarding the mean difference of skills between the two students in measuring the concentration of zinc. The result of the confidence interval procedure will be compared with the critical value (traditional) and P-value procedures.

The confidence interval procedure

Step 1: Specify the null and alternative hypotheses

The two hypotheses regarding the mean difference of skills between the two students are presented in Eq. (11.20).

$$H_0{:}\mu_d = 0 \text{ vs } H_1{:}\mu_d \neq 0 \tag{11.20}$$

We should make a decision regarding the null hypothesis as to whether the two students have equal skills in measuring the concentration of zinc or they are different.

Step 2: Select the significance level (α) for the study

The t critical value for the two-tailed test with a significance level of 0.01 ($\alpha = 0.01$), $\frac{\alpha}{2} = \frac{0.01}{2} = 0.005$ and $d.f = 10$ is 3.169 as appeared in the t table for critical values (Table B in the Appendix). Thus, the t critical value is $t_{(\frac{\alpha}{2},d.f)} = t_{(\frac{0.01}{2},10)} = 3.169$.

Step 3: Use the sample information to calculate the test statistic value

The hypothesis in Eq. (11.20) represents a two-tailed test. Thus, the two-tailed confidence interval formula presented in Eq. (11.17) should be used to calculate the confidence interval for the mean difference of skills between the two students because the alternative hypothesis of the form $\neq$, this formula represents the alternative hypothesis for a two-sided test.

The entries for the confidence interval formula as presented in Eq. (11.17) are already provided. The mean of the differences between the two students is 0.018, the standard deviation of the differences is 0.0098, and the sample size is 11.

$$\overline{d} - t_{(\frac{\alpha}{2},n-1)}\left(\frac{S_d}{\sqrt{n}}\right) \leq \mu_d \leq \overline{d} + t_{(\frac{\alpha}{2},n-1)}\left(\frac{S_d}{\sqrt{n}}\right)$$

We know from Example 9.10 that $\overline{d} = 0.018$ and $S_d = 0.0098$. Thus, the confidence interval is:

$$0.018 - 3.169\left(\frac{0.0098}{\sqrt{11}}\right) \leq \mu_d \leq 0.018 + 3.169\left(\frac{0.0098}{\sqrt{11}}\right)$$

$$0.018 - 0.009364602 \le \mu_d \le 0.018 + 0.009364602$$

$$0.009 \le \mu_d \le 0.027$$

Step 4: Make a decision using the confidence interval and interpret the results

Making a decision using the confidence interval procedure requires checking the confidence interval as to whether it contains the hypothesized (claimed) value or not. It can be seen that the hypothesized value ($\mu_d = 0$) is not in the calculated confidence interval $0.009 \le \mu_d \le 0.027$.

1. The decision reached by the confidence interval procedure: reject the null hypothesis and we believe that the two students have different skills in measuring the concentration of Zn.

   The result of the confidence interval is compared with the decisions reached by the critical value and *P*-value procedures:
2. The critical value (traditional) procedure: reject the null hypothesis as presented in Example 9.10.
3. The *P*-value procedure: reject the null hypothesis as presented in Example 9.10

We can say that the three procedures have reached the same decision: reject the null hypothesis at 1% and we believe that the two students have different skills in measuring the concentration of zinc.

Example 11.8: The concentration of lead in cockles: Example 9.11 is reproduced "A scientist wanted to determine the difference between two laboratories' equipment to measure the lead (Pb) concentration in cockles (mg/L). He believed that the mean concentration of lead measured by laboratory A is higher than the mean concentration of lead in cockles measured by laboratory B. He randomly selected 12 samples and half of each sample was sent to each laboratory. The data showed that the mean and standard deviation for the differences are 0.13 and 0.03, respectively. A significance level of $\alpha = 0.01$ is chosen to test the claim. Assume that the population is normally distributed."

The four steps for performing hypothesis testing employing the confidence interval procedure can be used to test the hypothesis regarding the difference between two laboratories' equipment to measure the lead concentration in cockles. The result of the confidence interval procedure will be compared with the critical value (traditional) and *P*-value procedures.

The confidence interval procedure

Step 1: Specify the null and alternative hypotheses

The two hypotheses regarding the difference between two laboratories' equipment to measure the lead concentration in cockles are presented in Eq. (11.21).

$$H_0: \mu_d \le 0 \text{ vs } H_1: \mu_d > 0 \quad (11.21)$$

We should make a decision regarding the null hypothesis as to whether the difference between the two laboratories is less than or equal to 0 (this means laboratory A provides lower records than laboratory B) or the difference is more than 0 (this means laboratory A provides higher records than laboratory B) regarding the concentration of lead in cockles.

Step 2: Select the significance level (α) for the study

The t critical value for a one-tailed test with a significance level of 0.01 ($\alpha = 0.01$) and $d.f = 12 - 1 = 11$ is 2.718 ($t_{(\alpha,d.f)} = t_{(0.01,11)} = 2.718$) as appeared in the t table for critical values (Table B in the Appendix). Thus, the t critical value for the right-tailed test is 2.718.

Step 3: Use the sample information to calculate the test statistic value

The hypothesis in Eq. (11.21) represents a one-tailed test. Thus, the lower one-tailed confidence interval formula presented in Eq. (11.18) should be used to calculate the confidence interval for the difference between two laboratories' equipment to measure the lead concentration in cockles because of the alternative hypothesis of the form $>$, this formula represents the alternative hypothesis for a one-sided test.

The entries for the confidence interval formula as presented in Eq. (11.18) are already provided. The mean of the differences between the two laboratories' equipment is 0.13, the standard deviation of the differences is 0.03, and the sample size is 12.

$$\bar{d} - t_{(\alpha,n-1)}\left(\frac{S_d}{\sqrt{n}}\right) \leq \mu_d$$

$$0.13 - 2.718\left(\frac{0.03}{\sqrt{12}}\right) \leq \mu_d$$

$$0.13 - 0.02689707 \leq \mu_d$$

$$0.103 \leq \mu_d$$

Step 4: Make a decision using the confidence interval and interpret the results

Making a decision using the confidence interval procedure requires checking the confidence interval as to whether it contains the hypothesized (claimed) value or not. It can be seen that the hypothesized value ($\mu_d = 0$) is not in the calculated confidence interval $0.103 \leq \mu_d$.

1. The decision reached by the confidence interval procedure: reject the null hypothesis and we believe that the two laboratories provide different results (laboratory A provides higher records than laboratory B).

   The result of the confidence interval is compared with the decisions reached by the critical value and *P*-value procedures:
2. The critical value (traditional) procedure: reject the null hypothesis as presented in Example 9.11.
3. The *P*-value procedure: reject the null hypothesis as presented in Example 9.11.

We can say that the three procedures have reached the same decision: reject the null hypothesis at 1% and we believe that the two laboratories provide different results, laboratory A provides higher records than laboratory B.

Example 11.9: The sales plan for a new coagulant: Example 9.12 is reproduced "An environmental company provided a new coagulant for treating wastewater. The company decided to start an intensive campaign to promote the new product. The sales before and after the intensive campaign covering 11 markets were recorded in thousands of dollars for a 1-month period, the data showed that the mean and standard deviation for the differences are -1.80 and 0.92, respectively. Can it be concluded that the sales before the advertising campaign are less than the sales after the campaign? A significance level of $\alpha = 0.01$ is chosen to test the claim. Assume that the population is normally distributed."

The four steps for performing hypothesis testing employing the confidence interval procedure can be used to test the hypothesis regarding the mean difference of sales of coagulant before and after the intensive campaign covering 11 markets. The result of the confidence interval procedure will be compared with the critical value (traditional) and *P*-value procedures.

The confidence interval procedure

Step 1: Specify the null and alternative hypotheses

The two hypotheses regarding the mean difference of sales of coagulants before and after the campaign covering 11 markets are presented in Eq. (11.22).

$$H_0{:}\mu_d \geq 0 \text{ vs } H_1{:}\mu_d < 0 \tag{11.22}$$

We should make a decision regarding the null hypothesis as to whether the sales before the campaign are more than or equal to the sales after the campaign regarding the new coagulant sales or the sales before the campaign are less than after the campaign.

Step 2: Select the significance level (α) for the study

The t critical value for a one-tailed test with a significance level of 0.01 ($\alpha = 0.01$) and $d.f = 11 - 1 = 10$ is 2.764 ($t_{(\alpha,d.f)} = t_{(0.01,10)} = 2.764$) as appeared in the t table for critical values (Table B in the Appendix).

Step 3: Use the sample information to calculate the test statistic value

The hypothesis in Eq. (11.22) represents a one-tailed test. Thus, the upper one-tailed confidence interval formula presented in Eq. (11.19) should be used to calculate the confidence interval for the mean difference in sales of the new coagulant before and after the campaign covering 11 markets because of the alternative hypothesis of the form >, this formula represents the alternative hypothesis for a one-sided test.

The entries for the confidence interval formula as presented in Eq. (11.19) are already provided. The mean of the differences in sales of new coagulant before and after the campaign is -1.80, the standard deviation of the differences is 0.92 and the sample size is 11.

$$\mu_d \leq \bar{d} - t_{(\alpha,n-1)}\left(\frac{S_d}{\sqrt{n}}\right)$$

$$\mu_d \leq -1.80 - 2.764\left(\frac{0.92}{\sqrt{11}}\right)$$

$\mu_d \leq -1.80 - 0.7666432$

$\mu_d \leq -1.033$

Step 4: Make a decision using the confidence interval and interpret the results

Making a decision using the confidence interval procedure requires checking the confidence interval as to whether it contains the hypothesized (claimed) value or not. It can be seen that the hypothesized value ($\mu_d = 0$) is not in the calculated confidence interval $\mu_d \leq -1.033$.

1. The decision reached by confidence interval procedure: reject the null hypothesis and we believe that the sales before the campaign were less than the sales after the campaign regarding the new coagulant.

   The result of the confidence interval is compared with the decisions reached by the critical value and *P*-value procedures:
2. The critical value (traditional) procedure: reject the null hypothesis as presented in Example 9.11.
3. The *P*-value procedure: reject the null hypothesis as presented in Example 9.11.

We can say that the three procedures have reached the same decision which is reject the null hypothesis at 1% and we believe that the mean sales before the campaign are less than after the campaign regarding the new coagulant.

11.6 Confidence interval for testing the difference between two proportions

Hypothesis testing for the difference between two population proportions can be tested using the confidence interval including one-tailed and two-tailed tests. A step-by-step procedure will be employed to illustrate the steps of making a decision concerning each hypothesis, including left-tailed, right-tailed, and two-tailed.

The two-tailed and one-tailed confidence intervals for the difference between two population proportions are:

- The two-tailed confidence interval for the difference between two population proportions is given in Eq. (11.23).

$$(\hat{p}_1 - \hat{p}_2) - Z_{\frac{\alpha}{2}}\sqrt{\frac{\hat{p}_1(1-\hat{p}_1)}{n_1} + \frac{\hat{p}_2(1-\hat{p}_2)}{n_2}} \leq p_1 - p_2 \leq (\hat{p}_1 - \hat{p}_2) + Z_{\frac{\alpha}{2}}\sqrt{\frac{\hat{p}_1(1-\hat{p}_1)}{n_1} + \frac{\hat{p}_2(1-\hat{p}_2)}{n_2}} \tag{11.23}$$

or it can be written as

$$\left[(\hat{p}_1 - \hat{p}_2) - Z_{\frac{\alpha}{2}}\sqrt{\frac{\hat{p}_1(1-\hat{p}_1)}{n_1} + \frac{\hat{p}_2(1-\hat{p}_2)}{n_2}}, (\hat{p}_1 - \hat{p}_2) + Z_{\frac{\alpha}{2}}\sqrt{\frac{\hat{p}_1(1-\hat{p}_1)}{n_1} + \frac{\hat{p}_2(1-\hat{p}_2)}{n_2}}\right].$$

Do not reject the null hypothesis if the hypothesized (claimed) value falls in the confidence interval and reject the null hypothesis if the hypothesized (claimed) value does not fall in the confidence interval which indicates that the result is statistically significant.

- The one-tailed confidence intervals are given in Eqs. (11.24) and (11.25).

The lower one-tailed confidence interval for the difference between two population proportions is given in Eq. (11.24).

$$(\hat{p}_1 - \hat{p}_2) - Z_\alpha \sqrt{\frac{\hat{p}_1(1-\hat{p}_1)}{n_1} + \frac{\hat{p}_2(1-\hat{p}_2)}{n_2}} \leq p_1 - p_2 \tag{11.24}$$

or it can be written as $\left[\left(\overline{Y}_1 - \overline{Y}_2\right) - Z_\alpha \sqrt{\frac{\sigma_1^2}{n_1} + \frac{\sigma_2^2}{n_2}}, \infty\right]$.

Reject the null hypothesis if the hypothesized (claimed) value does not fall in the confidence interval $(\hat{p}_1 - \hat{p}_2) - Z_\alpha \sqrt{\frac{\hat{p}_1(1-\hat{p}_1)}{n_1} + \frac{\hat{p}_2(1-\hat{p}_2)}{n_2}} \leq p_1 - p_2$, which means we reject the null hypothesis if $p_1 - p_2 < (\hat{p}_1 - \hat{p}_2) - Z_\alpha \sqrt{\frac{\hat{p}_1(1-\hat{p}_1)}{n_1} + \frac{\hat{p}_2(1-\hat{p}_2)}{n_2}}$.

The upper one-tailed confidence interval for the difference between two population proportions is given in Eq. (11.25)

$$p_1 - p_2 \leq (\hat{p}_1 - \hat{p}_2) + Z_\alpha \sqrt{\frac{\hat{p}_1(1-\hat{p}_1)}{n_1} + \frac{\hat{p}_2(1-\hat{p}_2)}{n_2}} \tag{11.25}$$

or it can be written as $[-\infty, (\hat{p}_1 - \hat{p}_2) + Z_\alpha \sqrt{\frac{\hat{p}_1(1-\hat{p}_1)}{n_1} + \frac{\hat{p}_2(1-\hat{p}_2)}{n_2}}]$.

Reject the null hypothesis if the hypothesized value does not fall in the confidence interval $p_1 - p_2 \leq (\hat{p}_1 - \hat{p}_2) + Z_\alpha \sqrt{\frac{\hat{p}_1(1-\hat{p}_1)}{n_1} + \frac{\hat{p}_2(1-\hat{p}_2)}{n_2}}$, which means we reject the null hypothesis if $p_1 - p_2 > (\hat{p}_1 - \hat{p}_2) + Z_\alpha \sqrt{\frac{\hat{p}_1(1-\hat{p}_1)}{n_1} + \frac{\hat{p}_2(1-\hat{p}_2)}{n_2}}$.

Example 11.10: The concentration of PM_{10} in air (dry and wet seasons): Example 9.13 is reproduced "A research team wants to investigate the claim regarding the proportion of PM_{10} in the air which says that PM_{10} is higher during the dry season than during the wet season. The collected data during the dry season showed that 14 samples out of 100 showed high PM_{10} and the collected data during the wet season showed that two out of 100 showed high PM_{10}. A significance level of $\alpha = 0.01$ is chosen to test the claim. Assume that the two populations are normally distributed."

The four steps for performing hypothesis testing employing the confidence interval procedure can be used to test the hypothesis regarding the difference in the proportion of PM_{10} in the air during the dry and wet seasons. The result of the confidence interval procedure will be compared with the critical value (traditional) and P-value procedures.

The confidence interval procedure

Step 1: Specify the null and alternative hypotheses

The two hypotheses (null and alternative) for the difference in the proportion of PM_{10} in the air during the dry and wet seasons are presented in Eq. (11.26).

$$H_0:\hat{p}_1 \le \hat{p}_2 \text{ vs } H_1:\hat{p}_1 > \hat{p}_2 \tag{11.26}$$

We should make a decision regarding the null hypothesis as to whether the proportion of PM_{10} in the air during the dry season ($\hat{p}_1$) is less than or equal to the proportion of PM_{10} in the air during the wet season ($\hat{p}_2$) or $\hat{p}_1$ is more than $\hat{p}_2$.

Step 2: Select the significance level (α) for the study

The Z critical value for a one-tailed test with a significance level of 0.01 ($\alpha = 0.01$) is $Z_\alpha = Z_{0.01} = 2.326$as appeared in the Z table (Table A in the Appendix).

Step 3: Use the sample information to calculate the test statistic value

The hypothesis in Eq. (11.26) represents a one-tailed test. Thus, the lower confidence interval formula presented in Eq. (11.24) should be used to calculate the confidence interval for the difference in the proportion of PM_{10} in the air during the dry season ($\hat{p}_1$) and the wet season ($\hat{p}_2$) because of the alternative hypothesis of the form $>$, this formula represents the alternative hypothesis for a one-sided test.

The entries for the confidence interval formula as presented in Eq. (11.24) are already provided. The proportion of PM_{10} in the air during the dry season ($\hat{p}_1$)is 0.14, while the proportion of PM_{10} in the air during the wet season ($\hat{p}_2$)is 0.02, and the sample size is 100 for each season.

$$(\hat{p}_1 - \hat{p}_2) - Z_\alpha \sqrt{\frac{\hat{p}_1(1-\hat{p}_1)}{n_1} + \frac{\hat{p}_2(1-\hat{p}_2)}{n_2}} \le p_1 - p_2$$

$$(0.14 - 0.02) - 2.326\sqrt{\frac{0.14(1-0.14)}{100} + \frac{0.02(1-0.02)}{100}} \le p_1 - p_2$$

$$0.12 - 0.08704397 \le p_1 - p_2$$

$$0.033 \le p_1 - p_2$$

Step 4: Make a decision using the confidence interval and interpret the results

Making a decision using the confidence interval procedure requires checking the confidence interval as to whether it contains the hypothesized (claimed) value or not. It can be seen that the hypothesized value ($p_1 - p_2 = 0$) is not in the calculated confidence interval $0.033 \le p_1 - p_2$.

1. The decision reached by the confidence interval procedure: reject the null hypothesis and we believe that the proportion of samples that exceed the permissible limits of PM_{10} in the air during the dry season ($\hat{p}_1$) is greater than the proportion of PM_{10} in the air during the wet season ($\hat{p}_2$).

The result of the confidence interval is compared with the critical value and *P*-value procedures:

2. The critical value (traditional) procedure: reject the null hypothesis as presented in Example 9.13.
3. The *P*-value procedure: reject the null hypothesis as presented in Example 9.13.

We can say that the three procedures have reached the same decision: reject the null hypothesis at 1% and we believe that the proportion of samples that exceed the permissible limits of PM_{10} in the air during the dry season ($\hat{p}_1$) is greater than the proportion of PM_{10} in the air during the wet season ($\hat{p}_2$).

Example 11.11: The level of turbidity in two locations: Example 9.14 is reproduced. "A research team wants to investigate the claim regarding the equality of the proportion of high turbidity in water of a dam and before the dam. The collected data for the water samples before the dam showed that 11 samples out of 80 showed high turbidity and collected data for the water samples from the dam showed that 10 out of 80 showed high turbidity. A significance level of $\alpha = 0.01$ is chosen to test the claim. Assume that the two populations are normally distributed."

The four steps for performing hypothesis testing employing the confidence interval procedure can be used to test the hypothesis regarding the difference in the proportion of sample that showed high turbidity of water before the dam ($\hat{p}_1$) and at the dam ($\hat{p}_2$). The result of the confidence interval procedure will be compared with the critical value (traditional) and *P*-value procedures.

The confidence interval procedure

Step 1: Specify the null and alternative hypotheses

The two hypotheses (null and alternative) for the difference in the proportion of samples that showed high turbidity of water before the dam ($\hat{p}_1$) and at the dam ($\hat{p}_2$). are presented in Eq. (11.27).

$$H_0{:}\hat{p}_1 = \hat{p}_2 \text{ vs } H_1{:}\hat{p}_1 \neq \hat{p}_2 \tag{11.27}$$

We should make a decision regarding the null hypothesis as to whether the proportion of high turbidity before the dam ($\hat{p}_1$) is equal to the proportion of high turbidity at the dam ($\hat{p}_2$), or $\hat{p}_1$ differs from $\hat{p}_2$.

Step 2: Select the significance level (α) for the study

The Z critical value for a two-tailed test with a significance level of $\alpha = 0.01$, $\frac{\alpha}{2} = \frac{0.01}{2} = 0.005$ is $Z_{\frac{\alpha}{2}} = Z_{0.005} = 2.58$ as appeared in the standard normal table (Table A in the Appendix).

Step 3: Use the sample information to calculate the test statistic value

The hypothesis in Eq. (11.27) represents a two-tailed test. Thus, the two-tailed confidence interval formula presented in Eq. (11.23) should be used to calculate the confidence interval for the difference in the proportion of high turbidity before the dam ($\hat{p}_1$) and the proportion of high turbidity at the dam ($\hat{p}_2$) because of the alternative hypothesis of the form $\neq$, this formula represents the alternative hypothesis for a two-sided test.

The entries for the confidence interval formula as presented in Eq. (11.23) are already provided. The proportion of high turbidity before the dam is 0.14, while the proportion of high turbidity at the dam is 0.12, and the sample size is 80 for each region.

$$(\hat{p}_1 - \hat{p}_2) - Z_{\frac{\alpha}{2}}\sqrt{\frac{\hat{p}_1(1-\hat{p}_1)}{n_1} + \frac{\hat{p}_2(1-\hat{p}_2)}{n_2}} \leq p_1 - p_2 \leq (\hat{p}_1 - \hat{p}_2)$$

$$+ Z_{\frac{\alpha}{2}}\sqrt{\frac{\hat{p}_1(1-\hat{p}_1)}{n_1} + \frac{\hat{p}_2(1-\hat{p}_2)}{n_2}}$$

$$(0.14-0.12)-2.58\sqrt{\frac{0.14(1-0.14)}{80}+\frac{0.12(1-0.12)}{80}} \leq p_1-p_2 \leq (0.14-0.12)$$

$$+ 2.58\sqrt{\frac{0.14(1-0.14)}{80} + \frac{0.12(1-0.12)}{80}}$$

$$0.02 - 0.1369072 \leq p_1 - p_2 \leq 0.02 + 0.1369072$$

$$-0.117 \leq p_1 - p_2 \leq 0.157$$

Step 4: Make a decision using the confidence interval and interpret the results

Making a decision using the confidence interval procedure requires checking the confidence interval as to whether it contains the hypothesized (claimed) value or not. It can be seen that the hypothesized value ($p_1 - p_2 = 0$) is in the calculated confidence interval $-0.117 \leq p_1 - p_2 \leq 0.157$.

1. The decision reached by the confidence interval procedure: fail to reject the null hypothesis and we believe that the proportion of high turbidity before the dam ($\hat{p}_1$) is equal to the proportion of high turbidity at the dam ($\hat{p}_2$).

   The result of the confidence interval is compared with the decisions reached by the critical value and P-value procedures:
2. The critical value (traditional) procedure: fail to reject the null hypothesis as presented in Example 9.14.
3. The P-value procedure: fail to reject the null hypothesis as presented in Example 9.14.

We can say that the three procedures have reached the same decision: Fail to reject the null hypothesis at 1% and we believe that the proportion of high turbidity before the dam ($\hat{p}_1$) is equal to the proportion of high turbidity at the dam ($\hat{p}_2$).

Example 11.12: The concentration of manganese in sediment: Example 9.15 is reproduced "A research team wants to investigate the claim regarding the proportion of samples that exceed the permissible limit of manganese (Mn) in sediment of two rivers. The collected data for sediment samples obtained from Juru River showed that two samples out of 100 exceeded the permissible limits, and the collected data for sediment samples obtained from Jejawi River showed that 15 out of 100 exceeded the permissible limits. A significance level of $\alpha = 0.01$ is chosen to test the claim that Juru River showed a lower proportion than Jejawi River. Assume that the two populations are normally distributed."

The four steps for performing hypothesis testing employing the confidence interval procedure can be used to test the hypothesis regarding the difference in the proportion of manganese samples in sediment of Juru and Jejawi Rivers. The result of the confidence interval procedure will be compared with the critical value (traditional) and P-value procedures.

The confidence interval procedure

Step 1: Specify the null and alternative hypotheses

The two hypotheses (null and alternative) for the proportion of manganese in the sediment of Juru and Jejawi Rivers are presented in Eq. (11.28).

$$H_0{:}\hat{p}_1 \geq \hat{p}_2 \text{ vs } H_1{:}\hat{p}_1 < \hat{p}_2 \tag{11.28}$$

We should make a decision regarding the null hypothesis as to whether the proportion of samples that exceeded the permissible limit regarding the concentration of manganese in sediment for Juru River ($\hat{p}_1$) is more than or equal to the proportion of samples that exceeded the permissible limit for Jejawi River ($\hat{p}_2$), or not.

Step 2: Select the significance level (α) for the study

The Z critical value for a one-tailed test with a significance level of 0.01 ($\alpha = 0.01$) is $Z_\alpha = Z_{0.01} = 2.326$ as appeared in the Z table (Table A in the Appendix).

Step 3: Use the sample information to calculate the test statistic value

The hypothesis in Eq. (11.28) represents a one-tailed test. Thus, the upper confidence interval formula presented in Eq. (11.25) should be used to calculate the confidence interval for the concentration of manganese in the sediment for Juru River ($\hat{p}_1$) and the concentration of manganese in the sediment for Jejawi River ($\hat{p}_2$) because of the alternative hypothesis of the form $<$, this formula represents the alternative hypothesis for a one-sided test.

The entries for the confidence interval formula as presented in Eq. (11.25) are already provided. The proportion of samples obtained from Juru River that exceeded the permissible limits of manganese is $\frac{2}{100} = 0.02$, while the proportion of samples obtained from Jejawi River that exceeded the permissible limits of manganese is $\frac{15}{100} = 0.15$, and the sample size is 100 for each river.

$$p_1 - p_2 \leq (\hat{p}_1 - \hat{p}_2) + Z_\alpha \sqrt{\frac{\hat{p}_1(1-\hat{p}_1)}{n_1} + \frac{\hat{p}_2(1-\hat{p}_2)}{n_2}}$$

$$p_1 - p_2 \leq (0.02 - 0.15) + 2.326\sqrt{\frac{0.02(1-0.02)}{100} + \frac{0.15(1-0.15)}{100}}$$

$$p_1 - p_2 \leq -0.013 + 0.08922386$$

$$p_1 - p_2 \leq -0.014$$

Step 4: Make a decision using the confidence interval and interpret the results

Making a decision using the confidence interval procedure requires checking the confidence interval as to whether it contains the hypothesized (claimed) value or not. It can be seen that the hypothesized value ($p_1 - p_2 = 0$) is not in the calculated confidence interval $p_1 - p_2 \leq -0.014$.

1. The decision reached by the confidence interval procedure: reject the null hypothesis and we believe that the proportion of samples that exceeded the permissible limit regarding the concentration of manganese in sediment for Juru River ($\hat{p}_1$) is less than the proportion of samples that exceeded the permissible limit for Jejawi River ($\hat{p}_2$).

   The result of the confidence interval is compared with the decisions reached by the critical value and P-value procedures:
2. The critical value (traditional) procedure: reject the null hypothesis as presented in Example 9.15.
3. The P-value procedure: reject the null hypothesis as presented in Example 9.15.

We can say that the three procedures have reached the same decision: reject the null hypothesis at 1% and we believe that the concentration of manganese in sediment for Juru River ($\hat{p}_1$) is less than the concentration of manganese in sediment for Jejawi River ($\hat{p}_2$).

11.7 Confidence interval for testing the ratio of two variances

Hypothesis testing for the ratio of two population variances can be tested using the confidence interval including one-tailed and two-tailed tests. A step-by-step procedure will be employed to illustrate the steps of making a decision concerning each hypothesis, including left-tailed, right-tailed, and two-tailed.

The confidence interval formulas for two-tailed and one-tailed tests for the ratio of two variances are:

- The two-tailed confidence interval for the ratio of two variances is given in Eq. (11.29).

$$\frac{1}{F_{(\frac{\alpha}{2},n_1-1,n_2-1)}}\frac{S_1^2}{S_2^2} \leq \frac{\sigma_1^2}{\sigma_2^2} \leq \frac{1}{F_{(1-\frac{\alpha}{2},n_1-1,n_2-1)}}\frac{S_1^2}{S_2^2} \quad (11.29)$$

  Do not reject the null hypothesis if the hypothesized (claimed) value falls in the confidence interval and reject the null hypothesis if the hypothesized (claimed) value does not fall in the confidence interval which indicates that the result is statistically significant.
- The one-tailed confidence interval for the ratio of two variances is given in Eqs. (11.30) and (11.31).

The one-tailed lower confidence interval for the ratio of two variances is given in Eq. (11.30).

$$\frac{1}{F_{(\alpha,n_1-1,n_2-1)}}\frac{S_1^2}{S_2^2} \le \frac{\sigma_1^2}{\sigma_2^2} \tag{11.30}$$

Reject the null hypothesis if the hypothesized (claimed) value does not fall in the confidence interval $\frac{1}{F_{(\alpha,n_1-1,n_2-1)}}\frac{S_1^2}{S_2^2} \le \frac{\sigma_1^2}{\sigma_2^2}$, which means we reject the null hypothesis if $\frac{1}{F_{(\alpha,n_1-1,n_2-1)}}\frac{S_1^2}{S_2^2} > \frac{\sigma_1^2}{\sigma_2^2}$.

The one-tailed upper confidence interval for the ratio of two variances is given in Eq. (11.31).

$$\frac{\sigma_1^2}{\sigma_2^2} \le \frac{1}{F_{(1-\alpha,n_1-1,n_2-1)}}\frac{S_1^2}{S_2^2} \tag{11.31}$$

Reject the null hypothesis if the hypothesized value does not fall in the confidence interval $\frac{\sigma_1^2}{\sigma_2^2} \le \frac{1}{F_{(1-\alpha,n_1-1,n_2-1)}}\frac{S_1^2}{S_2^2}$, which means we reject the null hypothesis if

$$\frac{\sigma_1^2}{\sigma_2^2} > \frac{1}{F_{(1-\alpha,n_1-1,n_2-1)}}\frac{S_1^2}{S_2^2}.$$

Example 11.13: The pH value of surface water before and after the dam: Example 9.16 is reproduced "A professor at an environmental section wanted to verify the claim that the variance of pH of surface water before Beris dam is equal to the pH value of surface water after the dam. He selected 11 samples from each region and tested for pH value. The collected data showed that the variance of pH before the dam is 0.26, while the collected data after the dam showed that the variance of pH is 0.71. A significance level of $\alpha = 0.01$ is chosen to test the claim. Assume that the two populations are normally distributed."

The four steps for performing hypothesis testing employing the confidence interval procedure can be used to test the hypothesis regarding the equality of variance of pH values of surface water before and after the dam. The result of the confidence interval procedure will be compared with the critical value (traditional) and P-value procedures.

The confidence interval procedure

Step 1: Specify the null and alternative hypotheses

The two hypotheses (null and alternative) for the equality of variance of pH values of surface water before and after the dam are presented in Eq. (11.32).

$$H_0{:}\sigma_1^2 = \sigma_2^2 \text{ vs } H_1{:}\sigma_1^2 \ne \sigma_2^2 \tag{11.32}$$

We should make a decision regarding the null hypothesis as to whether the variance of pH of surface water before the dam (σ_1^2) is equal to the variance of pH after the dam (σ_2^2), or the variance of pH before and after the dam are different ($\sigma_1^2 \neq \sigma_2^2$).

Step 2: Select the significance level (α) for the study

The F critical value for a two-tailed test with a significance level of 0.01 ($\alpha = 0.01$) and $d.f_1 = (n_1 - 1) = 11 - 1 = 10$ for numerator and $d.f_2 = (n_2 - 1) = 10$ for denominator can be computed as follows:

The left-tailed value: The F critical value for a left-tailed test for $d.f_1 = 10$, $d.f_2 = 10$, and a significance level of 0.01 ($\alpha = 0.01$), ($1 - \frac{\alpha}{2} = 0.995$) is 0.1710, $F_{(1-\frac{\alpha}{2}, n_1-1, n_2-1)} = F_{(0.995,10,10)} = 0.1710$ as shown in Table D in the Appendix or using software and presented in Fig. 9.13.

The right-tailed value: F critical value for the right-tailed test for $d.f_1 = 10$, $d.f_2 = 10$, and $\frac{\alpha}{2} = 0.005$ is 5.8467 $\left(F_{(\frac{\alpha}{2}, n_1-1, n_2-1)} = F_{(0.005,10,10)} = 5.8467\right)$ as shown in Table D in the Appendix or using software and presented in Fig. 9.13.

Step 3: Use the sample information to calculate the test statistic value

The hypothesis in Eq. (11.32) represents a two-tailed test. Thus, the two-tailed confidence interval formula presented in Eq. (11.29) should be used to calculate the confidence interval for pH values before the dam (σ_1^2) and the pH values after the dam (σ_1^2) because of the alternative hypothesis of the form $\neq$, this formula represents the alternative hypothesis for a two-sided test. The entries for the confidence interval formula as presented in Eq. (11.29) are already provided. The variance of pH values before the dam is 0.26, while the variance of pH values after the dam is 0.71, and the sample size is 11 for each region.

$$\frac{1}{F_{(\frac{\alpha}{2}, n_1-1, n_2-1)}} \frac{S_1^2}{S_2^2} \leq \frac{\sigma_1^2}{\sigma_2^2} \leq \frac{1}{F_{(1-\frac{\alpha}{2}, n_1-1, n_2-1)}} \frac{S_1^2}{S_2^2}$$

$$\frac{1}{5.8467} \frac{0.26}{0.71} \leq \frac{\sigma_1^2}{\sigma_2^2} \leq \frac{1}{0.1710} \frac{0.26}{0.71}$$

$$0.0626 \leq \frac{\sigma_1^2}{\sigma_2^2} \leq 2.1410$$

Step 4: Make a decision using the confidence interval and interpret the results

Making a decision using the confidence interval procedure requires checking the confidence interval as to whether it contains the hypothesized (claimed) value or not. It can be seen that the hypothesized value ($\frac{\sigma_1^2}{\sigma_2^2} = 1$) is in the calculated confidence interval $0.0626 \leq \frac{\sigma_1^2}{\sigma_2^2} \leq 2.1410$.

1. The decision reached by the confidence interval procedure: fail to reject the null hypothesis and we believe that the variance of pH values before the dam (σ_1^2) is not the same as the variance of pH values after the dam (σ_2^2).

The result of the confidence interval is compared with the decisions reached by the critical value and *P*-value procedures:

2. The critical value (traditional) procedure: fail to reject the null hypothesis as presented in Example 9.16.

3. The *P*-value procedure: fail to reject the null hypothesis as presented in Example 9.16.

We can say that the three procedures have reached the same decision: fail to reject the null hypothesis at 1% and we believe that the variance of pH values before the dam (σ_1^2) equals the variance of pH values after the dam (σ_2^2).

Example 11.14: The concentration of iron in sediment of two rivers: Example 9.17 is reproduced "A research group at a research center wishes to verify the claim that the variance of iron (Fe) concentration in sediment for the first river is less than the variance of iron in the second river. They selected 15 samples from each river and tested for iron concentration. The collected data showed that the variance of iron values for the first river is 0.29, while the collected data for the second river showed that the variance of iron is 1.42. A significance level of $\alpha = 0.01$ is chosen to test the claim. Assume that the two populations are normally distributed."

The four steps for performing hypothesis testing employing the confidence interval procedure can be used to test the hypothesis regarding the variance of iron concentration in the sediment of two rivers. The result of the confidence interval procedure will be compared with the critical value (traditional) and *P*-value procedures.

The confidence interval procedure

Step 1: Specify the null and alternative hypotheses

The two hypotheses (null and alternative) for the variance of iron concentration in the sediment of two rivers are presented in Eq. (11.33).

$$H_0 : \sigma_1^2 \geq \sigma_2^2 \text{ vs } H_1 : \sigma_1^2 < \sigma_2^2 \tag{11.33}$$

We should make a decision regarding the null hypothesis as to whether the variance of iron concentration in the sediment for the first river (σ_1^2) is greater than or equal to the variance of iron concentration in the sediment for the second river (σ_2^2), or the variance of iron concentration for the first river is less than the variance of iron concentration for the second river.

Step 2: Select the significance level (α) for the study

The F critical value for a one-tailed test with a significance level of 0.1 ($\alpha = 0.01$), $d.f_1 = 14$, and $d.f_2 = 14$ is 0.27045 ($F_{(\alpha,n_1-1,n_2-1)} = F_{(0.01,14,14)} = 0.27045$) as appeared in the F distribution table for critical values (using software or Table D in the Appendix).

Step 3: Use the sample information to calculate the test statistic value

The hypothesis in Eq. (11.33) represents a one-tailed test. Thus, the one-tailed upper confidence interval formula presented in Eq. (11.31) should be used to calculate the confidence interval for the variance of iron concentration in sediment of the first river (σ_1^2) and the variance of iron concentration in sediment of the second river (σ_2^2) because of the alternative hypothesis of the form $<$, this formula represents the alternative hypothesis for a one-sided test.

The entries for the confidence interval formula as presented in Eq. (11.31) are already provided. The variance of iron concentration in sediment for the first river is 0.29, while the variance of iron concentration in sediment for the second river is 1.42, and the sample size is 15 for each river.

$$\frac{\sigma_1^2}{\sigma_2^2} \leq \frac{1}{F_{(1-\alpha,n_1-1,n_2-1)}} \frac{S_1^2}{S_2^2}$$

$$\frac{\sigma_1^2}{\sigma_2^2} \leq \frac{1}{0.27045} \frac{0.29}{1.42}$$

$$\frac{\sigma_1^2}{\sigma_2^2} \leq 0.755$$

Step 4: Make a decision using the confidence interval and interpret the results

Making a decision using the confidence interval procedure requires checking the confidence interval as to whether it contains the hypothesized (claimed) value or not. It can be seen that the hypothesized value ($\frac{\sigma_1^2}{\sigma_2^2} = 1$) is not in the calculated confidence interval $\frac{\sigma_1^2}{\sigma_2^2} \leq 0.755$.

1. The decision reached by the confidence interval procedure: reject the null hypothesis and we believe that the variance of iron concentration in the sediment of the first river (σ_1^2) is less than the variance of iron concentration in the sediment of the second river (σ_2^2).

   The result of the confidence interval is compared with the decisions reached by the critical value and *P*-value procedures:
2. The critical value (traditional) procedure: reject the null hypothesis as presented in Example 9.17.
3. The *P*-value procedure: reject the null hypothesis as presented in Example 9.17.

We can say that the three procedures have reached the same decision: reject the null hypothesis at 1% and we believe that the variance of iron concentration in the sediment of the first river (σ_1^2) is less than the variance of iron concentration in the sediment of the second river (σ_2^2).

Example 11.15: The amount of solid waste for a palm oil mill: Example 9.18 is reproduced "An environmentalist wishes to investigate the claim about the amount of potash as solid waste for a palm oil mill. Two palm oil mills (A and B) were selected and the amount of potash was recorded (tons) for 12 months. The collected data showed that the variance of potash values for mill A is 5.54, while the collected data for mill B showed that the variance of potash values is 2.68. A significance level of $\alpha = 0.01$ is chosen to test the claim that the variance of mill A is higher than the variance of mill B. Assume that the two populations are normally distributed."

The four steps for performing hypothesis testing employing the confidence interval procedure can be used to test the hypothesis regarding the variance of potash of the two mills. The result of the confidence interval procedure will be compared with the critical value (traditional) and *P*-value procedures.

The confidence interval procedure

Step 1: Specify the null and alternative hypotheses

The two hypotheses (null and alternative) for the variance of potash of the two mills are presented in Eq. (11.34).

$$H_0{:}\sigma_1^2 \leq \sigma_2^2 \text{ vs } H_1{:}\sigma_1^2 > \sigma_2^2 \tag{11.34}$$

We should make a decision regarding the null hypothesis as to whether the variance of potash for mill A is less than or equal to the variance of potash for mill B, or the variance of potash for mill A is greater than the variance of potash for mill B.

Step 2: Select the significance level (α) for the study

The F critical value for a one-tailed test with a significance level of 0.01 ($\alpha = 0.01$), $d.f_1 = 11$, and $d.f_2 = 11$ is 4.4624 ($F_{(\alpha,n_1-1,n_2-1)} = F_{(0.01,11,11)} = 4.4624$) as appeared in the F distribution table for critical values (using a software or Table D in the Appendix).

Step 3: Use the sample information to calculate the test statistic value

The hypothesis in Eq. (11.34) represents a one-tailed test. Thus, the one-tailed lower confidence interval formula presented in Eq. (11.30) should be used to calculate the confidence interval for the variance of potash of mill A (σ_1^2) and the variance of potash of mill B (σ_2^2) because of the alternative hypothesis of the form $>$, this formula represents the alternative hypothesis for a one-sided test.

The entries for the confidence interval formula as presented in Eq. (11.30) are already provided. The variance of potash of mill A is 5.54, while the variance of potash of mill B is 2.68, and the sample size is 12 for each mill.

$$\frac{1}{F_{(\alpha,n_1-1,n_2-1)}}\frac{S_1^2}{S_2^2} \leq \frac{\sigma_1^2}{\sigma_2^2}$$

$$\frac{1}{4.4624}\frac{5.54}{2.68} \leq \frac{\sigma_1^2}{\sigma_2^2}$$

$$\frac{1}{4.4624}\frac{5.54}{2.68} \leq \frac{\sigma_1^2}{\sigma_2^2}$$

$$0.463 \leq \frac{\sigma_1^2}{\sigma_2^2}$$

Step 4: Make a decision using the confidence interval and interpret the results

Making a decision using the confidence interval procedure requires checking the confidence interval as to whether it contains the hypothesized (claimed) value or not. It can be seen that the hypothesized value ($\frac{\sigma_1^2}{\sigma_2^2} = 1$) is in the calculated confidence interval $0.463 \leq \frac{\sigma_1^2}{\sigma_2^2}$.

1. The decision reached by the confidence interval procedure: fail to reject the null hypothesis and we believe that the variance of potash for mill A (σ_1^2) is less than or equal to the variance of potash for mill B (σ_2^2).

   The result of the confidence interval is compared with the decisions reached by the critical value and *P*-value procedures:
2. The critical value (traditional) procedure: fail to reject the null hypothesis as presented in Example 9.18.
3. The *P*-value procedure: fail to reject the null hypothesis as presented in Example 9.18.

We can say that the three procedures have reached the same decision: fail to reject the null hypothesis at 1% and we believe that the variance of potash for mill A (σ_1^2) is less than or equal to the variance of potash for mill B (σ_2^2).

Further reading

Alkarkhi, A. F. M., Ahmad, A., Ismail, N., & Easa, A. M. (2008). Multivariate analysis of heavy metals concentrations in river estuary. *Environmental Monitoring and Assessment, 143*, 179–186.

Alkarkhi, A. F. M., & ALqaraghuli, W. A. A. (2020). *Applied statistics for environmental science with R* (1st ed.). Elsevier.

Alkarkhi, A. F. M., & Chin, L. H. (2012). *Elementary statistics for technologist* (1st ed.). Malaysia: Universiti Sains Malaysia.

Alkarkhi, A. F. M., Ismail, N., Ahmed, A., & Easa, Am (2009). Analysis of heavy metal concentrations in sediments of selected estuaries of Malaysia—A statistical assessment. *Environmental Monitoring and Assessment, 153*, 179–185.

Alkarkhi, F. M. A., Ismail, N., & Easa, A. M. (2008). Assessment of arsenic and heavy metal contents in cockles (*Anadara granosa*) using multivariate statistical techniques. *Journal of Hazardous Materials, 150*, 783–789.

Banch, T. J. H., Hanafiah, M. M., Alkarkhi, A. F. M., & Abu Amr, S. S. (2019). Statistical evaluation of landfill leachate system and its impact on groundwater and surface water in Malaysia. *Sains Malaysiana, 48*(11), 2391–2403.

Bluman, A. G. (1998). *Elementary statistics: A step by step approach* (3rd ed.). Boston: WCB/McGraw-Hill.

Eberly College of Science-Department of Statistics (2018). Two variances. In, Vol. 2020. Pennstate: The Pennsylvania State University. https://online.stat.psu.edu/stat414/node/206/.

Hossain, M.A., AlKarkhi, A.F.M., & AlKarkhi, N. N. (2008). Statistical and trend analysis of water quality data for the Baris Dam of Darul Aman river in Kedah, Malaysia. In *International Conference on Environmental Research and Technology (ICERT 2008)* (pp. 568–572). Penang-Malaysia: Universiti Sains Malaysia (USM).

Rahman, N. N. N. A., Chard, N. C., Al-Karkhi, A. F. M., Rafatullah, M., & Kadir, M. O. A. (2017). Analysis of particulate matters in air of palm oil mills – A statistical assessment. *Environmental Engineering and Management Journal, 16*(11), 2537–2543.

Weiss, N. A. (2012). *Introductory statistics* (9th ed.). Pearson.

Appendix

Table A Standard normal distribution area between 0 and z.

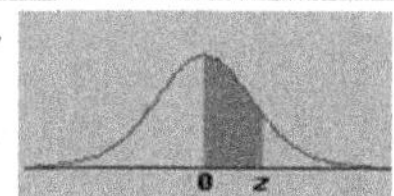

	0	0.01	0.02	0.03	0.04	0.05	0.06	0.07	0.08	0.09
0	0.5000	0.5040	0.5080	0.5120	0.5160	0.5199	0.5239	0.5279	0.5319	0.5359
0.1	0.5398	0.5438	0.5478	0.5517	0.5557	0.5596	0.5636	0.5675	0.5714	0.5753
0.2	0.5793	0.5832	0.5871	0.591	0.5948	0.5987	0.6026	0.6064	0.6103	0.6141
0.3	0.6179	0.6217	0.6255	0.6293	0.6331	0.6368	0.6406	0.6443	0.648	0.6517
0.4	0.6554	0.6591	0.6628	0.6664	0.67	0.6736	0.6772	0.6808	0.6844	0.6879
0.5	0.6915	0.695	0.6985	0.7019	0.7054	0.7088	0.7123	0.7157	0.719	0.7224
0.6	0.7257	0.7291	0.7324	0.7357	0.7389	0.7422	0.7454	0.7486	0.7517	0.7549
0.7	0.758	0.7611	0.7642	0.7673	0.7704	0.7734	0.7764	0.7794	0.7823	0.7852
0.8	0.7881	0.791	0.7939	0.7967	0.7995	0.8023	0.8051	0.8078	0.8106	0.8133
0.9	0.8159	0.8186	0.8212	0.8238	0.8264	0.8289	0.8315	0.834	0.8365	0.8389
1	0.8413	0.8438	0.8461	0.8485	0.8508	0.8531	0.8554	0.8577	0.8599	0.8621
1.1	0.8643	0.8665	0.8686	0.8708	0.8729	0.8749	0.8770	0.8790	0.8810	0.8830
1.2	0.8849	0.8869	0.8888	0.8907	0.8925	0.8944	0.8962	0.8980	0.8997	0.9015
1.3	0.9032	0.9049	0.9066	0.9082	0.9099	0.9115	0.9131	0.9147	0.9162	0.9177
1.4	0.9192	0.9207	0.9222	0.9236	0.9251	0.9265	0.9279	0.9292	0.9306	0.9319
1.5	0.9332	0.9345	0.9357	0.937	0.9382	0.9394	0.9406	0.9418	0.9429	0.9441
1.6	0.9452	0.9463	0.9474	0.9484	0.9495	0.9505	0.9515	0.9525	0.9535	0.9545
1.7	0.9554	0.9564	0.9573	0.9582	0.9591	0.9599	0.9608	0.9616	0.9625	0.9633
1.8	0.9641	0.9649	0.9656	0.9664	0.9671	0.9678	0.9686	0.9693	0.9699	0.9706
1.9	0.9713	0.9719	0.9726	0.9732	0.9738	0.9744	0.975	0.9756	0.9761	0.9767
2	0.9772	0.9778	0.9783	0.9788	0.9793	0.9798	0.9803	0.9808	0.9812	0.9817
2.1	0.9821	0.9826	0.983	0.9834	0.9838	0.9842	0.9846	0.9850	0.9854	0.9857
2.2	0.9861	0.9864	0.9868	0.9871	0.9875	0.9878	0.9881	0.9884	0.9887	0.9890
2.3	0.9893	0.9896	0.9898	0.9901	0.9904	0.9906	0.9909	0.9911	0.9913	0.9916
2.4	0.9918	0.992	0.9922	0.9925	0.9927	0.9929	0.9931	0.9932	0.9934	0.9936
2.5	0.9938	0.994	0.9941	0.9943	0.9945	0.9946	0.9948	0.9949	0.9951	0.9952
2.6	0.9953	0.9955	0.9956	0.9957	0.9959	0.996	0.9961	0.9962	0.9963	0.9964
2.7	0.9965	0.9966	0.9967	0.9968	0.9969	0.997	0.9971	0.9972	0.9973	0.9974
2.8	0.9974	0.9975	0.9976	0.9977	0.9977	0.9978	0.9979	0.9979	0.998	0.9981
2.9	0.9981	0.9982	0.9982	0.9983	0.9984	0.9984	0.9985	0.9985	0.9986	0.9986
3	0.9987	0.9987	0.9987	0.9988	0.9988	0.9989	0.9989	0.9989	0.9999	0.9999

Table B Critical values for t distribution.

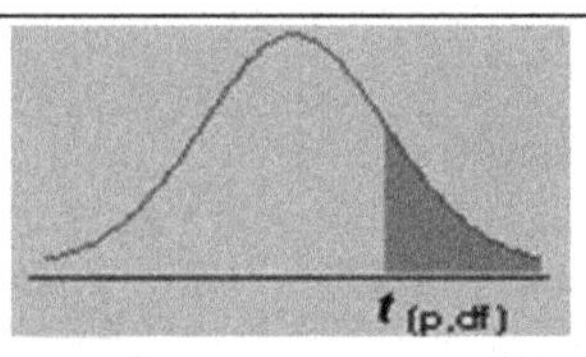

d.f.	One-tail α	0.25	0.10	0.05	0.025	0.01	0.005
	Two-tail α	0.50	0.20	0.10	0.05	0.02	0.01
1		1.000	3.078	6.314	12.706	31.821	63.657
2		0.816	1.886	2.920	4.303	6.965	9.925
3		0.765	1.638	2.353	3.182	4.541	5.841
4		0.741	1.533	2.132	2.776	3.747	4.604
5		0.727	1.476	2.015	2.571	3.365	4.032
6		0.718	1.440	1.943	2.447	3.143	3.707
7		0.711	1.415	1.895	2.365	2.998	3.499
8		0.706	1.397	1.860	2.306	2.896	3.355
9		0.703	1.383	1.833	2.262	2.821	3.250
10		0.700	1.372	1.812	2.228	2.764	3.169
11		0.697	1.363	1.796	2.201	2.718	3.106
12		0.695	1.356	1.782	2.179	2.681	3.055
13		0.694	1.350	1.771	2.160	2.650	3.012
14		0.692	1.345	1.761	2.145	2.624	2.977
15		0.691	1.341	1.753	2.131	2.602	2.947
16		0.690	1.337	1.746	2.120	2.583	2.921
17		0.689	1.333	1.740	2.110	2.567	2.898
18		0.688	1.330	1.734	2.101	2.552	2.878
19		0.688	1.328	1.729	2.093	2.539	2.861
20		0.687	1.325	1.725	2.086	2.528	2.845
21		0.686	1.323	1.721	2.080	2.518	2.831
22		0.686	1.321	1.717	2.074	2.508	2.819
23		0.685	1.319	1.714	2.069	2.500	2.807
24		0.685	1.318	1.711	2.064	2.492	2.797
25		0.684	1.316	1.708	2.060	2.485	2.787
26		0.684	1.315	1.706	2.056	2.479	2.779
27		0.684	1.314	1.703	2.052	2.473	2.771
28		0.683	1.313	1.701	2.048	2.467	2.763
29		0.683	1.311	1.699	2.045	2.462	2.756
30		0.683	1.310	1.697	2.042	2.457	2.750
INF.		0.674	1.282	1.645	1.960	2.326	2.576

Table C The critical values for chi-square distribution.

α / d.f.	0.995	0.990	0.975	0.950	0.900	0.750	0.500	0.250	0.100	0.050	0.025	0.010	0.005
1	0.00004	0.00016	0.00098	0.00393	0.01579	0.10153	0.45494	1.32330	2.70554	3.84146	5.02389	6.63490	7.87944
2	0.01003	0.02010	0.05064	0.10259	0.21072	0.57536	1.38629	2.77259	4.60517	5.99146	7.37776	9.21034	10.59663
3	0.07172	0.11483	0.21580	0.35185	0.58437	1.21253	2.36597	4.10834	6.25139	7.81473	9.34840	11.34487	12.83816
4	0.20699	0.29711	0.48442	0.71072	1.06362	1.92256	3.35669	5.38527	7.77944	9.48773	11.14329	13.27670	14.86026
5	0.41174	0.55430	0.83121	1.14548	1.61031	2.67460	4.35146	6.62568	9.23636	11.07050	12.83250	15.08627	16.74960
6	0.67573	0.87209	1.23734	1.63538	2.20413	3.45460	5.34812	7.84080	10.64464	12.59159	14.44938	16.81189	18.54758
7	0.98926	1.23904	1.68987	2.16735	2.83311	4.25485	6.34581	9.03715	12.01704	14.06714	16.01276	18.47531	20.27774
8	1.34441	1.64650	2.17973	2.73264	3.48954	5.07064	7.34412	10.21885	13.36157	15.50731	17.53455	20.09024	21.95495
9	1.73493	2.08790	2.70039	3.32511	4.16816	5.89883	8.34283	11.38875	14.68366	16.91898	19.02277	21.66599	23.58935
10	2.15586	2.55821	3.24697	3.94030	4.86518	6.73720	9.34182	12.54886	15.98718	18.30704	20.48318	23.20925	25.18818
11	2.60322	3.05348	3.81575	4.57481	5.57778	7.58414	10.34100	13.70069	17.27501	19.67514	21.92005	24.72497	26.75685
12	3.07382	3.57057	4.40379	5.22603	6.30380	8.43842	11.34032	14.84540	18.54935	21.02607	23.33666	26.21697	28.29952
13	3.56503	4.10692	5.00875	5.89186	7.04150	9.29907	12.33976	15.98391	19.81193	22.36203	24.73560	27.68825	29.81947
14	4.07467	4.66043	5.62873	6.57063	7.78953	10.16531	13.33927	17.11693	21.06414	23.68479	26.11895	29.14124	31.31935
15	4.60092	5.22935	6.26214	7.26094	8.54676	11.03654	14.33886	18.24509	22.30713	24.99579	27.48839	30.57791	32.80132
16	5.14221	5.81221	6.90766	7.96165	9.31224	11.91222	15.33850	19.36886	23.54183	26.29623	28.84535	31.99993	34.26719
17	5.69722	6.40776	7.56419	8.67176	10.08519	12.79193	16.33818	20.48868	24.76904	27.58711	30.19101	33.40866	35.71847
18	6.26480	7.01491	8.23075	9.39046	10.86494	13.67529	17.33790	21.60489	25.98942	28.86930	31.52638	34.80531	37.15645
19	6.84397	7.63273	8.90652	10.11701	11.65091	14.56200	18.33765	22.71781	27.20357	30.14353	32.85233	36.19087	38.58226
20	7.43384	8.26040	9.59078	10.85081	12.44261	15.45177	19.33743	23.82769	28.41198	31.41043	34.16961	37.56623	39.99685
21	8.03365	8.89720	10.28290	11.59131	13.23960	16.34438	20.33723	24.93478	29.61509	32.67057	35.47888	38.93217	41.40106
22	8.64272	9.54249	10.98232	12.33801	14.04149	17.23962	21.33704	26.03927	30.81328	33.92444	36.78071	40.28936	42.79565
23	9.26042	10.19572	11.68855	13.09051	14.84796	18.13730	22.33688	27.14134	32.00690	35.17246	38.07563	41.63840	44.18128
24	9.88623	10.85636	12.40115	13.84843	15.65868	19.03725	23.33673	28.24115	33.19624	36.41503	39.36408	42.97982	45.55851
25	10.51965	11.52398	13.11972	14.61141	16.47341	19.93934	24.33659	29.33885	34.38159	37.65248	40.64647	44.31410	46.92789
26	11.16024	12.19815	13.84390	15.37916	17.29188	20.84343	25.33646	30.43457	35.56317	38.88514	41.92317	45.64168	48.28988
27	11.80759	12.87850	14.57338	16.15140	18.11390	21.74940	26.33634	31.52841	36.74122	40.11327	43.19451	46.96294	49.64492
28	12.46134	13.56471	15.30786	16.92788	18.93924	22.65716	27.33623	32.62049	37.91592	41.33714	44.46079	48.27824	50.99338
29	13.12115	14.25645	16.04707	17.70837	19.76774	23.56659	28.33613	33.71091	39.08747	42.55697	45.72229	49.58788	52.33562
30	13.78672	14.95346	16.79077	18.49266	20.59923	24.47761	29.33603	34.79974	40.25602	43.77297	46.97924	50.89218	53.67196

Table D The critical values for F distribution.

d.f.D degrees of freedom: denominator	$\alpha = 0.10$																		
	d.f.N: Degrees of freedom, numerator																		
	1	2	3	4	5	6	7	8	9	10	12	15	20	24	30	40	60	120	INF
1	39.86	49.50	53.59	55.83	57.24	58.20	58.91	59.44	59.86	60.19	60.71	61.22	61.74	62.00	62.26	62.53	62.79	63.06	63.33
2	8.53	9.00	9.16	9.24	9.29	9.33	9.35	9.37	9.38	9.39	9.41	9.42	9.44	9.45	9.46	9.47	9.47	9.48	9.49
3	5.54	5.46	5.39	5.34	5.316	5.28	5.27	5.25	5.24	5.23	5.22	5.20	5.18	5.18	5.17	5.16	5.15	5.14	5.134
4	4.54	4.32	4.19	4.11	4.05	4.01	3.98	3.95	3.94	3.92	3.90	3.87	3.84	3.83	3.82	3.80	3.79	3.78	3.76
5	4.06	3.78	3.62	3.52	3.45	3.40	3.37	3.34	3.32	3.30	3.27	3.24	3.21	3.19	3.17	3.16	3.14	3.12	3.11
6	3.78	3.46	3.29	3.18	3.11	3.05	3.01	2.98	2.96	2.94	2.90	2.87	2.84	2.82	2.80	2.78	2.76	2.74	2.72
7	3.59	3.26	3.07	2.96	2.88	2.83	2.78	2.75	2.72	2.70	2.67	2.63	2.59	2.58	2.56	2.54	2.51	2.49	2.47
8	3.46	3.11	2.92	2.81	2.73	2.67	2.62	2.59	2.56	2.54	2.50	2.46	2.42	2.40	2.38	2.36	2.34	2.32	2.29
9	3.36	3.01	2.81	2.69	2.61	2.55	2.51	2.47	2.44	2.42	2.38	2.34	2.30	2.28	2.25	2.23	2.21	2.18	2.16
10	3.29	2.92	2.73	2.61	2.52	2.46	2.41	2.38	2.35	2.32	2.28	2.24	2.20	2.18	2.16	2.13	2.11	2.08	2.06
11	3.23	2.86	2.66	2.54	2.45	2.39	2.34	2.30	2.27	2.25	2.21	2.17	2.12	2.10	2.08	2.05	2.03	2.00	1.97
12	3.18	2.81	2.61	2.48	2.39	2.33	2.28	2.24	2.21	2.19	2.15	2.10	2.06	2.04	2.01	1.99	1.96	1.93	1.90
13	3.14	2.76	2.56	2.43	2.35	2.28	2.23	2.20	2.16	2.14	2.10	2.05	2.01	1.98	1.96	1.93	1.90	1.88	1.85
14	3.10	2.73	2.52	2.39	2.31	2.24	2.19	2.15	2.12	2.10	2.05	2.01	1.96	1.94	1.91	1.89	1.86	1.83	1.80
15	3.07	2.70	2.49	2.36	2.27	2.21	2.16	2.12	2.09	2.06	2.02	1.97	1.92	1.90	1.87	1.85	1.82	1.79	1.76
16	3.05	2.67	2.46	2.33	2.24	2.18	2.13	2.09	2.06	2.03	1.99	1.94	1.89	1.87	1.84	1.81	1.78	1.75	1.72
17	3.03	2.64	2.44	2.31	2.22	2.15	2.10	2.06	2.03	2.00	1.96	1.91	1.86	1.84	1.81	1.78	1.75	1.72	1.69
18	3.01	2.62	2.42	2.29	2.20	2.13	2.08	2.04	2.00	1.98	1.93	1.89	1.84	1.81	1.78	1.75	1.72	1.69	1.66
19	2.99	2.61	2.40	2.27	2.18	2.11	2.06	2.02	1.98	1.96	1.91	1.86	1.81	1.79	1.76	1.73	1.70	1.67	1.63
20	2.97	2.59	2.38	2.25	2.16	2.09	2.04	2.00	1.96	1.94	1.89	1.84	1.79	1.77	1.74	1.71	1.68	1.64	1.61
21	2.96	2.57	2.36	2.23	2.14	2.08	2.02	1.98	1.95	1.92	1.87	1.83	1.78	1.75	1.72	1.69	1.66	1.62	1.59
22	2.95	2.56	2.35	2.22	2.13	2.06	2.01	1.97	1.93	1.90	1.86	1.81	1.76	1.73	1.70	1.67	1.64	1.60	1.57
23	2.94	2.55	2.34	2.21	2.11	2.05	1.99	1.95	1.92	1.89	1.84	1.80	1.74	1.72	1.69	1.66	1.62	1.59	1.55
24	2.93	2.54	2.33	2.19	2.10	2.04	1.98	1.94	1.91	1.88	1.83	1.78	1.73	1.70	1.67	1.64	1.61	1.57	1.53
25	2.92	2.53	2.32	2.18	2.09	2.02	1.97	1.93	1.89	1.87	1.82	1.77	1.72	1.69	1.66	1.63	1.59	1.56	1.52
27	2.91	2.52	2.31	2.17	2.08	2.01	1.96	1.92	1.88	1.86	1.81	1.76	1.71	1.68	1.65	1.61	1.58	1.54	1.50
28	2.90	2.51	2.30	2.17	2.07	2.00	1.95	1.91	1.87	1.85	1.80	1.75	1.70	1.67	1.64	1.60	1.57	1.53	1.49
29	2.89	2.50	2.29	2.16	2.06	2.00	1.94	1.90	1.87	1.84	1.79	1.74	1.69	1.66	1.63	1.59	1.56	1.52	1.48
30	2.89	2.50	2.28	2.15	2.06	1.99	1.93	1.89	1.86	1.83	1.78	1.73	1.68	1.65	1.62	1.58	1.55	1.51	1.47
40	2.88	2.49	2.28	2.14	2.05	1.98	1.93	1.88	1.85	1.82	1.77	1.72	1.67	1.64	1.61	1.57	1.54	1.50	1.46
60	2.84	2.44	2.23	2.09	2.00	1.93	1.87	1.83	1.79	1.76	1.71	1.66	1.61	1.57	1.54	1.51	1.47	1.42	1.38
120	2.79	2.39	2.18	2.04	1.95	1.87	1.82	1.77	1.74	1.71	1.66	1.60	1.54	1.51	1.48	1.44	1.40	1.35	1.29
IFN.	2.75	2.35	2.13	1.99	1.90	1.82	1.77	1.72	1.68	1.65	1.60	1.55	1.48	1.45	1.41	1.37	1.32	1.26	1.19

Table D (Continued)

d.f.D degrees of freedom: denominator	$\alpha = 0.05$																		
	d.f.N: Degrees of freedom, numerator																		
	1	2	3	4	5	6	7	8	9	10	12	15	20	24	30	40	60	120	INF
1	161.45	199.50	215.71	224.58	230.16	233.99	236.77	238.88	240.54	241.88	243.91	245.95	248.01	249.05	250.10	251.14	252.20	253.25	254.31
2	18.51	19.00	19.16	19.25	19.30	19.33	19.35	19.37	19.38	19.40	19.41	19.43	19.45	19.45	19.46	19.47	19.48	19.49	19.50
3	10.13	9.55	9.28	9.12	9.01	8.94	8.89	8.85	8.81	8.79	8.74	8.70	8.66	8.64	8.62	8.59	8.57	8.55	8.53
4	7.71	6.94	6.59	6.39	6.26	6.16	6.09	6.04	6.00	5.96	5.91	5.86	5.80	5.77	5.75	5.72	5.69	5.66	5.63
5	6.61	5.79	5.41	5.19	5.05	4.95	4.88	4.82	4.77	4.74	4.68	4.62	4.56	4.53	4.50	4.46	4.43	4.40	4.37
6	5.99	5.14	4.76	4.53	4.39	4.28	4.21	4.15	4.10	4.06	4.00	3.94	3.87	3.84	3.81	3.77	3.74	3.70	3.67
7	5.59	4.74	4.35	4.12	3.97	3.87	3.79	3.73	3.68	3.64	3.57	3.51	3.44	3.41	3.38	3.34	3.30	3.27	3.23
8	5.32	4.46	4.07	3.84	3.69	3.58	3.50	3.44	3.39	3.35	3.28	3.22	3.15	3.12	3.08	3.04	3.01	2.97	2.93
9	5.12	4.26	3.86	3.63	3.48	3.37	3.29	3.23	3.18	3.14	3.07	3.01	2.94	2.90	2.86	2.83	2.79	2.75	2.71
10	4.96	4.10	3.71	3.48	3.33	3.22	3.14	3.07	3.02	2.98	2.91	2.85	2.77	2.74	2.70	2.66	2.62	2.58	2.54
11	4.84	3.98	3.59	3.36	3.20	3.09	3.01	2.95	2.90	2.85	2.79	2.72	2.65	2.61	2.57	2.53	2.49	2.45	2.40
12	4.75	3.89	3.49	3.26	3.11	3.00	2.91	2.85	2.80	2.75	2.69	2.62	2.54	2.51	2.47	2.43	2.38	2.34	2.30
13	4.67	3.81	3.41	3.18	3.03	2.92	2.83	2.77	2.71	2.67	2.60	2.53	2.46	2.42	2.38	2.34	2.30	2.25	2.21
14	4.60	3.74	3.34	3.11	2.96	2.85	2.76	2.70	2.65	2.60	2.53	2.46	2.39	2.35	2.31	2.27	2.22	2.18	2.13
15	4.54	3.68	3.29	3.06	2.90	2.79	2.71	2.64	2.59	2.54	2.48	2.40	2.33	2.29	2.25	2.20	2.16	2.11	2.07
16	4.49	3.63	3.24	3.01	2.85	2.74	2.66	2.59	2.54	2.49	2.42	2.35	2.28	2.24	2.19	2.15	2.11	2.06	2.01
17	4.45	3.59	3.20	2.96	2.81	2.70	2.61	2.55	2.49	2.45	2.38	2.31	2.23	2.19	2.15	2.10	2.06	2.01	1.96
18	4.41	3.55	3.16	2.93	2.77	2.66	2.58	2.51	2.46	2.41	2.34	2.27	2.19	2.15	2.11	2.06	2.02	1.97	1.92
19	4.38	3.52	3.13	2.90	2.74	2.63	2.54	2.48	2.42	2.38	2.31	2.23	2.16	2.11	2.07	2.03	1.98	1.93	1.88
20	4.35	3.49	3.10	2.87	2.71	2.60	2.51	2.45	2.39	2.35	2.28	2.20	2.12	2.08	2.04	1.99	1.95	1.90	1.84
21	4.32	3.47	3.07	2.84	2.68	2.57	2.49	2.42	2.37	2.32	2.25	2.18	2.10	2.05	2.01	1.96	1.92	1.87	1.81
22	4.30	3.44	3.05	2.82	2.66	2.55	2.46	2.40	2.34	2.30	2.23	2.15	2.07	2.03	1.98	1.94	1.89	1.84	1.78
23	4.28	3.42	3.03	2.80	2.64	2.53	2.44	2.37	2.32	2.27	2.20	2.13	2.05	2.01	1.96	1.91	1.86	1.81	1.76
24	4.26	3.40	3.01	2.78	2.62	2.51	2.42	2.36	2.30	2.25	2.18	2.11	2.03	1.98	1.94	1.89	1.84	1.79	1.73
25	4.24	3.39	2.99	2.76	2.60	2.49	2.40	2.34	2.28	2.24	2.16	2.09	2.01	1.96	1.92	1.87	1.82	1.77	1.71
27	4.23	3.37	2.98	2.74	2.59	2.47	2.39	2.32	2.27	2.22	2.15	2.07	1.99	1.95	1.90	1.85	1.80	1.75	1.69
28	4.21	3.35	2.96	2.73	2.57	2.46	2.37	2.31	2.25	2.20	2.13	2.06	1.97	1.93	1.88	1.84	1.79	1.73	1.67
29	4.20	3.34	2.95	2.71	2.56	2.45	2.36	2.29	2.24	2.19	2.12	2.04	1.96	1.91	1.87	1.82	1.77	1.71	1.65
30	4.18	3.33	2.93	2.70	2.55	2.43	2.35	2.28	2.22	2.18	2.10	2.03	1.94	1.90	1.85	1.81	1.75	1.70	1.64
40	4.17	3.32	2.92	2.69	2.53	2.42	2.33	2.27	2.21	2.16	2.09	2.01	1.93	1.89	1.84	1.79	1.74	1.68	1.62
60	4.08	3.23	2.84	2.61	2.45	2.34	2.25	2.18	2.12	2.08	2.00	1.92	1.84	1.79	1.74	1.69	1.64	1.58	1.51
120	4.00	3.15	2.76	2.53	2.37	2.25	2.17	2.10	2.04	1.99	1.92	1.84	1.75	1.70	1.65	1.59	1.53	1.47	1.39
IFN.	3.92	3.07	2.68	2.45	2.29	2.18	2.09	2.02	1.96	1.91	1.83	1.75	1.66	1.61	1.55	1.50	1.43	1.35	1.25

(*Continued*)

Table D (Continued)

d.f.D degrees of freedom: denominator	$\alpha = 0.025$																		
	d.f.N: Degrees of freedom, numerator																		
	1	2	3	4	5	6	7	8	9	10	12	15	20	24	30	40	60	120	INF
1	647.79	799.50	864.16	899.58	921.85	937.11	948.22	956.66	963.28	968.63	976.71	984.87	993.10	997.25	1001.41	1005.60	1009.80	1014.02	1018.26
2	38.51	39.00	39.17	39.25	39.30	39.33	39.36	39.37	39.39	39.40	39.41	39.43	39.45	39.46	39.47	39.47	39.48	39.49	39.50
3	17.44	16.04	15.44	15.10	14.88	14.73	14.62	14.54	14.47	14.42	14.34	14.25	14.17	14.12	14.08	14.04	13.99	13.95	13.90
4	12.22	10.65	9.98	9.60	9.36	9.20	9.07	8.98	8.90	8.84	8.75	8.66	8.56	8.51	8.46	8.41	8.36	8.31	8.26
5	10.01	8.43	7.76	7.39	7.15	6.98	6.85	6.76	6.68	6.62	6.52	6.43	6.33	6.28	6.23	6.18	6.12	6.07	6.02
6	8.81	7.26	6.60	6.23	5.99	5.82	5.70	5.60	5.52	5.46	5.37	5.27	5.17	5.12	5.07	5.01	4.96	4.90	4.85
7	8.07	6.54	5.89	5.52	5.29	5.12	4.99	4.90	4.82	4.76	4.67	4.57	4.47	4.42	4.36	4.31	4.25	4.20	4.14
8	7.57	6.06	5.42	5.05	4.82	4.65	4.53	4.43	4.36	4.30	4.20	4.10	4.00	3.95	3.89	3.84	3.78	3.73	3.67
9	7.21	5.71	5.08	4.72	4.48	4.32	4.20	4.10	4.03	3.96	3.87	3.77	3.67	3.61	3.56	3.51	3.45	3.39	3.33
10	6.94	5.46	4.83	4.47	4.24	4.07	3.95	3.85	3.78	3.72	3.62	3.52	3.42	3.37	3.31	3.26	3.20	3.14	3.08
11	6.72	5.26	4.63	4.28	4.04	3.88	3.76	3.66	3.59	3.53	3.43	3.33	3.23	3.17	3.12	3.06	3.00	2.94	2.88
12	6.55	5.10	4.47	4.12	3.89	3.73	3.61	3.51	3.44	3.37	3.28	3.18	3.07	3.02	2.96	2.91	2.85	2.79	2.73
13	6.41	4.97	4.35	4.00	3.77	3.60	3.48	3.39	3.31	3.25	3.15	3.05	2.95	2.89	2.84	2.78	2.72	2.66	2.60
14	6.30	4.86	4.24	3.89	3.66	3.50	3.38	3.29	3.21	3.15	3.05	2.95	2.84	2.79	2.73	2.67	2.61	2.55	2.49
15	6.20	4.77	4.15	3.80	3.58	3.41	3.29	3.20	3.12	3.06	2.96	2.86	2.76	2.70	2.64	2.59	2.52	2.46	2.40
16	6.12	4.69	4.08	3.73	3.50	3.34	3.22	3.12	3.05	2.99	2.89	2.79	2.68	2.63	2.57	2.51	2.45	2.38	2.32
17	6.04	4.62	4.01	3.66	3.44	3.28	3.16	3.06	2.98	2.92	2.82	2.72	2.62	2.56	2.50	2.44	2.38	2.32	2.25
18	5.98	4.56	3.95	3.61	3.38	3.22	3.10	3.01	2.93	2.87	2.77	2.67	2.56	2.50	2.45	2.38	2.32	2.26	2.19
19	5.92	4.51	3.90	3.56	3.33	3.17	3.05	2.96	2.88	2.82	2.72	2.62	2.51	2.45	2.39	2.33	2.27	2.20	2.13
20	5.87	4.46	3.86	3.51	3.29	3.13	3.01	2.91	2.84	2.77	2.68	2.57	2.46	2.41	2.35	2.29	2.22	2.16	2.09
21	5.83	4.42	3.82	3.48	3.25	3.09	2.97	2.87	2.80	2.73	2.64	2.53	2.42	2.37	2.31	2.25	2.18	2.11	2.04
22	5.79	4.38	3.78	3.44	3.22	3.05	2.93	2.84	2.76	2.70	2.60	2.50	2.39	2.33	2.27	2.21	2.15	2.08	2.00
23	5.75	4.35	3.75	3.41	3.18	3.02	2.90	2.81	2.73	2.67	2.57	2.47	2.36	2.30	2.24	2.18	2.11	2.04	1.97
24	5.72	4.32	3.72	3.38	3.15	2.99	2.87	2.78	2.70	2.64	2.54	2.44	2.33	2.27	2.21	2.15	2.08	2.01	1.94
25	5.69	4.29	3.69	3.35	3.13	2.97	2.85	2.75	2.68	2.61	2.51	2.41	2.30	2.24	2.18	2.12	2.05	1.98	1.91
27	5.66	4.27	3.67	3.33	3.10	2.94	2.82	2.73	2.65	2.59	2.49	2.39	2.28	2.22	2.16	2.09	2.03	1.95	1.88
28	5.63	4.24	3.65	3.31	3.08	2.92	2.80	2.71	2.63	2.57	2.47	2.36	2.25	2.19	2.13	2.07	2.00	1.93	1.85
29	5.61	4.22	3.63	3.29	3.06	2.90	2.78	2.69	2.61	2.55	2.45	2.34	2.23	2.17	2.11	2.05	1.98	1.91	1.83
30	5.59	4.20	3.61	3.27	3.04	2.88	2.76	2.67	2.59	2.53	2.43	2.32	2.21	2.15	2.09	2.03	1.96	1.89	1.81
40	5.57	4.18	3.59	3.25	3.03	2.87	2.75	2.65	2.57	2.51	2.41	2.31	2.20	2.14	2.07	2.01	1.94	1.87	1.79
60	5.42	4.05	3.46	3.13	2.90	2.74	2.62	2.53	2.45	2.39	2.29	2.18	2.07	2.01	1.94	1.88	1.80	1.72	1.64
120	5.29	3.93	3.34	3.01	2.79	2.63	2.51	2.41	2.33	2.27	2.17	2.06	1.94	1.88	1.82	1.74	1.67	1.58	1.48
IFN.	5.15	3.80	3.23	2.89	2.67	2.52	2.39	2.30	2.22	2.16	2.05	1.95	1.82	1.76	1.69	1.61	1.53	1.43	1.31

Table D (Continued)

d.f.D degrees of freedom: denominator	$\alpha = 0.01$																		
	d.f.N: Degrees of freedom, numerator																		
	1	**2**	**3**	**4**	**5**	**6**	**7**	**8**	**9**	**10**	**12**	**15**	**20**	**24**	**30**	**40**	**60**	**120**	**INF**
1	4052.18	4999.50	5403.35	5624.58	5763.65	5858.99	5928.36	5981.07	6022.47	6055.85	6106.32	6157.29	6208.73	6234.63	6260.65	6286.78	6313.03	6339.39	6365.86
2	98.50	99.00	99.17	99.25	99.30	99.33	99.36	99.37	99.39	99.40	99.42	99.43	99.45	99.46	99.47	99.47	99.48	99.49	99.50
3	34.12	30.82	29.46	28.71	28.24	27.91	27.67	27.49	27.35	27.23	27.05	26.87	26.69	26.60	26.51	26.41	26.32	26.22	26.13
4	21.20	18.00	16.69	15.98	15.52	15.21	14.98	14.80	14.66	14.55	14.37	14.20	14.02	13.93	13.84	13.75	13.65	13.56	13.46
5	16.26	13.27	12.06	11.39	10.97	10.67	10.46	10.29	10.16	10.05	9.89	9.72	9.55	9.47	9.38	9.29	9.20	9.11	9.02
6	13.75	10.93	9.78	9.15	8.75	8.47	8.26	8.10	7.98	7.87	7.72	7.56	7.40	7.31	7.23	7.14	7.06	6.97	6.88
7	12.25	9.55	8.45	7.85	7.46	7.19	6.99	6.84	6.72	6.62	6.47	6.31	6.16	6.07	5.99	5.91	5.82	5.74	5.65
8	11.26	8.65	7.59	7.01	6.63	6.37	6.18	6.03	5.91	5.81	5.67	5.52	5.36	5.28	5.20	5.12	5.03	4.95	4.86
9	10.56	8.02	6.99	6.42	6.06	5.80	5.61	5.47	5.35	5.26	5.11	4.96	4.81	4.73	4.65	4.57	4.48	4.40	4.31
10	10.04	7.56	6.55	5.99	5.64	5.39	5.20	5.06	4.94	4.85	4.71	4.56	4.41	4.33	4.25	4.17	4.08	4.00	3.91
11	9.65	7.21	6.22	5.67	5.32	5.07	4.89	4.74	4.63	4.54	4.40	4.25	4.10	4.02	3.94	3.86	3.78	3.69	3.60
12	9.33	6.93	5.95	5.41	5.06	4.82	4.64	4.50	4.39	4.30	4.16	4.01	3.86	3.78	3.70	3.62	3.54	3.45	3.36
13	9.07	6.70	5.74	5.21	4.86	4.62	4.44	4.30	4.19	4.10	3.96	3.82	3.67	3.59	3.51	3.43	3.34	3.26	3.17
14	8.86	6.52	5.56	5.04	4.70	4.46	4.28	4.14	4.03	3.94	3.80	3.66	3.51	3.43	3.35	3.27	3.18	3.09	3.00
15	8.68	6.36	5.42	4.89	4.56	4.32	4.14	4.00	3.90	3.81	3.67	3.52	3.37	3.29	3.21	3.13	3.05	2.96	2.87
16	8.53	6.23	5.29	4.77	4.44	4.20	4.03	3.89	3.78	3.69	3.55	3.41	3.26	3.18	3.10	3.02	2.93	2.85	2.75
17	8.40	6.11	5.19	4.67	4.34	4.10	3.93	3.79	3.68	3.59	3.46	3.31	3.16	3.08	3.00	2.92	2.84	2.75	2.65
18	8.29	6.01	5.09	4.58	4.25	4.02	3.84	3.71	3.60	3.51	3.37	3.23	3.08	3.00	2.92	2.84	2.75	2.66	2.57
19	8.19	5.93	5.01	4.50	4.17	3.94	3.77	3.63	3.52	3.43	3.30	3.15	3.00	2.93	2.84	2.76	2.67	2.58	2.49
20	8.10	5.85	4.94	4.43	4.10	3.87	3.70	3.56	3.46	3.37	3.23	3.09	2.94	2.86	2.78	2.70	2.61	2.52	2.42
21	8.02	5.78	4.87	4.37	4.04	3.81	3.64	3.51	3.40	3.31	3.17	3.03	2.88	2.80	2.72	2.64	2.55	2.46	2.36
22	7.95	5.72	4.82	4.31	3.99	3.76	3.59	3.45	3.35	3.26	3.12	2.98	2.83	2.75	2.67	2.58	2.50	2.40	2.31
23	7.88	5.66	4.77	4.26	3.94	3.71	3.54	3.41	3.30	3.21	3.07	2.93	2.78	2.70	2.62	2.54	2.45	2.35	2.26
24	7.82	5.61	4.72	4.22	3.90	3.67	3.50	3.36	3.26	3.17	3.03	2.89	2.74	2.66	2.58	2.49	2.40	2.31	2.21
25	7.77	5.57	4.68	4.18	3.86	3.63	3.46	3.32	3.22	3.13	2.99	2.85	2.70	2.62	2.54	2.45	2.36	2.27	2.17
27	7.72	5.53	4.64	4.14	3.82	3.59	3.42	3.29	3.18	3.09	2.96	2.82	2.66	2.59	2.50	2.42	2.33	2.23	2.13
28	7.68	5.49	4.60	4.11	3.79	3.56	3.39	3.26	3.15	3.06	2.93	2.78	2.63	2.55	2.47	2.38	2.29	2.20	2.10
29	7.64	5.45	4.57	4.07	3.75	3.53	3.36	3.23	3.12	3.03	2.90	2.75	2.60	2.52	2.44	2.35	2.26	2.17	2.06
30	7.60	5.42	4.54	4.05	3.73	3.50	3.33	3.20	3.09	3.01	2.87	2.73	2.57	2.50	2.41	2.33	2.23	2.14	2.03
40	7.56	5.39	4.51	4.02	3.70	3.47	3.30	3.17	3.07	2.98	2.84	2.70	2.55	2.47	2.39	2.30	2.21	2.11	2.01
60	7.31	5.18	4.31	3.83	3.51	3.29	3.12	2.99	2.89	2.80	2.67	2.52	2.37	2.29	2.20	2.11	2.02	1.92	1.81
120	7.08	4.98	4.13	3.65	3.34	3.12	2.95	2.82	2.72	2.63	2.50	2.35	2.20	2.12	2.03	1.94	1.84	1.73	1.60
IFN.	6.85	4.79	3.95	3.48	3.17	2.96	2.79	2.66	2.56	2.47	2.34	2.19	2.04	1.95	1.86	1.76	1.66	1.53	1.38

(Continued)

Table D (Continued)

d.f.D degrees of freedom: denominator	$\alpha = 0.995$																	
	d.f.N: Degrees of freedom, numerator																	
	1	2	3	4	5	6	7	8	9	10	12	15	20	24	30	40	60	120
1	0.00	0.01	0.02	0.03	0.04	0.05	0.06	0.07	0.07	0.08	0.09	0.09	0.10	0.11	0.01	0.11	0.11	0.12
2	0.00	0.01	0.02	0.04	0.06	0.07	0.08	0.10	0.10	0.11	0.12	0.13	0.14	0.15	0.16	0.17	0.17	0.18
3	0.00	0.01	0.02	0.04	0.06	0.08	0.09	0.10	0.12	0.12	0.14	0.15	0.17	0.18	0.19	0.20	0.21	0.22
4	0.00	0.01	0.02	0.04	0.06	0.08	0.10	0.11	0.13	0.14	0.15	0.17	0.19	0.21	0.22	0.23	0.24	0.26
5	0.00	0.01	0.02	0.05	0.07	0.09	0.11	0.12	0.13	0.15	0.17	0.19	0.21	0.22	0.24	0.25	0.27	0.28
6	0.00	0.01	0.02	0.05	0.07	0.09	0.11	0.13	0.14	0.15	0.17	0.20	0.22	0.24	0.25	0.27	0.29	0.30
7	0.00	0.01	0.02	0.05	0.07	0.09	0.11	0.13	0.15	0.16	0.18	0.21	0.24	0.25	0.27	0.29	0.30	0.32
8	0.00	0.01	0.02	0.05	0.07	0.10	0.12	0.13	0.15	0.16	0.19	0.21	0.25	0.26	0.28	0.30	0.32	0.34
9	0.00	0.01	0.02	0.05	0.07	0.10	0.12	0.14	0.15	0.17	0.19	0.22	0.25	0.27	0.29	0.31	0.33	0.36
10	0.00	0.01	0.02	0.05	0.07	0.10	0.12	0.14	0.16	0.17	0.20	0.23	0.26	0.28	0.30	0.32	0.34	0.37
11	0.00	0.01	0.02	0.05	0.07	0.10	0.12	0.14	0.16	0.17	0.20	0.23	0.27	0.29	0.31	0.33	0.36	0.38
12	0.00	0.01	0.02	0.05	0.08	0.10	0.12	0.14	0.16	0.18	0.20	0.24	0.27	0.29	0.32	0.34	0.37	0.39
13	0.00	0.01	0.02	0.05	0.08	0.10	0.12	0.14	0.16	0.18	0.21	0.24	0.28	0.30	0.32	0.35	0.37	0.40
14	0.00	0.01	0.02	0.05	0.08	0.10	0.13	0.15	0.16	0.18	0.21	0.24	0.28	0.30	0.33	0.35	0.38	0.41
15	0.00	0.01	0.02	0.05	0.08	0.10	0.13	0.15	0.17	0.18	0.21	0.25	0.29	0.31	0.33	0.36	0.39	0.42
16	0.00	0.01	0.02	0.05	0.08	0.10	0.13	0.15	0.17	0.18	0.21	0.25	0.29	0.31	0.34	0.37	0.40	0.43
17	0.00	0.01	0.02	0.05	0.08	0.10	0.13	0.15	0.17	0.19	0.22	0.25	0.29	0.32	0.34	0.37	0.40	0.44
18	0.00	0.01	0.02	0.05	0.08	0.10	0.13	0.15	0.17	0.19	0.22	0.25	0.30	0.32	0.35	0.38	0.41	0.44
19	0.00	0.01	0.02	0.05	0.08	0.10	0.13	0.15	0.17	0.19	0.22	0.26	0.30	0.32	0.35	0.38	0.41	0.45
20	0.00	0.01	0.02	0.05	0.08	0.10	0.13	0.15	0.17	0.19	0.22	0.26	0.30	0.33	0.35	0.39	0.42	0.46
21	0.00	0.01	0.02	0.05	0.08	0.11	0.13	0.15	0.17	0.19	0.22	0.26	0.30	0.33	0.36	0.39	0.42	0.46
22	0.00	0.01	0.02	0.05	0.08	0.11	0.13	0.15	0.17	0.19	0.22	0.26	0.31	0.33	0.36	0.39	0.43	0.47
23	0.00	0.01	0.02	0.05	0.08	0.11	0.13	0.15	0.17	0.19	0.23	0.26	0.31	0.34	0.36	0.40	0.43	0.47
24	0.00	0.01	0.02	0.05	0.08	0.11	0.13	0.15	0.18	0.19	0.23	0.26	0.31	0.34	0.37	0.40	0.44	0.48
25	0.00	0.01	0.02	0.05	0.08	0.11	0.13	0.15	0.18	0.19	0.23	0.27	0.31	0.34	0.37	0.40	0.44	0.48
27	0.00	0.01	0.02	0.05	0.08	0.11	0.13	0.16	0.18	0.20	229.00	0.27	0.32	0.34	0.37	0.41	0.45	0.49
28	0.00	0.01	0.02	0.05	0.08	0.11	0.13	0.16	0.18	0.20	0.23	0.27	0.32	0.35	0.38	0.41	0.45	0.50
29	0.00	0.01	0.02	0.05	0.08	0.11	0.13	0.16	0.18	0.20	0.23	0.27	0.32	0.35	0.38	0.41	0.45	0.50
30	0.00	0.01	0.02	0.05	0.08	0.11	0.13	0.16	0.18	0.20	0.23	0.27	0.32	0.35	0.38	0.42	0.46	0.50
40	0.00	0.01	0.02	0.05	0.08	0.11	0.14	0.16	0.18	0.20	0.24	0.28	0.33	0.36	0.40	0.44	0.48	0.54
60	0.00	0.01	0.02	0.05	0.08	0.11	0.14	0.16	0.19	0.21	0.24	0.29	0.34	0.38	0.41	0.46	0.51	0.57
120	0.00	0.01	0.02	0.05	0.08	0.11	0.14	0.17	0.19	0.21	0.25	0.30	0.36	0.39	0.44	0.49	0.55	0.62
IFN.	0.00	0.01	0.02	0.05	0.08	0.11	0.14	0.17	0.19	0.22	0.26	0.31	0.37	0.41	0.46	0.52	0.59	0.70

Table D (Continued)

d.f.D degrees of freedom: denominator	$\alpha = 0.005$																	
	d.f.N: Degrees of freedom, numerator																	
	1	2	3	4	5	6	7	8	9	10	12	15	20	24	30	40	60	120
1	16210.72	19999.50	21614.74	22499.58	23055.80	23437.11	23714.57	23925.41	24091.00	24224.49	24426.37	24630.21	24835.97	24939.57	25043.63	25148.15	25253.14	25258.57
2	198.50	199.00	199.17	199.25	199.30	199.33	199.36	199.37	199.39	199.40	199.42	199.43	199.45	199.46	199.47	199.48	199.48	199.49
3	55.55	49.80	47.47	46.19	45.39	44.84	44.43	44.13	43.88	43.69	43.39	43.08	42.78	42.62	42.47	42.30	42.15	41.99
4	31.33	26.28	24.26	23.15	22.46	21.97	21.62	21.35	21.14	20.97	20.70	20.44	20.17	20.03	19.89	19.75	19.61	19.47
5	22.78	18.31	16.53	15.56	14.94	14.51	14.20	13.96	13.77	13.62	13.38	13.15	12.90	12.78	12.66	12.53	12.40	12.27
6	18.64	14.54	12.92	12.03	11.46	11.07	10.79	10.57	10.39	10.25	10.03	9.81	9.59	9.47	9.36	9.24	9.12	9.00
7	16.24	12.40	10.88	10.05	9.52	9.16	8.89	8.68	8.51	8.38	8.18	7.97	7.75	7.65	7.53	7.42	7.31	7.19
8	14.69	11.04	9.60	8.81	8.30	7.95	7.69	7.50	7.34	7.21	7.01	7.97	6.61	6.50	6.40	6.29	6.18	6.06
9	13.61	10.11	8.72	7.96	7.47	7.13	6.88	6.69	6.54	6.42	6.23	6.03	5.83	5.73	5.62	5.52	5.41	5.30
10	12.83	9.43	8.08	7.34	6.87	6.54	6.30	6.12	5.97	5.85	5.66	5.47	5.27	5.17	5.07	4.97	4.86	4.75
11	12.23	8.91	7.61	6.88	6.42	6.10	5.86	5.68	5.54	5.42	5.24	5.05	4.86	4.76	4.65	4.55	4.45	4.37
12	11.75	8.51	7.23	6.52	6.07	5.76	5.52	5.35	5.20	5.09	4.91	4.72	4.53	4.43	4.33	4.23	4.12	4.01
13	11.37	8.19	6.93	6.23	5.79	5.48	5.25	5.08	4.94	4.82	4.64	4.46	4.27	4.17	4.07	3.97	3.87	3.76
14	11.06	7.92	6.68	6.00	5.56	5.26	5.03	4.86	4.72	4.60	4.43	4.25	4.059	3.96	3.86	3.76	3.66	3.55
15	10.80	7.70	6.48	5.80	5.37	5.07	4.85	4.67	4.54	4.42	4.30	4.07	3.88	3.79	3.69	3.59	3.48	3.37
16	10.58	7.51	6.30	5.64	5.21	4.91	4.69	4.52	4.38	4.27	4.10	3.92	3.73	3.64	3.54	3.44	3.33	3.22
17	10.38	7.35	6.16	5.50	5.07	4.78	4.56	4.39	4.25	4.14	3.97	3.79	3.61	3.51	3.41	3.31	3.21	3.10
18	10.22	7.21	6.03	5.37	4.96	4.66	4.44	4.28	4.14	4.03	3.86	3.68	3.50	3.40	3.30	3.20	3.10	2.99
19	10.07	7.09	5.92	5.27	4.85	4.56	4.34	4.18	4.04	3.93	3.76	3.59	3.40	3.31	3.21	3.11	3.00	2.89
20	9.94	6.99	5.82	5.17	4.76	4.47	4.26	4.09	3.96	3.85	3.68	3.50	3.32	3.22	3.12	3.02	2.92	2.81
21	9.83	6.89	5.73	5.09	4.68	4.39	4.18	4.01	3.88	3.77	3.60	3.43	3.24	3.15	3.05	2.95	2.84	2.73
22	9.73	6.81	5.65	5.02	4.61	4.32	4.11	3.94	3.81	3.70	3.54	3.36	3.18	3.08	2.98	2.88	2.77	2.66
23	9.63	6.73	5.58	4.95	4.54	4.26	4.05	3.88	3.75	3.64	3.47	3.30	3.12	3.02	2.92	2.82	2.71	2.60
24	9.55	6.66	5.52	4.89	4.49	4.20	3.99	3.83	3.69	3.59	3.42	3.25	3.06	2.97	2.87	2.77	2.66	2.55
25	9.48	6.60	5.46	4.84	4.43	4.15	3.94	3.78	3.64	3.54	3.37	3.20	3.01	2.92	2.82	2.72	2.61	2.50
27	9.34	6.49	5.36	4.74	4.34	4.06	3.85	3.69	3.56	3.45	3.28	3.11	2.93	2.83	2.73	2.63	2.52	2.41
28	9.28	6.44	5.32	4.70	4.30	4.02	3.81	3.65	3.52	3.41	3.25	3.07	2.89	2.79	2.69	2.59	2.48	2.37
29	9.23	6.40	5.28	4.66	4.26	3.98	3.77	3.61	3.48	3.38	3.21	3.04	2.86	2.76	2.66	2.56	2.45	2.33
30	9.18	6.3547	5.24	4.62	4.23	3.95	3.74	3.58	3.45	3.34	3.18	3.01	2.82	2.73	2.63	2.54	2.42	2.30
40	8.83	6.07	4.98	4.37	3.99	3.71	3.51	3.35	3.22	3.12	2.95	2.78	2.60	2.50	2.40	2.30	2.18	2.06
60	8.49	5.80	4.73	4.14	3.76	3.49	3.29	3.13	3.01	2.90	2.74	2.57	2.39	2.29	2.19	2.08	1.96	1.83
120	8.18	5.54	4.50	3.92	3.55	3.28	3.09	2.93	2.81	2.71	2.54	2.37	2.19	2.09	1.98	1.87	1.75	1.61
IFN.	7.88	5.30	4.28	3.72	3.35	3.09	2.90	2.74	2.62	2.52	2.36	2.19	1.99	1.89	1.79	1.67	1.53	1.36

Index

Note: Page numbers followed by "*f*" and "*t*" refer to figures and tables, respectively.

Printed in the United States
By Bookmasters